THE ENGINEER AND SOCIETY

Already published

ENGINEERING HERITAGE (VOLUME I)

ENGINEERING HERITAGE (VOLUME II)

MECHANICAL ENGINEERING

ENGINEERING MATERIALS AND METHODS

MANAGEMENT FOR ENGINEERS

THE ENGINEER
AND SOCIETY

Edited by

E. G. SEMLER, BSc, CEng, FIMechE

The Institution of Mechanical Engineers

LONDON

First published as a series of articles in *The Chartered Mechanical
Engineer* between 1962 and 1971. Reprinted by photolitho in this form
by J. W. Arrowsmith Ltd, Bristol, on cartridge paper and bound by
G. and J. Kitcat Ltd, London, England

Made and printed in Great Britain

CONTENTS

PREFACE

by H. G. Conway, CBE, MA, CEng, FIMechE
President of The Institution of Mechanical Engineers, 1967-68

The benefits that the Engineer and Scientist have brought to civilisation have also produced problems – many of them severe, especially in this country that nurtured the Industrial Revolution 200 years ago. But if the Engineer must carry his share of responsibility for the harms he has caused to Society he can take heart in the certainty that the good he has done far outpaces the damage.

If some of the pollution – in the broadest sense of the word – we see around us is at least in part the problem of the Engineer, it is certain also that it is he who will find the answers. The man who creates the noisy machine can be the one to produce the silent one, or the smokeless one, or the wasteless one.

Perhaps the most significant development is that the Engineer wants to debate the issues that he finds around him. Conscious – perhaps more than others – of his strength he wants to debate his place in the society around him. He wants to debate it with his professional colleagues from other fields of endeavour.

And this brings in turn to his consciousness the problem of his own status, complicated by an unsatisfactory terminology and a certain lack of prestige which this entails. While it may be good to debate his status from time to time, let him not spend too much of this time on the debate. There is plenty to do in the disciplining of the advancing frontiers of Engineering which will automatically bring with it the enhancement of his status by his own achievements.

In this book the problem surrounding the engineer are discussed from all angles, bringing light to dark areas, and provoking thought in the most praiseworthy way.

Technology and the Quality of Life

by the Rt Hon A. Wedgwood Benn, PC, MP*

People want technology—but technology with a human face. The ex-Minister of Technology gives his views on the directions of future technical advance which will yield benefits to society as a whole, without turning us into robots. Public accountability is one solution but first we must persuade ordinary people that they ought to influence decisions and are qualified to do so.

The sixties saw a greatly quickened public interest in technology. More and more people have come to see how the rapid application of scientific methods could improve their performance and help to raise the standard of living. Mintech has been encouraging the idea that technology can and ought to be put into service in industry and society more quickly and the growing acceptance of it is most welcome. We have had to give economic growth a high priority.

At the same time public awareness about the broader effects of technology on human beings has also been growing. As economic standards rise, there is an increasing fear that technology is being misused, that we are pursuing increased production of goods and services too singlemindedly and without regard for the general quality of life. People are becoming increasingly concerned that this general quality of life will be damaged because the wider effects of technological change are either not thought about or disregarded.

A balanced assessment

In its most extreme form this reaction sometimes seems to be in danger of becoming a campaign against technology. To read some of the wilder comments made one would imagine that technology had produced a steady reduction in the quality of life over the last 50 years; and that we were all being actively poisoned to the point where the immediate survival of the species was now in question. This of course is a ludicrous overstatement of the nature of the problem. And it leaves out of account the fact that it is only by the application of technology that we can cope with the consequences of technological change that has already occurred. In fact, the application of technology has noticeably increased not only the standard and quality of life but the average length of life itself.

For example, the generation of electricity by nuclear or hydro-power is immensely cleaner than any previous method of providing light and heat. Domestic technology has liberated millions of women from hideous drudgery. Communications technology has given us all a wider range of experience, either by extending our capacity to move freely or to enjoy a fuller range of experience in our own home. Medical technology—in the form of drugs and appliances, surgical and other equipment—has not only lengthened the expectation of life but is making life itself more tolerable for the old, the sick and the disabled. Computer technology, too, has opened up possibilities of managing complex systems

without which we could not hope to control the power we have created.

No one wants to turn his back on all this. But what is now, quite rightly, under direct challenge is the tendency for an unthinking acceptance of everything that is scientifically exciting and technically within our capability; regardless of its social consequences.

Technology, like all power, is neutral and the question is how do we use it. It is the decision-making process that we are concerned with and how it can be improved. This is indeed one reason why the Ministry of Technology itself was set up.

What people want is a better method of decision making which takes into account the wider considerations that have in the past been left out of account. They are dissatisfied with narrow assessments that do not take sufficient account of the wider social and human consequences which may, in the end, turn out to be more important than any immediate short-term and purely economic gain. They also want the information and the time necessary to allow these wider factors to be publicly discussed before final decisions are made.

The time scale of technological development is so long and the costs so high that, if there is no discussion during the formative stage, people may wake up one day and find that major technological changes are well advanced and it is too late to stop them because so much money and effort has already been committed. This, more than almost anything else, gives people the feeling that their views count for nothing and drives them to the extremes of obscurantist opposition to technology itself or, worse still, to scientists and engineers in general. It is therefore to the decision-making process that we must look for improvements.

In fact the position is not as bad as it may sound. A great deal of progress has been made, but there is a great deal more to be done. As far as Government is concerned, the process of analysis must be widened at the very outset. Everyone working on, or promoting, new projects or processes must be actively encouraged to think much more widely about the social implications of what they are doing while they are actually engaged in their work, and so must all those in responsible positions throughout industry and society, who are concerned with technological change.

* *From an Address to the Manchester Technology Association at the Royal Society.*

The Rt Hon. Anthony Wedgwood Benn, *PC, MP, was Minister of Technology from 1966 to 1970, and since then opposition spokesman on trade and industry, and chairman of the Labour Party 1971/72. Educated at Westminster School and New College, Oxford, Mr Wedgwood Benn served as a pilot officer with the RAFVR from 1943-45 and in the RNVR from 1945-46 with the rank of sub-lieutenant. He joined the BBC North American Service in 1949 and a year later was elected to Parliament as the Member for Bristol South-East. After 10 years with the constituency he was debarred from sitting because he had succeeded to a peerage. His long fight to disclaim his title, which he won in 1963, enabled him to return to the House in that year. In 1964 he was made Postmaster General and he served in this capacity until his move to Mintech in 1971.*

These wider considerations must be independently examined by interdisciplinary groups to be composed of those with a wider range of human experience who can identify the people most likely to be affected. This whole exercise of initial proposition, evaluation and assessment must be made public so that everyone can join in, before the final decision is made.

Though these proposals sound quite mild when simply stated, they may require further fundamental changes in our decision-making machinery. All this must apply increasingly in industry as well.

The wider responsibilities which need to be opened up for managers, engineers and scientists could help to liberate them from the restrictions which may have hitherto limited their scope. Most of them are highly qualified, with a deep knowledge of, and concentration upon, their field of work or responsibility. No one wants to restrict or limit the specialist. Quite the reverse.

What we want is what Dr Paine, the NASA administrator, calls 'T-shaped men'. The vertical stem of the T reflects the deep knowledge of a special subject in which a man may study and work throughout his life. The horizontal cross-bar at the top of the T symbolises his wide and general interest over the whole range of human activities.

Interrogating the initiators

We shall have to devise some more comprehensive interrogation to which we can subject advocates of new industrial processes, methods or projects as they emerge from Government and industry. In recent years propositions of this kind have been investigated by increasingly sophisticated analysis as regards economic costs and benefits. But in addition to that we shall now have to push some fairly basic questions down the line and require them to be answered at the point of initiation, by the initiators. In this way, we may hope to establish an early warning system, which will widen the discussion at the outset.

The sort of questions we might want to ask in the case of a major development could include:

1. Would your project—if carried through—promise benefits to the community? if so, what are they, how will they be distributed and to whom and when would they accrue?

2. What disadvantages would you expect from your work? who would experience them? what, if any, remedies would correct them? and is the technology sufficiently advanced for the remedies to be available when the disadvantages began to accrue?

3. What demands would the development of your project make upon our resources of skilled manpower and are these likely to be available?

4. Is there a cheaper, simpler and less sophisticated way of achieving at least a part of your objective?

5. What new skills would have to be acquired by people using the innovation and how could these be created?

6. What skills would be rendered obsolete? how serious a problem would this create for the people who had them?

7. Is the work being done, or has it been done, in other parts of the world and what experience is available from abroad?

8. What disadvantages will accrue to the community and what are the alternatives?

9. If your proposition is accepted, what other work should be set in hand to cope with the consequences or to prepare for the next stage?

10. If a start is made, for how long will the option to stop remain open and how reversible will the decision be at various stages?

The real value of such an interrogation would be that, in order to answer the questions, the whole research team, board of management or sponsoring agency would have to engage in an intense discussion of a kind that does not always take place now. At least it doesn't take place in the widest context, even though in better-managed organisations the internal implications will be debated.

I am now considering whether such a procedure could be introduced into the process of evaluation within the Ministry of Technology itself, with a view to ensuring that major submissions, whether they come from outside or inside, should include an annexe containing answers to these or similar questions.

But this of course will not be sufficient by itself. The answers provided would, in many cases, have to be considered by a wider, inter-disciplinary group, including specialists in all the fields that would be affected by the innovation. These may require more information than has been furnished or identify greater benefits or more far-reaching side-effects than have been pin-pointed by the originators themselves. They would also have the task of identifying all those groups or interests which should be specifically consulted.

They may find further studies necessary and recommend that these be set in hand, either before the project is approved, or in parallel with it.

A think tank is not enough

This group would not be a 'Think Tank' into which to off-load the full responsibility for assessment. What is wrong to-day is that too few people are encouraged to think for themselves outside their own field. To seek to correct that by establishing 'think tanks' as new centres of specialisation could have the effect of encouraging even more people to believe that they need not think because thinking had been made the responsibility of others.

The assessment group, to be effective, has therefore to be made up of people who are themselves regularly engaged in their own work and who have been drawn out of it part-time to work with others for the purpose of the assessment. What we want is 'sabbatical' groups, made up of people who do not lose their contact with reality. Membership would be selected according to the problem that is to be thrown to the group, always changing while maintaining its inter-disciplinary character. I am considering how best we might organise such studies of major projects now coming forward.

But these two stages of interrogation and independent assessment are not enough. There is a third essential process —public discussion. Without full public discussion, decision-making would still become inbred.

People today are just not prepared to accept their own exclusion from the process of assessment. They do not accept that anyone should be able to secure the commitment of large chunks of scarce resources of qualified people and money without a public debate. And they are absolutely right.

For this whole process that I have been describing has, in fact, a much wider significance than may at first appear. It represents the demand by an ever-growing number of thinking people that the power of technology—by whomever exercised—be brought more effectively into the arena of public affairs and made subject to democratic decision-making.

Just as in earlier centuries the power of kings and feudal landowners was made subject to the crude and imperfect popular will as expressed in our primitive parliamentary system; and just as the new power created by the Industrial Revolution was tamed and shaped by the public which demanded universal franchise; so now the choices we make as between the alternatives opened up by technology have got to be exposed to far greater public scrutiny and subjected more completely to public decision, especially by those whose interests are most intimately affected. The case can best be demonstrated by considering the effect of choosing the opposite course.

If Parliament and the electorate were solemnly to decide that decisions involving technological judgments were so intrinsically complicated and specialised that it would be best to leave them to people who understood these subjects, we could reduce democracy to the discussion of those matters with which it has hitherto been preoccupied, the election of MPs and their work in economic and social fields where the choices are thought to be within the range of understanding of ordinary simple people.

This would throw away in a single decade all the gains that generations of people have struggled to achieve for public accountability. But if this is not to happen, the discussion process must be stimulated and made more real.

Public accountability

The development of the Ministry of Technology which brought most industrial sponsorships and the control of many public research resources together under a single Minister, accountable to Parliament, represented a significant shift in the right direction. The establishment of a Select Committee on Science and Technology in the House of Commons in recent years marked another important step forward and it has significantly altered the balance of power in favour of elected MPs. Quite properly it has kept my old Department's work under close examination. It has helped to educate Parliament in the new problems that confront us all and has brought the House more directly into contact with the decision-making processes in this field.

Another important development has been the Green Papers through which the last Government shared its thinking with the public before it committed itself to a firm policy decision. The Green Paper on Industrial Research was a perfect example of this completely new approach to the role of the public in our national policy for technology.

The Government actually invited people to give their views, and invited Parliament—whether through the Select Committee or not—to participate in policy making.

If this development is to be carried further there are two obstacles to be overcome. The first will be opposition from those who believe that technical decisions require such specialist knowledge that it would be foolish, dangerous and wrong to allow ordinary people to have a say in them. They argue that it would be disastrous if a nation, the majority of whom, by definition, have failed the 11 plus, should be allowed to decide things which can only be understood by PhDs, the chairmen of big corporations or their highly qualified teams of economists and technologists.

However superficially persuasive this argument may seem, it is in fact exactly the same argument as was used in the last century—and in this—against both universal suffrage and votes for women. Our technological policy is now the stuff of government and it must either be under democratic control or not: there is no middle course. To argue for the exclusion of these issues from popular control is not only fundamentally undemocratic but is also completely impracticable.

As the implication of scientific and technological decisions becomes more and more the subject of public interest, people will insist upon having a greater say in these decisions. If you were to try to shut them out from participation, public dissatisfaction could reach the same explosive proportions as it would have done if the vote had not been given to everyone.

But there is another reason too. The doctrine of control by experts is based upon an under-estimation of the general level of public intelligence and knowledge. Even with all their present, unacceptable defects, the educational system and the mass media have enormously raised the level of public understanding in the course of a single generation. The genie has got out of the bottle and he cannot ever be put back in again and the cork replaced.

The next obstacle—and it is a far more formidable one— lies in the minds of ordinary people themselves. Far too many still believe that they have not got the knowledge to make independent judgments on these matters or that, if they tried to do so, their efforts would be doomed to failure, because nobody really cares what they think.

This combination of lack of self-confidence and defeatism —both self-generated—must be overcome. You may have to be a brilliant surgeon before you can do a heart-transplant but you don't have to have any scientific qualifications at all to be able to reach a view as to whether the large sums of money and the medical research teams involved would be better employed in developing a better industrial health service to cut down the thousands of preventable deaths and disabilities that occur each year.

You may have to be a brilliant engineer to design a space capsule that will land on the moon, but you don't have to have any qualifications before you express the view that some of the money spent in space research might be better employed in improving the quality of public transport and the development of quicker, quieter, cleaner and more comfortable bus services or commuter trains.

Judgments of this kind may be difficult to reach but, if sufficient information about alternative strategies and their costs is more available, the choice between objectives can be made by anyone.

Anyone is perfectly qualified and fully entitled to contribute his opinion as to the purposes that technology should serve, even though he may know nothing about the first law of thermodynamics, nor be able to mend a blown fuse.

But if we are able to persuade people that they ought to be able to influence decisions and are qualified to do so, we still face the much more difficult job of overcoming their suspicion that, even if they were to make the attempt, it would be bound to fail because nobody cares two hoots what they think. This defeatism is borne out of past frustration. As people realise the significance of what has been done to give them more influence, confidence will slowly return.

The right to opt out — and in

Of course, there will always be people who genuinely don't want to participate, either because the issues don't interest them or because they don't feel strongly enough. The right to opt out, like the right to abstain in an election, is a fundamental human right and, if persuasion fails to change that view, no one should be compelled to join in. But if the right to decide is exercised, it carries with it the duty to become informed and study the evidence.

If people who want to join in effective discussion and decision-making are not able to do so, then they either become apathetic or they are driven into a frenzy of protest. Apathy and protest are both evidence of alienation. No society can be stable unless it provides the machinery for peaceful change and institutions capable of reflecting the desires of ordinary people.

In a democratic society like our own, the future must be shaped by the wishes of the community at large. Indeed to a remarkable extent it is. The social values which we have developed and the national temperament and character which have built up over the years will offer to the serious student a far better and more accurate explanation of what and why we are than can be obtained from the incessant scrutiny of the policies of those in authority.

This structure of values may be changed by political and public discussion, bringing new objectives into the forefront to replace old. The process by which people change their minds is at least as important as the shelf of printed statutes which record the legislative work of Parliament. We would, I think, all see the future more clearly if we studied the way that community thought has developed, rather than focusing exclusively on what scientists are capable of giving us, or what the big corporations put into their five-year plans, or even what Ministers and Shadow Ministers are putting before us by way of proposals.

But the task must not be limited to the vetting of technological proposals submitted to the community. We have also got to identify the main problems facing society and find ways and means of converting needs into real demands which can be met best by the use of technology.

There are already many thousands of human pressure groups or action groups in existence with proposals to do just this. Unless we can provide better facilities for these people to have access to those who might be able to solve their problems, we could miss one of the most important ways in which technology could be used for human benefit.

Indeed I suspect that the political leadership that has the most lasting effect exists not in the confrontations so beloved in Fleet Street, but in the stream of analysis, exploration interpretation and argument that slowly but surely changes the collective will.

What I am really saying is not at all new. It is no more than that the method we use to reach our decisions is at least as important as, if not more important than, the decisions themselves, and the expertise that lies behind them.

This, of course, is exactly what the Parliamentary System is all about. It is based upon the belief that how you govern yourself—by argument, election and accountability instead of thought-control, civil war and dictatorship—is what really matters.

I am, therefore, simply arguing that the methodology of self-government, based on the concept of talking our way through to decisions, must now be clearly extended to cover the whole area, at all levels, of technology which is in our century the source of new power, just as ownership of the land or the ownership of factories was in earlier times.

If we don't succeed in doing this, we shall run the risk of becoming robots. It won't be the machines themselves that make us robots but the fact that we have subcontracted our future to huge organisations. backed by their own resources of managerial, scientific or professional talent. Whether we do this or not is, I believe, one of the most important single issues in the whole area of public policy.

And before we can really make sense of it we need more national debate. Now, as we enter the 70s, with great decisions lying ahead of us, now is the time when that debate should begin.

People want technology—but it must be technology with a human face.

Our Responsibility to Mankind

by M. W. Thring, ScD, CEng, FIEE, FIMechE

The affluent society with ever-increasing waste and pollution— or the creative society which can maximise human happiness? The choice is ours and the engineer is better able to influence the result than others.

Criticism

The work of the engineer* is coming under increasing criticism in the press and public discussion for certain, often unintended, consequences of his actions, which affect adversely the life of the individual. Some of these consequences are the following.

Noise is increasing, especially near airports, roads and railways. In general, the machine-made noise with which the ordinary citizen has to live rises steadily year by year and is already almost intolerable in many places.

Chemical pollution of air is caused by partially-burnt fuels (soot and tarry, evil-smelling and carcinogenic by-products from hydrocarbons) by SO_2 and SO_3, by dusts, fumes and aerosols from many processes. Seas and rivers are polluted by oil-water emulsions, industrial effluents and now undecomposed insecticides, sewerage and even escaped poisons and radioactive wastes. Land, too, is polluted (eg, agricultural, parkland and beaches) by the fallout from air and water pollution, by refuse of all kinds, including derelict motor cars and refrigerators, and by the use of insecticides and fertilisers with harmful side-effects.

Accidents to humans occur in transport, sea-fishing, farm, factory and home; a large proportion of these could have been avoided if more engineering precautions had been included.

The development of mechanical, chemical and biological weapons of war of ever-increasing destructiveness to human life cannot take place without the assistance of engineers.

The continual thoughtless destruction of natural beauty, wildlife and beautiful old buildings and other historical relics is often associated with technology, and so is the rapid exhaustion of readily obtained mineral resources.

Until recently the engineer has replied to such criticisms by saying that he is obeying the dictates of the political or economic system of his country (doing what he is paid to do) or he may blame it on the population explosion. However, it is rapidly becoming clear that some group of people must call a halt to these processes before human life becomes intolerable. I shall try to show why I believe the engineer must give a lead in doing this and to indicate how he can set about it.

The first industrial revolution

In the last 200 years, the average industrial output of consumer goods produced as a result of one hour's work of one man or woman has gone up by a factor of the order of 10 in those countries which have been mechanised. Even on

*For the purposes of this paper, I propose to use the term 'engineer' in the broadest possible sense to include everyone concerned with the application of scientific and technical knowledge to the purposes of man's daily life.

farms, the increase is of the same order so that we now produce considerably more food per acre although a great proportion of labour has migrated to the towns. The result of this mechanisation has been that the hours of work in the lifetime of an ordinary person have roughly been halved and that the remaining factor of 5 has enabled the factory worker to buy 5 times as much in terms of quantity or quality. He can now have facilities that even princes could not have had before the first industrial revolution, such as telephones, television, cars, central heating, washing machines and good medical services.

Man has achieved this tenfold increase in output essentially by the production of primary and secondary power devices, machine tools and mass production lines. Unfortunately, at the same time, the work of the craftsmen who were personally responsible for building a whole windmill, coach or chest of drawers and so could take a pride in the finished article, has been largely replaced by work which is essentially repetitive on a short time-cycle. This work is so boring that people necessarily only do it for money and various devices have to be developed to prevent it driving them mad. Even here, therefore, the engineer's gift to the ordinary man has been accompanied by a loss of pleasure in craft work.

Professional ethics

It is certainly the engineer who was basically responsible for the first industrial revolution—his machines were quite essential to it whereas it could have happened with entirely different political, economic or legal systems. Moreover, the engineer can foresee more clearly than anyone else what types of machine can be developed in the future. For these reasons it seems clear that he, more than any other professional man, must give the lead in deciding what machines are to be developed in the next generation. He must accept that the results of his work affect the life of every individual human being so closely that he can by no means opt out of responsibility for choosing what he develops.

One can draw a parallel with the medical profession who accepted more than two thousand years ago that their work was so important to their patient that they must have a moral responsibility to him. They have been struggling with the problems of a professional ethic ever since and new and extremely complex problems are arising from medical and surgical technology, eg, heart transplants and genetic control. They have also had, and doubtless still have, members of their profession who have broken their Hippocratic oath, but nevertheless the idea of a medical professional ethic is so strong that a general would hesitate even in the

6

middle of a battle before he attempted to order his doctors not to care for enemy wounded.

The engineer affects the life of the individual not only in sickness but in all his activities day and night and, therefore, the need for a professional ethic is at least as great for him as it is for the doctor.

I suggest that we can take it as axiomatic that the professional ethic of the engineer must be to use his skills to serve the best interests of the individual man or woman, the world citizen. I shall now discuss how best to convert this broad principle into concrete action.

The affluent society

The present political and economic thinking is based on the assumption that the success of a man is judged primarily by his possessions. The assumption that the profit motive is the primary driving force of society leads logically and inevitably to the Affluent Society conditions illustrated in Fig. 1. The ordinate is the production of goods per man-hour, plotted on a logarithmic scale (with an approximate time scale) and the full curve represents the wealth, possessions and standard of living of the average man or woman measured in non-inflating money. In our affluent society the attempt is made to absorb the ever-increasing output per man-hour by using publicity to persuade people to buy all the goods produced, many of them gadgets for keeping up with the Jones's.

The solid curve can be extrapolated to the future. We can suppose that the engineer is likely to increase the production per man hour by a further factor of the order of 10 by the year 2020 and that, in the Affluent Society, the paid hours worked in a lifetime will stay at about their present figure. It follows that each individual will have to consume ten times as many manufactured goods in the year 2020 as he does at present.

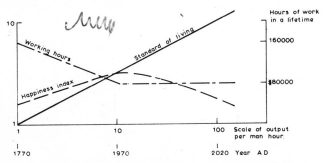

Fig. 1. Consumption stimulated by publicity does not necessarily increase happiness beyond an optimum value

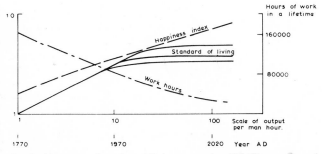

Fig. 2. His essential needs satisfied, man needs self-fulfilment as much as a higher standard of living. But creativity usually requires leisure

This brings us to the great defect of the Affluent Society: not only does it imply that all the bad consequences of the first industrial revolution will be accentuated but that two even worse ones will inevitably become more obvious.

Of the problems discussed above, ugliness, noise and exhaustion of mineral resources and wildlife will be particularly acute. Pollution and accidents may be checked by strong laws but these laws will be resisted by powerful lobbies, using the argument of international competitiveness and the techniques of advertising which are basic to the Affluent Society.

The two new difficulties which are only now becoming visible are, first, that however hard you advertise you cannot persuade everyone to buy many more goods than a certain optimum; you cannot eat too much meat and cannot drive two cars at once. In the USA there are even groups of people who reject all individual possessions. Thus, unless defence budgets or space races are enormously inflated, there inevitably comes a surplus of production and hence a steadily rising unemployment figure. The Affluent Society is in fact based on an untenable hypothesis and leads to chaos.

The second difficulty is even more serious. It is rapidly becoming quite clear that none of the developed countries have found a satisfactory way of giving real help to the developing countries. The inevitable consequence is that the latter fall steadily further behind in mechanisation, standard of living, health and education. This is largely due to the fact that the profit motive is completely inimical to such help. If this problem cannot be solved before the end of this century, it must inevitably lead to Armageddon.

The dotted curve in Fig. 1 is an attempt to express these consequences by supposing that we can find a numerical measure of the happiness of the individual in a society. It might be measured by his tendency to whistle or burst into song, to enjoy looking at trees, flowers or birds or, negatively, by suicides, deaths due to drugs and alcoholism or the breakup of family life. Probably this happiness index rose somewhat during the period of the first industrial revolution because constant fear of starvation, cold and disease are certainly detrimental to happiness, but it is clearly going to fall right back if society pursues the path of affluence to the bitter end.

One can therefore sum up the Affluent Society by saying that the attempt to base a system solely on the profit motive, more or less restrained by Government, leads inevitably to the enslavement of man by machines, money, politics and law.

The creative society

The only hope of controlling these forces is to base our society as much on a motive which is more real and important to man than his desire to acquire possessions. This is the motive of doing a self-imposed task really well because one feels it is worth doing. It feels worth doing, both because it uses one's talents to the full in a creative way and because one is convinced that it is of real value to other people.

Such a society can be called the Creative Society and its parameters are plotted in Fig. 2. It would be bi-motival because people would be judged and respected at least as much for the value of their creative work as for their wealth.

In this society the engineer's great gift to mankind (a further tenfold increase in output/man-hour) would be used, not primarily to give him more possessions than he wants but to shorten the working period so much that the bulk of man's energy is freed for creative activities he undertakes for their own sake.

The amount of work needed for the manufacture of consumer goods must go down by a factor of about five if the consumption of such goods is to reach equilibrium at not much more than twice the present level. This work will consist of the maintenance, repair and development of the robots and automatic machinery doing the manufacturing instead of the repetitive work so prevalent at the moment.

Since we must assume that normal people need to feel that their work is valuable to the community, we cannot pay most of them a dole to do nothing and hence the hours of work for everyone must come down to less than half the present, as illustrated in Fig. 2.

There are many other ways in which the extra productive capacity of the machines could be used in the Creative Society to give the individual a more satisfactory life in addition to freeing him from long hours of drudgery and toil. Some of these are discussed below.

Instead of making consumer goods for a short life, these would be made really well to last a lifetime (to conserve raw materials and reduce refuse), to be as safe as possible, to use as little fuel as possible, to avoid pollution of all kinds and to minimise the rape of the countryside. The engineer could certainly also develop a transport system combining all the advantages of private transport with those of a really comfortable convenient public system (rail and air), if he were free to use his inventive talents without the constant cry of the Affluent Society, 'we can't afford it'.

Ways could be found of using effectively all waste materials such as metals, glass, plastics and ceramics, by means of a pipeline distribution system for delivering goods to every house and collecting and sorting refuse. Sorting, concentration and utilisation of combustible refuse as fuels could all be done automatically. All these are essential if we are to preserve our environment in a world with increasing population, but all are blocked at the moment by short-sighted economic policies.

Not only would the engineer relieve people completely of repetitive work, by the development of programmable and adaptive robots, but also of all work in dangerous environments, such as underground, under the sea, in hot or radioactive places or in danger from accidental fires. In these cases 'telechiric' machines would enable the operator to sit in a safe, comfortable place, receive all the sensory impressions he would need from the danger spot, and then operate mechanical hands by remote control.

Human services, such as teaching, nursing, visiting sick or lonely people would be put at the top of the job-value scale instead of at the bottom as at present. Thus far more people would do them and more thought and effort would be given to the ways in which the life of the individual could be made more attractive than it has become in our overcrowded cities.

The engineer can produce machines to take all the menial and back-breaking tasks from nursing and machines to leave the human teachers free from all routine tasks so that they can concentrate on inspiring and helping pupils individually, which is the real task of teaching. He can also make devices to enable the elderly and crippled to do much more for themselves and he can help doctors and surgeons in many ways, even to the extent of solving their moral problems (eg, by providing artificial organs so that they do not have

Prof. M. W. Thring entered Trinity College, Cambridge, with a scholarship in Mathematics in 1934 and was awarded a Senior Scholarship in 1937. He worked on combustion of coal and coke and on producer gas for BCURA from 1937-46; was concerned with the development of the down jet combustion process and awarded the Students Medal of the Institute of Fuel in 1938. Appointed Head of the Physics Department, BISRA, in 1946, he established this Department in the Battersea Laboratories. He was instrumental in setting up the International Flame Research Foundation in 1948 of which he was appointed General Superintendent of Research in 1951—a post he held until 1971 when he was made Hon. Gen. Supt. From 1935 to 1964, he was Professor of Fuel Technology and Chemical Engineering at Sheffield and has since been Professor of Mechanical Engineering at Queen Mary College, London. He is now working on machines to help cripples and invalids, robots and telechine machines.

to take them from corpses). Great progress has already been made in this direction, for instance, an artificial kidney small enough to be worn externally.

However, this type of engineering development is stifled in our Affluent Society by the low priority given to such work by the Government, and by the fact that there is little or no industrial profit to be made out of it.

The Creative Society would be based on the fundamental satisfaction that every normal being can obtain from the full exercise of his or her talents in a really creative activity, whether done alone, as in the fine arts, or in a group or team. Many people, such as engineers developing new machines which improve the life of the individual, will find sufficient outlet for their creative urges through their paid work. Others, however, will be educated to enable them to find a creative hobby which they will do for its own satisfaction, ranging from gardening through hand-making of furniture to playwriting or acting. Here again the engineer could help to give everyone opportunities or facilities if we had our work priorities right, by providing tools, travel and transport devices and workshops.

The way in which man's natural desire for self-fulfilment through useful creative work can be realised in a Creative Society is a subject far greater than can be worked out in a short article. Enough has been said, however, to indicate that there are two possible extremes of the future of mankind: one that will make life for the individual steadily less tolerable and the other which will completely reverse the trend. What is equally certain is that we shall automatically drift into the Affluent Society with all its horrors, unless millions of people all over the world consciously decide to make the large and continual effort necessary to remotivate society.

Clearly all thoughtful people have a contribution to make, especially those professional people who organise the four masters of mankind who have to be turned into his slaves: machines money, politics and law.

Equally, all those concerned with education and with the moulding of public opinion have to strengthen and canalise the idealism which is natural to all normal people and teach them to respect creative activity above possessions. However, at some point it is necessary to break into the automatic feedback of inadequate motives.

I submit that it is the engineering profession that can most easily make this break because the engineer has the most clear-cut decision to make between the consequences of his work which are harmful and others which are beneficial.

The moral scale of machines

The engineer, indeed anyone concerned with applying science to human needs, is inevitably brought up against the fact that his work may have consequences which are actively harmful to some. He can range most of his work according to its effects on individuals according to a moral scale of machines.

Right at the top must come all those machines primarily designed to increase the possibilities of a satisfactory, self-fulfilling creative life for the individual. In this category come all the machines that help education, communication of ideas and travel; those that help people in creative work and also those that help to restore normal physical abilities to sick, maimed or elderly people. These might be called the human service machines and it should be the aim of every engineer to devote as much of his working life as possible to this and the next category.

Still high on the scale of human service come all the machines designed to provide human beings with an adequate standard of living without their having to do excessively long hours, repetitive or dangerous work. In the developed countries this means robots and telechiric machines. In the developing countries, the engineer has special responsibility for inventing machines that really can give everyone a decent standard of living without introducing the bad features of the industrial revolution that are causing so much misery here. I would stress the need to avoid destroying the heritage of craft skills that one finds among peasant populations all over the world.

Examples of such machines are: a mechanical elephant to carry heavy loads across rough country with small fuel consumption; machines using solar energy for lifting fresh water or distilling salt water; and machines for punching holes in the ground to aerate it without ploughing.

The third category comes just above the line that divides beneficial from harmful machines. These machines are of value to the individual, not because they increase his possibilities for a self-fulfilling life, but because it is fashionable for him or her to have them. It includes all advertising machines, fashion and cosmetic machines, unnecessary gadgets and unnecessarily large or flashy machines. None of these are harmful except that they tend to increase the harmful consequences below the dividing line.

Table 1—A Hippocratic Oath for Engineers

I vow to strive to apply my professional skills only to projects which, after conscientious examination, I believe to contribute to the goal of co-existence of all human beings in peace, human dignity and self-fulfilment.

I believe that this goal requires the provision of an adequate supply of the necessities of life (good food, air, water, clothing and housing, access to natural and man-made beauty), education and opportunities to enable each person to work out for himself his life objectives and to develop creativeness and skill in the use of the hands as well as the head.

I vow to struggle through my work to minimise danger, noise, strain or invasion of privacy of the individual: pollution of earth, air or water, destruction of natural beauty, mineral resources and wildlife.

Machines specifically designed to hurt or kill humans come below the dividing line and so do all harmful by-products: pollution, ugliness, premature exhaustion of mineral resources, accidents and noise.

What the individual engineer can do

Ultimately, all the engineers and applied scientists of the world will have to agree so strongly on a professional ethic that, if any one of them insists on applying it, he cannot be penalised because all will back him. If they cannot do so, the machines they make will destroy man. Such a professional ethic might be along the lines of the draft 'Hippocratic' oath in Table 1.

However, that day is still far away and in the meantime each individual engineer has to face his own problems with only his own conscience to help him. This means he will be penalised if he says to his employer 'I refuse to make such a machine because I believe its net effect on the ordinary human being will be harmful (eg, by noise or pollution)' or 'I insist on building a more expensive machine which reduces harmful by-product effects or dangers'. If he says 'I believe we should develop a new system of public transport because it will alleviate the discomforts of the city rush hour travel but it is quite uneconomic', he cannot get the necessary money, whereas quite different criteria are applied to defence projects.

Basically, therefore, he has to apply the dictates of his conscience in three ways. First, he can choose the job he enters according to his estimate of how far up the moral scale he will be working. Everyone has moments of choice between different jobs. These are not easy; for example, one man will feel that the development of war machines is genuinely necessary for the good of his fellow countrymen, while the conscience of another will not allow him to work on such machines under any circumstances. Nevertheless, each man must clearly face the fact that his choice involves him in an overall moral judgment which he must allow his conscience to decide.

Second, in any job there are smaller but still very important moments of choice where the engineer can exert influence according to the decisions of his own conscience. For example, he can struggle to find ways of minimising atmospheric pollution or noise which are not too costly, or he can make special efforts to foresee all possible types of accident and to build in safety devices to minimise these. The higher he rises on the professional ladder, the greater his opportunity will be to look after the interests of the ordinary human being. This is why it is important that we should make it very clear to young people choosing their careers that the engineer has unique opportunities to serve these interests.

Thirdly, I submit that it is the responsibility of every thoughtful engineer, whose conscience drives him to feel that these matters are vitally important, to take every possible opportunity to try to rectify the whole use to which engineering is being put. If we are to save civilisation from destruction by its own machines, we must re-think the whole basis on which we choose what machines we will develop. So we have to publish papers, give lectures and discuss the ethical basis of engineering with the younger generation, strive for a greater influence in long term politico-engineering planning and use publicity media to develop a whole new philosophy ensure that machines remain our slaves rather than become our masters.

Why We Need a Code of Ethics

by K. W. Matthews, BA

The question of a Hippocratic code for engineers is raised every time a technical failure involves loss of life; it also arises in connection with the engineer's standing in society. Is it their code of conduct which gives doctors and lawyers their prestige? Perhaps we can see the pros and cons more clearly through the eyes of a man who, while not an engineer himself, has long been associated with our profession.

Mr K. W. Matthews was educated at Tonbridge School, Kent, and Trinity College, Oxford, where he graduated with first class honours in the School of Modern History in 1937. Passing the first division Civil Service examination in 1937, he joined the Admiralty as Assistant Principal, Military and Air Branches, and from 1944 to 1946 worked in the Commission and Warrant Branch. He served in the Forces from 1940–44, becoming a Lieutenant in the Royal Artillery. From 1946 until 1952 he worked with the Control Commission for Germany as Deputy Chief, Displaced Persons Division. After a spell at the Ministry of Supply, he became Secretary of the Radar Research Establishment in 1953 and was Assistant Secretary of the Air Division, Ministry of Supply, 1957–58. Since then he has been Head of Administration, at the Dounreay Establishment of the UKAEA. His chief hobbies are farming and ceramics.

Should the profession have an overall code of ethics? A few argue the point passionately but does the majority care? Yet, a great deal may depend on the answer which must eventually be found to this problem.

The case against a code

The viewpoint of the average engineer is perhaps something like this. What is good or bad engineering is well enough understood, let alone the general rules of honest conduct. There is no evidence that the country is not excellently served by its engineers or that there exist among them more than the usual number of rogues and weaklings to be found in any section of the community. The institutions who all profess ethical aims have done a good job in promoting and defending standards.

Besides, any code of ethics can be no more than a few pious generalities, not often applicable to the individual situation. Engineers work in every kind of circumstance and for all sorts of people. The profession is not like that of medicine where the doctor–patient relationship and its difficulties are universal and so general rules of conduct may be framed and uniformly applied. Nor have engineers the power of the legal profession which, by denying backsliders the right to practice before the Courts, can exercise a real sanction and enforce it by its own machinery.

It is true that some other countries have codes and find

them of enduring value. But their history is quite different from that of Britain. They are or have been emergent nations, compressing their development within comparatively few generations; frontier countries which have taken in large numbers of immigrants from all over the World to whom they have had to teach much within a short time. These are the sort of circumstances that call for written constitutions and statements of aims to guide and convince those with ideas on ethics still deriving from their different places of origin; who must be encouraged, even forced, into accepting the unifying ethics of their adopted countries.

By contrast the United Kingdom's development has been slower and more stable. Ethics have been bred into the nation, are well known and any attempt at codification is unnecessary, maybe even dangerous. The rigidity of a code could easily become rigor mortis. It is alien to a living and successful tradition.

Furthermore, there is no evidence that the engineers of countries with written codes are more upright or more highly regarded than those of Britain.

Protective aspects

Unfortunately this line of thinking seems to be reinforced by the fact that much counter-argument in favour of a code is necessarily protective and thus, to many ears, negative and uninspiring. The public need protection against charlatans and incompetents. Engineers need protection against one another lest, for example, they poach each others' jobs. Engineers also need protection against employers whose advantage may not always be served by too nice an integrity among their engineers.

All these things are true but they are only limited to odd cases. Nor, in all probability, would a code help very much even in these cases, for it is often difficult to establish the full facts. Any decision given under a code must rest on a value judgement that will seem to many to be arbitrary and of little use for general guidance.

Moreover, the critic continues to reason, engineering is a practical world where, if a man does badly, he suffers an immediate penalty in his personal circumstances. It is quite wrong to put the profession on trial in the person of some allegedly errant member of it. If an engineer is oppressed by his employer, he can always find another or, in an acute matter, he has, if he deserves it, the protection of the law. His institution or the Engineer's Guild will help him fight if his cause is good.

There is no gain in superimposing on him a defensive code

Table 1. A Tentative Code for Engineers

1. Whether actively engaged in engineering or not, a chartered engineer will regard the integrity of the profession as his primary loyalty: his conduct will always be regulated accordingly.

2. In no matter touching on the profession will he conceal truth or lend himself to any deception of other engineers or anyone engaged in, or making use of, engineering services.

3. His constant endeavour will be to keep himself abreast of discoveries and developments, especially in his particular field, so that his own work may advance engineering knowledge and its application for the benefit of humanity.

4. He will accept no post, position or assignment unless he is himself satisfied that his capacity and experience are equal to the responsibilities proposed.

5. He will seek to understand the duties and responsibilities of other professions as he may be concerned with them, so that any joint responsibility he may have with them can be properly weighed: conversely he will try to impart to them an equal understanding of his own work.

6. He should promote the corporate life and fellowship of the profession, especially by participation in the work of his own institution and, where possible and appropriate, in that of other institutions.

7. He should allow no criticism of the profession made in his presence which he regards as unjust to pass without challenge.

8. He will neither accept nor give any technical instruction that in his own professional opinion is not based on sound engineering principles or practice.

9. He will pass no drawing or specification that is not in his judgement complete and, where appropriate, authenticated.

10. He should always be alert for any risk to structures and installations or threat to the safety or health of anyone working in, or affected by, any project on which he is engaged. He must not allow himself to be party to any decision that in his view involves an unacceptable engineering hazard in the project concerned.

11. He will ensure that the provisions of all statutes and regulations pertinent to whatever work he is doing are met in full.

12. When working under superiors who are not engineers he must endeavour to put necessary engineering considerations to them in terms they can understand.

13. He will uphold the right of any professional engineer—including his own subordinates—to express an unfettered professional opinion on a technical subject within such engineer's competence and his further right to have this opinion formally recorded if he so wishes.

14. He has the responsibility to ensure that all professional engineers serving under him are properly employed on duties appropriate to their status and, where necessary, trained and guided to the full use of their professional knowledge for the benefit of society.

15. He will also be concerned for the efficiency, utilisation and development of all other persons serving under him, whether technicians, supervisors, craftsmen or unskilled; in particular he will promote the highest standards of craftsmanship and general engineering work.

16. He will strive for the continual betterment of industrial relations and will make himself fully conversant with the agreements and working practices accepted within any area of work on which he is engaged.

17. In pursuing personal claims for advancement, remuneration, posts or other improvements to his own circumstances he shall not deny or infringe upon the rights of his fellow engineers or conduct himself in any way such as to bring the profession into disrepute.

18. He shall not use his professional standing to promote his private interests in any non-professional matter save for the normally accepted purposes of everyday life.

19. He shall always positively endeavour to uphold this code himself and by his example to encourage his fellow professional engineers to keep it inviolate and alive.

20. If any instance of serious unworthy behaviour by another professional engineer comes to his notice or he has reason to believe that observance of this code may be endangered by outside pressures or, in his opinion, any part of this code is inadequate or out of date, he will consider whether he has not a duty to bring his view to the attention of the appropriate institution or its' Council.

that will subject him to an arbitrary yardstick in an attempt to secure some kind of uniformity at the expense of the flexibility needed for an infinite number of variable circumstances.

The wider issues

What, however, this reasoning fails to recognise is that the deepest problems of today's life are ethical. At a time when every accepted standard and established belief are under fundamental challenge, can the engineer stand aside? It is no longer the comforting fact that the morality which once underlay the ethics of all professions remains an instinct of the British people.

The reverence for exact craftsmanship, the acceptance that an engineer must build the best and the most enduring, these are no longer prime considerations. The dominance of economics, the struggle for exports can mean grievous assaults on professional integrity. No-one now believes that British workmanship has anything particular to commend it. Engineers must engage in management, in sales, in advertising and in other activities which certainly each have their own ethics but which, equally certainly, are sometimes in conflict with the engineer s professional standards.

"—allow no criticism of the profession . . . which he regards as unjust to pass without challenge"

The opportunities provided by modern technology are immense but in every advance of human knowledge lie temptations and dangers. In the great uprush of new professions and quasi-professions lies another risk that engineering may be depreciated.

The difficulty in recruiting to engineering and science may stem from failure to meet competition from other professions, rather than from alleged faults in the educational system. There is even fear at the other extreme that technology and democracy are incompatible.

Does not all this impose a responsibility on engineers that they can only meet corporately? In the Middle Ages, when social and political life were chronically disrupted, the growth of civilisation was significantly protected by the ethical traditions of the craft guilds: we have a similar responsibility to society today.

If this is true reasoning, there is indeed a need for an ethical code, but a positive one, a restatement of the profession's aims and its relationship with society. Such a restatement should centre on the search for truth and emphasise the concept of service.

Even when personally engaged in non-engineering matters, a professional engineer should regard the profession's integrity as his first loyalty. When in charge of men he should seek to promote human dignity and an increasing harmony of relationships throughout industry. He should always seek to attract the youth of the country to engineering, professional or other sections, and regard training as among his most important preoccupations.

The improvement of standards, his own as well as others, should equally be his responsibility. It might be prudent to enjoin upon him, for all other professions, the respect which he rightfully demands in return from them. To this, of course, must be added the more technical injunctions that each institution may deem necessary.

While the idea of such a code is widely respected, the crucial question remains whether it is practicable. Prolonged discussion and expert drafting will be needed to arrive at a final version but the 20 rules proposed in Table 1 are an attempt to set the ball rolling.

They are deliberately drawn in wide terms to cover the whole field and are each intended to be reasonably self evident.

I have differentiated between what a man can and must do, as opposed to what he can only try to do: this point is fundamental. I hope that, in my proposed code, a tolerable freedom for individual conscience has, in principle, been allowed for.

The guiding intention has been to demand no more from the professional engineer than must be imposed on all professions if service to, and the progress of, the whole community are to be the aims.

Conclusions

These are only outlines of ideas. Some of them may be mistaken; there are many others that could be developed on these lines.

The important point is that a positive code could be drawn up as a permanent inspiration for engineers. It would also make their place in the community challengingly clear to all. It would seek to establish simple principles; it must be neither obscure nor legalistic, which are the attributes of the protective mind at work. It should be aimed at guidance and encouragement, not punishment.

Could it nevertheless be enforced where necessary? The only requisite is that the general body of professional engineers shall wish it to be enforced. No law is enforceable if the consensus of those who should uphold it is against it. There should be no need to bother with small infractions and minor penalties; no-one always lives up to his standards. Only in cases of a major breach or a clear, continuing failure would the institutions resort to warning or expulsion. If they and the code have the confidence of the country's engineers, they will know when and how to act.

As a safeguard against mistakes or loss of touch, the code should be regularly reviewed with the aid of wide consultation. If the tradition is sound and alive, majority opinion in the profession can make itself felt. Wherever else democracy may fail it should never be in the case of the engineering profession.

Much is said and written, these days, about the community's image of the engineer. There does seem to be evidence to suggest that trust is more readily given to a body with clearly defined aims and methods, with pride and self-confidence in its bearing and self-discipline enforcing a high standard of which the general public as well as the profession itself can understand.

Otherwise, however, important and benevolent such a body of men may be, it tends to be regarded as something of a secret society and this can be dangerous.

Automatic Control—
What about People?

by HRH Prince Philip,
Duke of Edinburgh (Hon. MIMechE)

*This speech was made at the opening of the
third Congress of the International
Federation of Automatic Control in London.
It is included here because it shows clearly how
someone well-placed to observe society as a whole sees
the activities of engineers.*

One of the hazards of inviting an amateur to open the proceedings of such a highly technical conference as this, is that members may have to endure a lot of elementary rubbish, a string of excruciating platitudes or, at best, a brave attempt by the speaker to make it sound as if he understood the technical jargon he has extracted from his brief. You will probably find all three in what I am going to say.

The International Federation of Automatic Control exists to further the development and application throughout the world of automatic control. The assumption being that automatic control improves industrial efficiency and industrial efficiency makes an important contribution to national prosperity.

I think this is a perfectly reasonable assumption but there is one glaring omission, there is no mention of people and in the long run every step in technological progress has got to be related to human existence.

As it is, the industrialised countries of the world are just beginning to face the human consequences of the industrial age with its problems of mass employment, urban sprawl, population explosion, mobility and insecurity. I freely admit that all this must be set against the very obvious benefits which modern technology and industry have brought to the individual but merely to exchange one lot of awkward problems for another cannot be described as progress.

Automatic control is part of the general development of the industrial system but in itself it has a direct impact on people and the industrial community. Popular opinion, of course, has it that automatic control merely reduces the number of people employed and therefore it is a cause of unemployment. Figures bear this out. In this country the labour employed for the same output has been reduced to between one tenth and one fifteenth since the turn of the century.

The point, of course, is in the words 'for the same output'. Without technological **advance** and automatic control we would need a labour **force** of hundreds of millions to produce the volume of consumer **goods** which we now take for granted.

In fact if we look at it from the opposite point of view, only technological advance and increasing productivity in industry make it possible to have a world population limited to a reasonable size and able to make and enjoy the products of modern industry and technology.

Automatic control has a large element of decision making in relation to the operation which is being controlled. In many other areas computers are being used increasingly to arrive at what is hoped will be rational decisions. This is to be welcomed provided we don't make the existence of computers and scientific research an excuse to give up thought and judgment.

Policy decisions have still to be made by people for people. It is the relationship between people with particular responsibilities and special interests which eventually controls the ability of any organisation to arrive at the necessary decisions.

The growing use of automatic control will obviously affect many other areas of human activity but I think one of the most important will be the effect it has on the sort of work people will be required to do in industry and the programme training which this will require.

The automatic control industry itself, and all the commercial and industrial organisations which use its techniques, demands a whole new range of technical skills and intellectual qualifications. If automatic control is to prosper the programme of basic training, the structure of qualification and the opportunities for retraining cannot be ignored or neglected.

As I said, I am an amateur, but I have some rather special opportunities as a spectator in the industrial field. From all I have seen it would be my guess that automatic control is still in its infancy in comparison with what lies ahead. The path it follows and the way in which it is integrated into our human civilisation will depend on the discussions and the general guidance given by this international organisation.

I hope the third Congress of the International Federation of Automatic Control is a great success and that all the delegates will return home refreshed and inspired by new ideas, new contacts and friendships.

The Professions in our National Life*

by K. H. Platt, CBE, BSc, CEng, FIMechE

What distinguishes the professions from other occupations is the nature of the services they render. Unlike most other professions, qualified engineers in Britain do not enjoy a monopoly, nor even a legally protected title. Is there a case for state registration or other action to remedy this?

In their book *Professional People*† Lewis and Maude wrote:

> The history of the professions is the history of specialization—and of the realization that breadth of experience, liberality of education, and an understanding of fundamentals must somehow be preserved in professional training if the specialist is to be adequate to his task.

In the early Middle Ages, when the professions of Medicine and Law emerged, the Church was the dominating influence and every human activity was related to the spiritual theme; but it is doubtful today whether it is possible to find a common theme which binds all professional people together, imparting unity and purpose to their varied efforts.

Historically, the professions came to be identified with those people who provided the skilled intellectual services which society required. Prior to the eighteenth century they consisted of physicians, surgeons, ecclesiastics, civil servants, lawyers, teachers and apothecaries. Then came the architects. In the late eighteenth century, scientific knowledge had reached such proportions that its application to practical problems became more general. This development was reinforced by favourable social and industrial conditions and, as a consequence, new professions arose.

In this development engineering was basic. Engineers made large-scale industrial organization possible and this, in turn, gave rise to other professions founded on new intellectual techniques; for example, accountants, secretaries, surveyors, and so on.

Learned societies

The practitioners of these new techniques desired social intercourse with those who were doing the same work and facing the same problems. Dining clubs were formed where 'shop' was talked; the earliest example of this in engineering was the Smeatonian Society. Formal meetings for the presentation and discussion of papers followed. One of the first objects of such groups, therefore, was the promotion of study activities; they became 'learned societies'.

However, meeting for such purposes, they found that they also had other things in common. They adopted titles which, to themselves, implied competence in their own sphere.

But the general public was not so knowledgeable and so the skilled practitioners wanted the competent to be distinguished in the public mind, and protected, from those not qualified. Admission to the professional associations was therefore restricted, to gain prestige for the members.

* Secretary, Institution of Mechanical Engineers.

†*Professional People*, by Roy Lewis and Angus Maude, 1952 Phoenix House, London, on which much of the subsequent information is based.

At the same time standards of competence were raised and methods of testing them introduced or improved. Eventually the examination system was adopted.

But, just as the general public had failed to distinguish between the competent and the incompetent, so they also failed to distinguish honourable from dishonourable practitioners. This led the former to draw up a code of ethics. Inevitably interest spread to remuneration, for a certain level of income implied public recognition of status and the best way to obtain the latter was to press for the former; thus protection of members interests became another activity.

More activities followed; concern about standards of competence led to an interest in educational policy, for example, and so the associations became active in this field. They were not only ready to help and advise, they wanted to take the initiative by presenting their views and influencing public policy in such matters as affected them.

The history of the growth of a typical professional association since the industrial revolution may thus be summarized by the following five stages:

1. Study activities, following 'social' meetings
2. Standards of competence
3. Code of ethics
4. Protection of interests
5. Public activities.

The General Medical Council

The importance attached to each of these five activities has varied from profession to profession. In the medical profession there was an early segregation of specialists from general practitioners, the former trying to keep the latter subservient. Each group however, eventually came to understand its true place, the result being that the doctors became centralized, achieving unity and monopoly in their profession.

Today the General Medical Council consists of 47 members of which 8 are nominated by Her Majesty, 28 appointed by universities and 11 elected by fully registered persons. Only 3 are non-registered. The duties of this Council are to register practitioners with adequate experience and to supervise and report to the Privy Council on the instruction and examination of medical students. This is done by inspectors who attend but may not interfere.

There is a right of appeal to the Privy Council against refusal by the General Council to register.

Registered practitioners receive certain privileges in that no person is entitled to recover fees unless registered; certain appointments are reserved for them; and they are exempted from liability to certain offices.

There is a Disciplinary Committee which may direct the

name of a registered person to be erased if he or she has been convicted of any crime or offence, or if found guilty of infamous conduct in any professional respect. Again there is a right of appeal against erasure to the Judicial Committee of the Privy Council.

The architects

By way of comparison, the Royal Institute of British Architects was founded in 1834 and the R.I.B.A. Board of Architectural Education was created in 1910. Even after this date there still remained practising architects who were not members, who had not passed its examination, and who were not bound by its Code of Conduct. Eventually the Architects Registration Acts were passed in 1931 and 1938 to prevent any person practising under the title 'architect' unless his name was on the National Register of Architects. Over 20 000 are on this Register, and of those nearly 17 000 are corporate members of the R.I.B.A.

The present distribution of the profession by occupation is 50 per cent in private practice (30 per cent being principals and 20 per cent salaried assistants); 40 per cent in central and local government and in nationalized industries; 6 per cent in private firms; and 4 per cent in teaching and other architectural employment. Thus, 70 per cent of registered architects are salaried employees.

Under the Architects Registration Council there are three statutory bodies: the Board of Architectural Education whose duty is to recommend exempting qualification to the Council; the Admission Committee, which considers applicatons for registration and reports to the Council; and the Discipline Committee with the duty to inquire, when directed by the Council, into any case of alleged disgraceful conduct by an architect.

Removal of a name from the Register is in the power of the Registration Council which is composed of 3 nominees of the Council of the R.I.B.A., 10 of other architectural associations; one of the Association of Building Technicians; 5 'unattached' architects; 5 in Government Departments; and 3 from engineering bodies (Municipals, Structurals and the Society of Engineers); and one each from four other bodies. Not more than a quarter of the Council can consist of non-architects.

The engineers

The engineers, like the doctors, had to face the problem of specialization which arose for them because of the variety of applications of the basic sciences involved in all of them. Concentrating, as they have done, on 'learned society' activities and the hall-marking of their members for competence in the sub-divisions concerned, protection of status and interests were neglected and unity and monopoly have not been achieved.

Is there a lesson to be learned here: that, whilst preserving the many-sided learned society activities, unity should be sought for the profession as a whole in its approach to other activities?

In their book *The Professions*,* Carr-Saunders and Wilson pointed out that while no simple test can be devised to distinguish clearly those vocations which are professional from those which are not, certain characteristic features are common to the acknowledged professions. These are: a technique founded upon a basic field of enquiry and acquired by long specialized intellectual training, enabling them

The Professions, by A. M. Carr-Saunders and P. A. Wilson, 1933, Clarendon Press.

to render a particular service to the community; a sense of responsibility, manifested in a concern for competence and integrity; machinery for imposing tests and enforcing standards of conduct; and service rendered for a fixed fee or salary.

Material considerations of income and status are not neglected, but they are subservient to the needs of society. There is an implied contract to serve society over and beyond all other specific duty. The outstanding feature is, however, the possession of a specialized technique.

It is interesting to compare with this the definitions of a professional engineer and of an engineering technician, produced by the engineering institutions of Western Europe and the U.S. (EUSEC) and issued by our Council for the guidance of those sponsoring applications for Corporate Membership of this Institution.

> A professional engineer is competent by virtue of his fundamental education and training to apply the scientific method and outlook to the analysis and solution of engineering problems. He is able to assume personal responsibility for the development and application of engineering science and knowledge, notably in research, designing, construction, manufacturing, superintending, managing and in the education of the engineer. His work is predominantly intellectual and varied, and not of a routine mental or physical character. It requires the exercise of original thought and judgment and the ability to supervise the technical and administrative work of others.
>
> His education will have been such as to make him capable of closely and continuously following progress in his branch of engineering science by consulting newly published work on a world-wide basis, assimilating such information and applying it independently. He is thus placed in a position to make contributions to the development of engineering science or its applications.
>
> His education and training will have been such that he will have acquired a broad and general appreciation of engineering science as well as a thorough insight into the special features of his own Branch. In due time he will be able to give authoritative technical advice, and to assume responsibility for the direction of important tasks in his branch.

and, regarding technicians:

> An engineering technician is one who can apply in a responsible manner proven techniques which are commonly understood by those who are expert in a branch of engineering, or those techniques specially prescribed by professional engineers.
>
> Under general professional engineering direction, or following established engineering techniques, he is capable of carrying out duties which may be found among the list of examples set out below.
>
> In carrying out many of these duties, competent supervision of the work of skilled craftsmen will be necessary. The techniques employed demand acquired experience and knowledge of a particular branch of engineering, combined with the ability to work out the details of a task in the light of well-established practice.
>
> An engineering technician requires an education and training sufficient to enable him to understand the reasons for and purposes of the operations for which he is responsible.

A list of typical types of work for technicians is then given.

The image of our profession

The EUSEC definition of a professional engineer is excellent as far as it goes. It is understood and accepted by those who think carefully about engineering, and it forms the basis for entry to corporate membership of many engineering institutions. But does it go far enough? Is it understood clearly by the general public? Is it accepted by those who employ the services of professional engineers?

Two prominent features of engineering today pose problems which call for remedial action. First there is confusion among the general public and in the lay press about what a professional engineer is and does. This leads to a false image of the engineering profession, one evil consequence of which is that it prevents us from attracting a fair share of the best young brains. Another consequence is that credit which often should fairly be given to the engineer goes in fact to the scientist.

Secondly, there is great need to clarify the position of technicians in relation to professional engineers; their education and training, and the corporate facilities which they need. A rising standard of living and a corresponding expansion of industry have called for more and more experts of various kinds. Not all of these specialists have been able to qualify for the older professional institutions, so some of them have set up their own associations, often patterned on the older bodies, in an effort to give their activities the formal status of a profession.

Many who are qualified to judge consider that the remedy is for professional engineers to seek the legal status provided by state registration.

State registration

In the Middle Ages all vocations, whether professional or not, were regulated by the State, as was social life generally. The objects were: to ensure competence, honest dealing with clients, fair dealing with those of inferior status, maintenance of discipline, and proper supervision (e.g., in medicine, in connexion with drugs). The State was not always successful in achieving all this. In any case, by the sixteenth century, more individual freedom was developing and gradually State intervention ceased.

Much later, in the nineteenth century, the State had again to intervene though reluctantly and with hesitation. Intervention took the form of registration to provide a list of practitioners which included only those of proved competence: to those certain professional functions were reserved.

Professions which have been thus regulated fall into five groups:
1. where the service rendered is vital (e.g., Medicine);
2. where the service is fiduciary in a marked sense (e.g., Law);
3. when public safety is intimately concerned (e.g., Merchant Navy and mine management);
4. when the State is an employer of professional people and sets up tests of competency. Those eligible for employment constitute in effect a registered group (civil servants);
5. in certain cases (e.g. architects) practitioners have themselves sought regulation, expecting thereby self-government and certain privileges. In the case of the architects, Parliament, with doubt and hesitation, conceded their demand.

Registration has had profound effects. It implies statutory authority for the body which makes the rules governing admission and expulsion. Such a body virtually becomes an organ of the State which governs the profession concerned. The registration authority may take one of three forms: a Government Department; a professional association; or a special 'ad hoc' body set up by Parliament. Thus, for patent agents the authority is the Board of Trade which, however, delegates some of its authority to the Chartered Institute. For the legal profession it is the Law Society. In the medical profession a special 'ad hoc' body was created, the General Medical Council; and architects, too, have a special body, as has been shown above.

The object of the register is to distinguish the qualified from the unqualified. Prohibition of the unqualified may be general or specific. In the case of the patent agents, for example, the functions of the practitioner are not specified; but whatever they may be, unqualified persons must not perform any of them; it is in fact a closed profession with a valuable monopoly. Even if the profession is not closed, however, monopolistic advantages accrue as a result of registration, for this carries with it prestige and standing.

Protection of title helps towards monopoly, but can only be obtained by statute. An Act of Parliament is required to prevent an unqualified person using a particular title to which he is not entitled.

What professions should be registered? If the public can easily make the distinction between qualified and unqualified, it may be said that State intervention is not necessary. The State is only justified in assuming responsibility in those cases where the public is unable, or has not the opportunity, to distinguish this for itself, or where the consequences of consulting an unqualified man are very serious; that is to say, in those professions where the service is vital or fiduciary to a marked degree, and demands a prolonged intellectual training of a specialized kind.

Other countries

Many other countries have found it desirable to introduce some form of protection for the engineering profession. In Canada, U.S.A., Belgium, Greece, Italy and Spain the professional practice of engineering is, to some extent, regulated by law. In Austria, France, Norway and Switzerland the special title awarded to recognized professional engineers is protected by law.

In Canada legal registration is carried out by the Provincial Government. The title 'engineer' is fully protected, and may be used only by those who are registered by the licensing authority. Certain functions may only be exercised by licensed persons.

In the U.S.A. legal registration is carried out in each state, by a board of professional engineers. The principle of registration has been generally accepted only in the last 30 years. About 227 000 out of 400 000 eligible engineers are registered. The legal definition of a 'professional engineer' varies from state to state but, in general, the principles of the EUSEC definitions given above are accepted. Generally, the legal requirements apply only to those responsible for services involving the health and welfare of the community. Plans which require the approval of public authorities must be prepared by, or under the supervision of, registered professional engineers. The title 'professional engineer' is restricted to those who are licensed.

There is a federation of national engineering organizations called the Engineers' Council for Professional Development (ECPD) one of whose main functions is the accreditation of engineering curricula. Graduation from an accredited college is an important element in the acquisition of a licence.

In Belgium the title 'Ingénieur civil' is protected by law and conferred only by certain educational establishments. There are penalties for violating the law. An Engineer in a Government Department must be an ingénieur civil, and this condition also applies to most provincial bodies and large towns, and to some private bodies. There is no such legal restriction for appointments in private industry.

In Greece the right to practice as a professional engineer is granted only to graduates of recognized engineering schools. These persons are registered by the Greek Technical Chamber which issues licenses. Official engineering documents such as projects for Government authorities must be signed by licensed persons.

In Italy the title 'Ingegnero' is granted by State examination and both the profession and the title are strictly controlled. Registration is in the 'ALBO professionale' and only those registered may freely practise. Registration is not however required for industrial, commercial or teaching practice.

In Spain the title 'Ingeniero' is restricted to those who have graduated from an official school of engineering. The professional practice of engineering is confined to those who possess the title, but there are people without the title in private industry and elsewhere who practise engineering in the EUSEC definition sense.

Historically the idea of regulation has nearly always been associated with the idea of public benefit. Unless the community will clearly gain from the exercise of the powers granted to a particular profession, the State has been reluctant to grant them.

If, as professional engineers, we desire entry to our profession to be more closely controlled, we must first of all establish that control is clearly in the public interest. This should not be difficult to do. Judging from Lord Hailsham's remarks in the House of Lords during the debate on Scientific Policy, there will be a sympathetic reception at Ministerial level of anything which tends to raise the status of engineering.

By making complex consumer equipment available to almost every family in the country, the engineer is creating a Domestic Revolution and repaying part of the debt incurred in human misery during the Industrial Revolution. Today the direct contribution of the engineer to the economy—great though it is, particularly in the vital export field—constitutes only a fraction of his total influence. For he keeps the entire manufacturing industry and many other trades and services equipped with the modern devices they need to function efficiently. This survey of the principal productive industries by the Institution of Mechanical Engineers' new Honorary Member shows that their progress in recent years has been largely due to engineering and that further advances will depend on their ability to recruit enough engineers of the right calibre. This article was published in 1964 but the statements made are little affected by the passage of time.

The Contribution of Engineering to the British Economy*

by Sir Harold Hartley, GCVO, FRS, Hon MIMechE

Although I shall be dealing, in the main, with the present and the future, I cannot omit any reference to the great engineers of the Industrial Revolution; the men who led the world and laid the foundations of our present British economy. Watt's steam engine gave the critical impetus but it was men like Murdock, Wilkinson, Trevithick, Stephenson, Rennie, Telford and others who engineered the great forward movement that changed the way of life.

The background

When L. J. Henderson said that science owed more to the steam engine than the steam engine owed to science, he forgot that its birth depended on the discovery of latent heat by Watt's friend, Joseph Black. However, by and large, these great pioneers worked by the light of their experience and intuition and science played little part in their designs. They had that intuitive sense of rightness that has guided engineers since the design of the Roman aqueducts and which, according to the Feilden Report, is still the prime arbiter in the art of engineering design. The bridges built by these men, before the days of soil mechanics and the study of creep and fatigue, still stand and carry traffic, not only in this country but on the Continent where British engineers established the tradition of overseas contracts.

When the French Government offered to pay for the repair of the first railway from Rouen to Paris after a wash-out, Thomas Brassey sent his proud reply: "I undertook to build and maintain the railway and no man shall ever say that Thomas Brassey is not as good as his word".

That was the spirit that won confidence overseas. I suppose British engineering reached its competitive climacteric soon after 1851, when the whole world came to the Great Exhibition to see how it was done and establish its own industries. Some countries, like Germany, coming late into the field,

[Courtesy: U.K.A.E.A.]

Fig. 1. The advanced gas-cooled reactor at Windscale, operating at a gas outlet temperature 200°above that at Calder Hall, promises considerable economies. This is a view of the pile cap with the charging machine in the background and two of the heat exchungers on the right

*This condensed version of the first Maurice Lubbock Lecture is published by permission of the Trustees of the Maurice Lubbock Memorial Fund.

Sir Harold Hartley *has combined a distinguished career in the academic world with a lifetime of public service. Educated at Dulwich College, he was first an undergraduate at Balliol College, Oxford, then the Natural Science Tutor and Bedford Lecturer in Physical Chemistry, and finally a Research Fellow of the College. His association with Balliol lasted over 60 years, though his time there was interrupted by service in the 1914–18 war. During that war he was mentioned in despatches three times and was awarded an M.C. In public service he has held numerous appointments; he was Chairman of the Fuel Research Board for 15 years from 1932. His other interests in power have included Chairmanship of the Electricity Supply Research Council from 1949–52; Presidency of the World Power Conference from 1950–56; and Chairmanship of the Energy Commission of the European Economic Community from 1955–56. He has also held high office with both the British Airways Corporations, with the Railways and, from 1954–56, he was Chairman of Council of the Duke of Edinburgh's Study Conference on Human Problems in Industrial Communities. The author of many scientific papers, Sir Harold has received honorary degrees from universities in Britain and the U.S.A. He became a Fellow of the Royal Society in 1926 and was knighted in 1928. In April 1964 he was elected an Hon. Member of the Institution of Mechanical Engineers.*

looked to industries where science could help and, in spite of great British inventors like Kelvin and Parsons and the influence of men like Mond, Siemens and Marconi, who brought their ideas to Britain to be developed, reliance here on traditional methods allowed other countries to forge ahead. Our weakness was revealed by the 1914–18 war and the remedy was found in the establishment of the Department of Scientific and Industrial Research, the Fuel Research Board, research associations and by industrial research laboratories which were staffed from the expanding scientific departments of the universities.

The year 1914 marks the watershed between the old and the new. The past 50 years have seen that amazingly rapid development resulting from the closer association of engineering with scientific research and the successive emergence of new specializations. I would assign the greatest influence to the science and practice of metallurgy. Metals are the limiting factor in engineering design. The study of chemical and physical metallurgy has provided the engineer with new materials that have made new ventures possible and has given him data about those insidious enemies, creep and fatigue, that enable him to design for a projected life with economy.

Then there has come the great upsurge in our knowledge of electronics with its outcome in communications, in new measuring instruments, in the transistor, the maser and the laser, and in analogue and digital computers. The computer has now become an immensely powerful instrument, assisting the engineer in research, in the design office and in management.

The study of the theory of automatic control has brought into being a new primary technology, control engineering, applicable over the whole range of engineering projects from rockets and satellites to steel mills and chemical processes. The control engineer is an essential member of the team of specialists engaged in the design study of any great engineering project.

However, control engineering is only the most recent of the new techniques to emerge. Going back to the 1920's there was chemical engineering, with its immense influence on the process industries; aeronautical engineering, promoting the development of new prime movers; and structural engineering, with its new techniques of soil mechanics and relaxation. Then, with the impact of another world war, came nuclear engineering and space engineering, made possible by the control engineer.

The Domestic Revolution

That is the background against which I shall try to answer the question posed in the title. Obviously, I shall have to rely largely on statistics. However, when I studied the *Census of Production* and the *Annual Abstract of Statistics*, invaluable as they are, I found myself hedged in by their definition of what constitutes the engineering industries of this country. Accepting that definition, the engineering industries account for 35 per cent of the contribution of manufacturing industry to the gross national product and for nearly half of U.K. total exports. About one-third of the engineering industries' output is exported.

Now, these figures, impressive as they are, convey a completely inadequate idea of the contribution of engineering to our economy because they neglect the importance of engineering in other industries. I hope to convince you of this by means of an analysis of the organic structure of British industry.

We are today in the middle of a Domestic Revolution (the use of power and heat in the home in countless contrivances) that has eased the work of the housewife (and also of her husband) and has added to health and comfort. It may be regarded as the repayment by the technologist of a debt long overdue for the social consequences of the Industrial Revolution.

The mechanized production of standardized products with the aid of quality control has fundamentally changed the nature of most manual work, substituting for the fitter the more skilled tool-room, maintenance or instrument worker. The control engineer, by using automatic devices much more rapid and accurate than the human senses, is further changing the types of work available in industry and elsewhere.

All these new techniques depend on the availability of energy in the form of power and heat. I need not remind you of the close relation of energy consumption with national income, though it is not always easy to decide which is the cause and which the effect.

I start my analysis, therefore, by examining the energy industries of Britain and their contribution to our gross national product. In terms of output, the G.N.P. is the measure of the net outputs of all the individual industries, trades and professions, plus net income from abroad, while, in terms of income, the G.N.P. is the sum of personal incomes arising from employment, rent, dividends and interest and undistributed company profits. So, on the one hand, it is a measure of additional value created and, on the other, of what we, as a nation have to spend.

I shall, throughout, use figures from the 1958 Census of Production, the latest available, as the percentage make-up of the G.N.P. today does not differ materially from that in recent years. For 1958 it was £20,000 m. and the details will be found in Table 1. Table 2 shows the figures obtained by partial regrouping according to the classifications set out in this article. The good agreement is accidental as there is some overlapping and omission.

The energy industries

The energy industries, coal, oil, electricity and gas, have all undergone revolutionary changes since 1945. These affect their production costs and productivity and contribute to the changing pattern of energy consumption. For all those changes the engineer, with the help of the scientist, has been responsible.

The nationalization of coal, electricity and gas, unified each of those industries and made possible a co-ordinated programme of development, with the concentration of production in large and more economic units. In the coalfields, practically no new shafts had been sunk during the 20 years before 1947 so that shaft sinking techniques had to be brought into line with modern technology and the speed of sinking was quickly trebled. A systematic search with new borings, both on land and on the sea bed, together with geophysical techniques, added some 10 000 million tons to our reserves. Rapid progress was also made in the reconstruction of existing collieries and in the development of new mines—60 per cent of the present output is from reconstructed or new collieries and in a few years it will be 80 per cent.

The most spectacular advance has been made at the coal face itself. In 1947 the standard form of mechanization was to undercut the seam, then to bring down the coal with explosives and hand-load it on to conveyors. Today*, 70 per cent of the coal is power-loaded by machines developed in Britain, which both cut and load the coal on to conveyors. The substitution of hydraulic supports for wooden pit-props has made possible the remotely controlled 'walking' supports which play an important part in fully-mechanized mining. An important new break-through in the working of thin coal seams below 2 ft 6 in has been made by the Collins Miner, equipped with a nucleonic coal-sensing probe.

Thanks to these developments the productivity of coal-mining rose by 40 per cent between 1957 and 1963 and this rise will continue. Concentration of production in the new and reconstructed pits has made British coal the cheapest in Europe. The progress made in the development of these new techniques has also resulted in a large sale of British machinery in overseas markets.

Electricity is the next largest energy industry in terms of output and there again spectacular progress has been made since nationalization. The size of the generating units has increased from 30 MW to 500 MW with an increase in thermodynamic efficiency from 27 to 39 per cent, gained by higher steam temperatures and pressures and by reheat, while the capital cost per kW sent out has fallen from £67 to £37. If account were taken of the lower value of money, this reduction would be even more striking.

The siting of the power stations near coal supplies and cooling water involves the large-scale transmission of power to the load centres. This has been made possible by reinforcing the original national grid and raising the voltage from 132 kV to 400 kV. The unification of the industry and better telecommunications have enabled the load to be carried in the most economical manner. Thanks to all this the average cost of electricity has hardly risen in five years.

New engineering problems had to be solved in the commercial utilization of nuclear power. Here the Atomic Energy Authority led the way with Calder Hall, the first commercial reactors in the world. The manufacture of nuclear fuel elements was a triumph for the chemical engineers and metallurgists while the design of the reactor demanded

* This article was published in 1964.

Table 1—The G.N.P. as given in the 'Annual Abstract of Statistics'

Industry, trade etc.	£ m.
Agriculture, forestry and fishing	873
Mining and quarrying	709
Manufacturing	6,977
Construction	1,183
Gas, electricity and water	525
Transport and communication	1,575
Total production	11,842
Distributive trades	2,490
Insurance, banking and finance (inc. real estate)	577
Other services	1,981
Total production and trade	16,890
Public administration and defence	1,226
Public health and educational services	719
Ownership of dwellings	681
Domestic services to households	92
Services to private non-profit-making bodies	101
less Stock appreciation	6
Residual error	275
Gross domestic product at factor cost	19,990
Net property income from abroad	294
Gross National Product	20,284

Table 2—The first section of Table 1 rearranged

Industry, trade etc.	£ m.
The Energy Industries	1,160
Food and raw materials	950
Processing Industries	2,700
Structures	1,630
Capital and consumer goods	3,830
Transport and communication	1,575
Total	11,845
Remainder as in Table 1	

entirely new standards of precision and reliability. Much research and ingenuity has been built into the next generation of gas-cooled reactors and the advanced gas-cooled reactor shown in Fig. 1 promises a considerable increase in efficiency.

The gas industry too is undergoing revolutionary changes. The conventional carbonizing plants producing gas, coke and chemical by-products have become uneconomic, except for the metallurgical coke in coke-ovens, and no more will be built. The gas engineer has become a chemical engineer and developed a number of new processes, based on petroleum feedstocks ranging from waste refinery gases to heavy oil. Apart from two Lurgi plants, built for the complete gasification of cheaper coal in the Midlands and in Scotland, all the new installations will use petroleum products, either for reforming or enrichment. In addition liquid methane, imported from the Sahara, will be distributed in a high-pressure, 18 in national pipeline, running from Canvey Island to Leeds. Another important development has been the process for reforming light distillate, of which there are ample supplies, to produce an almost non-toxic, lean gas, rich in hydrogen, which can then be further enriched with methane from the Sahara or elsewhere. These plants have a capital cost one eighth that of a conventional carbonizing plant; they occupy less space, need less labour, can be started up quickly to meet peak loads and produce gas at a pressure sufficient to transmit it 100 miles without any further compression. In addition, two processes have been developed to make rich gas for blending from light distillate.

Orders are already coming in from abroad for these new plants and it is no exaggeration to say that they have changed the entire future prospects of the gas industry.

Most of these developments in the gas industry have been made possible by the great expansion of oil refining in Britain, from 1·8m. tons in 1946 to 59½m. tons in 1964. The building and operation of these great refineries have presented many problems for the chemical engineer. The striking changes in the density spectrum of British oil product consumption has had to be met by great flexibility in operation to secure the right balance of output. While the total consumption of oil products has increased sixfold since 1946, the proportion of gasoline has been halved and that of fuel oil doubled. Fig. 2 shows an experimental boiler installed in one of the new British refineries for doing fuel oil research, part of the engineering effort which has promoted this remarkable change of proportions. There have also been large increases both in the consumption and quality of aviation turbine fuels, derv and diesel oil. The improvement in the octane number of motor gasoline has enabled the compression ratios of engines to be raised with corresponding gains in acceleration and fuel economy. Catalytic cracking and reforming have played the main parts in meeting these changing demands. Alongside these developments in the refineries have come the economies in the transport of oil products due to the use of pipelines and super-tankers. All this represents a great technological achievement by the oil industry which has been of material benefit to our balance of trade.

In 1958 the net output of the four energy industries was £1,160 m., representing nearly 6 per cent of the G.N.P., but their indirect contribution was much greater, as all industrial activity and much else besides was dependent on the availability of heat and power.

[Courtesy: The British Petroleum Co. Ltd]

Fig. 2. Since the war oil-refining capacity in the U.K. has expanded enormously: the total consumption of oil products has increased sixfold and the proportion of heavy fuel oil has doubled. This experimental boiler is part of the modern equipment at a Kent Oil Refinery used for fuel oil combustion research

Food and raw materials

For half her food and most of her raw materials Britain has to rely on imports. She has few minerals except coal (dealt with already under energy) and iron ore. In 1958 the gross output of home agriculture was £1,470 m., making a contribution of about 5 per cent to the G.N.P. The value of imports of food (including drink, tobacco and animal feeding stuffs) was £1,491 m.

What contribution can I claim for the engineer in our food production? Something substantial, I suggest, as the agricultural engineer has changed the face of British farming in recent years and, since 1956, the productivity of labour has increased by 5·5 per cent each year; an important point, in view of the big increase in farming wages. The horse has almost disappeared and been replaced by over 400 000 tractors. Crops are sown, cultivated and harvested by over a million machines of various types. Rural electrification has made possible over 300 000 milking machines, as well as crop drying and various mechanized aids. Meanwhile the yield of nearly all crops has shown a very substantial increase during the past 10 years, for part of which the engineer should take credit. I believe that the productivity of the U.K. farm worker and his wages are the highest in Europe. This is a useful example of the increased living standards that can be provided by applying modern engineering developments to long-established industries.

Our export of agricultural tractors is the largest in the world and the export of other agricultural machinery considerably exceeds home consumption. So the agricultural engineer helps our balance of trade in a dual manner, both by increasing home production and by exporting the products of his design. This is a good example of the need of a strong home market to promote exports.

As regards raw materials for industry, in 1958 we imported agricultural and mining products (omitting oil) costing £889 m. while our home production of iron ore, quarry materials, wool and hides was worth about £200 m.

British engineers were the pioneers of mining development in many overseas countries from which we draw supplies of ore. Flotation, which has revolutionized ore-dressing, was a British invention. Today, when much of this work is done by nationals of the countries concerned, British mining engineers are still playing an important part in the Commonwealth and in such major projects as the exploitation of the bauxite deposits in N. Australia, the iron ore in Mauretania and the lignite beds of India.

The processing industries

The processing of raw materials is a much longer story, due to the infinite variety of products. Chemical engineering has been the predominant factor, particularly in developing synthetic materials.

The chemist, the metallurgist and the engineer have combined in developing new techniques for the winning of those metals on which our modern civilization is so largely based, for their purification and for the production of alloys. Then comes their fabrication by casting, rolling, pressing and heat treatment into the forms in which they are used in industry to produce the machines, the structures and the gadgets that make possible our modern way of life.

All this is the province of the engineer and vast metallurgical plants are the manifestation of his creative genius. Perhaps the most impressive is the modern continuous strip mill in which slabs are hot-rolled automatically until they emerge in the form of steel strip at up to 3500 ft/min. Thanks

[Courtesy: Steel Co. of Wales Ltd]

Fig. 3. The increasing continuity and automatic control of steel finishing processes has been the outstanding feature of their post-war development. This furnace is part of a strip galvanizing line at the Abbey Works

to automatic gauge control, a tolerance of ±1 or 2 per cent can now be achieved with a modern mill on all except the thinnest gauges. Fig. 3 shows part of a modern continuous tinplate processing line. Other major developments in the steel industry include the introduction of oxygen steelmaking on a large scale, continuous casting, and vacuum de-gassing.

Next in importance—actually its output is of greater financial value—is the chemical industry, itself largely the product of chemical engineering during the last 50 years. Heavy chemicals, acids, alkalis, explosives and fertilizers, originally its main objectives, are now balanced by the organic chemicals, synthetic dyes and pharmaceuticals, dependent originally on coal tar and fermentation, now re-inforced by the immense range of synthetics, largely based on petrochemicals.

Is there any limitation to their ubiquity? I was once dis-cussing the competition between synthetic chemicals and tropical agriculture with Carl Bosch, who engineered the Haber process for the fixation of atmospheric nitrogen—probably the most significant application of chemical engin-eering as regards its impact on human existence. He told me of a conversation with Steinmetz, the great electrical engineer: "Bosch", said Steinmetz, "I know you can make indigo cheaper than God, some day you may make rubber cheaper than God, you will never make cellulose cheaper than God".

That penetrating comment exposes the Achilles heel of man-made synthetics and the importance of the latest devel-opment, biochemical engineering, in which the living cell is harnessed to provide complex structures like the antibiotics which nature produces so cleanly by her template methods.

It is on nature, directly or indirectly, that we must rely for those complicated molecules on which, in the long course of evolution, she has decided we must feed. Notwithstanding the triumphs of engineering, its final purpose must be partnership with nature. Of this there is no better example than the impact of engineering on the food industries in the conservation of perishable matter subject to pathogenic contamination. Refrigerated and air-conditioned transport and storage have diversified the range of cheap foods, prevented deteriora-tion and made large savings. Mechanization has revolution-ized the canning and food packaging industries with great hygienic advantages. The pasteurization and bottling of milk have become engineering operations. Much food is now cooked and packaged in the factory and distributed with the

help of deep freeze equipment. As a sequel to all this, British food machinery is making an increasing contribution to our export trade.

Textiles constitute another industry of great importance in the processing stage and my last example is the manufacture of paper and board from imported cellulose and fibre in mills which have much in common with steel strip mills.

These processing industries form part of the chain of manu-facture and the five I have mentioned contributed, in 1958, nearly 40 per cent of the value added to materials by manu-facture and 13 per cent of the G.N.P.

Structures and capital goods

Structures include buildings of all kinds in which the structural engineer now plays so important a part with the use of steel, pre-stressed concrete and light-weight materials. These great modern buildings are bringing into existence yet another type of specialist, the environmental engineer, to deal with heating and ventilation, air-conditioning, lighting, sound-proofing and all the other amenities of our busy life.

Structures also include roads, bridges, tunnels, railways, dams, hydraulic works, drainage and pipelines. In all these, the engineer has made great progress in recent years in providing the facilities that modern life demands. Here again the work at home is the proving ground for techniques and men and leads to contracts overseas, many of which take the form of technical aid to the emergent countries.

These overseas contracts are important sources of revenue, their value in the past nine years being over £1,000 m. They include a variety of exciting engineering projects all over the world, such as hydro-electric projects in Turkey and Brazil, the largest power scheme in S. America, steel works in India, a fertilizer factory in Iran, a new township in Bahrein, port works in the Sudan, a geothermal steam generating station in New Zealand, hundreds of miles of roads in mountainous districts of Ethiopia and pipelines in several countries. Many of these works are designed and administered by consulting

[Courtesy: Westland Aircraft Ltd]

Fig. 4. In the transport field Britain continues to lead in new ideas, if not in operation: the 37·5 ton hovercraft, under construction for the Ministry of Aviation, exemplifies progress made with this invention in recent years

[Courtesy: British Motor Corp.]

Fig. 5. Modern production methods have brought personal transport and consumer durables within the reach of most families. Even where manual production skills are still needed, the degree of mechanization is high

engineers who still enjoy the confidence that Britain won during the Industrial Revolution, in spite of great political changes.

Among structures I also include ships and aircraft because of the various problems of structural design that they involve, together with their hydrodynamic and aerodynamic problems. They require many specialized forms of prime movers and today over half the engines of all civil aircraft in service or on order in the free world are of British design. The latest British invention in this field is the Hovercraft and Fig. 4 shows how much progress has already been made in a few years.

All these structures together make a sizeable contribution to the G.N.P., amounting to between 7 and 8 per cent.

First in productive importance among capital goods are the machine tools on which all stages of production ultimately depend. They must serve an immense range of duties and their design, the responsibility of mechanical and electrical engineers, is a major factor in determining the efficiency of industry as a whole. One interesting development is the accurate automatic control of machine tools by means of moiré fringes made with almost perfect optical gratings.

Next come the prime movers, steam and diesel, and the turbo-generators of the power stations with the electric motors, large and small, that supply the drive of almost all modern machinery.

Motor vehicles, including tractors, account for the largest net output of any industry and there are substantial contributions from the production of locomotives and railway rolling stock, mechanical handling plant and mining machinery. The last 20 years have also seen an immense increase in the range of consumer goods, particularly in domestic appliances and in radio and television sets. New materials like plastics have invaded many old consumer industries and stimulated some new ones. All this has involved many fresh manufacturing techniques.

It is not always possible to draw a clear line between capital and consumer goods: their joint contribution to the G.N.P. amounted to 19 per cent. Engineering, heavy or light, enters into this at every stage and even where manual skill is involved in assembly, a mechanized production line is the general rule. This is clearly seen in Fig. 5.

A further useful contribution of 8 per cent to the G.N.P. is made by the operation of British transport by land, sea and air and by postal, cable and wireless communication. These services, essential to our commercial activity and the mobility of modern life, all depend on engineering.

The continuous technical improvement of the trunk-line network has enabled it to meet rapidly rising demands, including those of television, with greater efficiency, and much reduced capital cost per trunk circuit. Automatic dialling is now being extended to overseas calls and the system is being strengthened by microwave lines while the telex system is meeting the new need to transmit digital data with speed and accuracy.

One outstanding achievement stands to the credit of British communication engineers; the new type of deep-sea cable in which the tension member is transferred to the centre of a cable, consisting of two coaxial groups of high-tensile steel wires. They have opposite lays and are torsionally balanced, with a polythene outer sheath to protect them. This new light-weight cable was first used in the high-capacity system between Britain and Canada and was adopted for the round-the-world Commonwealth cable system which has just been completed.

The challenge

The part that engineering plays in each of these productive industries is unmistakable. Between them, they account for 59 per cent of the G.N.P. Another sector of the economy, the distribution trades, finance and services, contribute 25 per cent of the G.N.P. and, in many ways, they too are dependent on engineering for transport and communications, mechanical handling, office machinery, and for electronic computers, analysers and recorders.

Engineering was responsible for the great upsurge of our economy in the 19th century, and it is no exaggeration to say that it is the mainstay of that economy today and will become even more important in the future. Many of our great industries have, however, already reached a very advanced stage in the exploitation of engineering, and their further progress, though essential to our economy in a highly competitive world, will be possible only through highly sophisticated advances in technology. We must find the solutions to the exciting problems that now confront us; the direct conversion of heat and nuclear energy to electricity; the development of new materials with super-properties; the design of supersonic aircraft, colour television, telecommunication by satellites and other new means; the design and use of computers as everyday tools; and the optimization of production processes, including machine tools, under computer control.

These are just a few of the fascinating problems of the near future. To solve them we need in engineering men of the highest ability and training, not only to work on design but also to engage in the research that is today a vital first step in most engineering advances. The intellectual and financial level of engineering research is sometimes not realized: it already accounts for 60 per cent of our total research expenditure and there is no doubt that it presents some of the most difficult and challenging problems that now face us.

However, the recent Report of the Oxford Education Department and other documents show that engineering has failed so far to attract its share of talent, both in ability and number; recent admissions to universities show no sign of a change of heart; rather the reverse. This is among the most vital problems that Britain is facing since, as I have tried to show, the basis of our wealth lies in engineering. Sir George Pickering recognized this: our relative poverty in research endowments, compared with the U.S.A., would only be overcome, he said, when we, as a nation, became once more interested in the creation of wealth, rather than in its distribution and consumption.

Let us face this issue: the future quality of our engineering will depend largely on the decisions of youth about their choice of a career. In this they will be influenced inevitably by the climate of their education. In my own case, I refused a promising job in industry in 1900 because I had watched my chemistry master's investigation of dry reactions and my mind was set on research. Those days were easier for the lone wolf, operating in uncharted areas of knowledge; today the tendency towards specialization in narrower fields, each requiring immensely expensive equipment, leads inevitably to planned team work and, in today's school curricula, the scales are weighted in favour of a career in pure science.

What can be done to convince young people of the satisfaction that comes from taking part in the creation of some great engineering work and of the happiness that derives from being part of the human chain of responsibility in some co-operative productive enterprise? Memories of Sanderson, of Dulwich and Oundle, make me wonder whether the school workshop equipped for wood and metal working is not part of the answer.

Science and engineering

Education is asking for greatly increased expenditure; should it not give more thought to redressing the imbalance between the recruitment of scientists and engineers which,

if it persists, will imperil the British economy?

In the future, Britain must rely on producing goods with a high technological content, with the maximum of added value built into them. Success will depend on a close alliance between engineering and science. Prince Philip put it very neatly in his Graham Clark Lecture of 1961: "The engineer is in fact the means by which the people are enabled to enjoy the fruits of science".

It is not an easy partnership as the basic outlooks are so different. The scientist is interested in discovery, time doesn't enter much into his programme. The engineer is working on a definite project and time is the essence of the contract; he needs information at the right moment to build it into his design, and the scientist is unwilling to part with a half-baked answer. But the scientist can and will work to a timetable if he sees the reason for it. There is no better example of this than the building of the first commercial reactor at Calder Hall, when the engineer was dependent on the scientist for his design data.

I believe that this co-operation will be greatly helped by the new Oxford School of Engineering and Economics. Analysis of costs will, I am sure, provide valuable common ground between the scientist and the engineer, and I know from experience the scientist's feeling of satisfaction when you know that your work has been applied and are told the measure of its practical value. Moreover, economic studies will help the engineer to decide where the research effort is needed.

Oxford is also considering the possibility of having one of the much-needed postgraduate schools of engineering design, the success of which will depend on close relationship with those who are engaged in industry. Visiting Fellows are becoming fashionable in Oxford and I hope that colleges will offer such fellowships to engineers who might spare time to come, say, once a fortnight, to take some part in the work of this School and help to keep it in touch with the problems of industry. This should be of special value in design.

Technology and the Developing Countries

by Lord Jackson of Burnley, FRS, CEng, FIMechE*

The gap between the 'haves' and the 'have-nots' among the nations is, if anything, increasing despite the efforts made by the underdeveloped countries themselves and the international aid provided for them. The reasons are partly the population explosion but also the failure to apply the available aid in the most effective way. This article was published in 1969 but the position has not greatly changed since then.

We in this country are extremely fortunate and privileged people compared with a very large proportion of the world's population. We are among the most favoured beneficiaries of the progress of science and technology, no doubt justifiably so, since we have been major contributors to this progress. It is natural that we should wish to remain in the forefront of further progress and we are therefore wise to analyse why we lag behind other highly industrialised countries in our ability to exploit the potentialities of recent scientific work. The strengthening of our national technological achievement will not only further our own standard of living but is essential if we are to make a much greater contribution towards solving the problems of the developing countries. I wish the people of Britain could be brought to a clearer realisation of the importance and urgency of this objective.

The size of the problem

· I ought first, no doubt, to define a developing country. In fact there is nothing approaching a clear definition. In some aspects of technology, such as space, of microelectronics and computers, and in average standard of living, this country may perhaps be regarded as underdeveloped, relative to the USA.

Usually the distinction is based on strictly economic considerations such as Gross National Product per head of population, with all the limitations this involves, including the somewhat misleading consequences of presenting comparative figures in American dollars, calculated on international currency exchange rates. On this basis there are countries in Europe, such as Spain and Greece, with a diversified production and export structure, rich natural resources, satisfactory levels of education and reasonably developed social and economic institutions, which have a lower GNP per capita than some countries of Asia and South America.

The limitations of this criterion, however, do not greatly affect the gross differences shown approximately in Table 1 for the year 1967, and it is differences of this order which are the cause of my concern.

* This is a condensed version of his Earl Grey Memorial Lecture given at the University of Newcastle upon Tyne.

Table 1—Gross National Product / capita ($)

USA	3600
United Kingdom	1800
India	90
Tanzania	70
Indonesia	70

Table 2—Average Annual Growth rates of GNP and population

Regions	GNP	Population	GNP/capita
Europe	7·4	1·5	5·9
Africa	3·1	2·2	0·9
Latin America	4·7	2·9	1·8
Asia	5·1	2·6	2·5
Total developing regions	5·0	2·5	2·4

The primary aim of overseas aid is to help establish situations within the developing countries which will lead to a reduction in these discrepancies—discrepancies of which the ordinary people in these countries are being made increasingly aware through world-wide telecommunications and air transport, which are bringing them into contact with the standards and conditions of life, and with the modes of behaviour —bad as well as good—in countries such as ours. Within a few years many such countries have been taken politically through evolutionary phases which with us have been a gradual progress, and for which they are educationally ill prepared. No wonder that their powers of adaptation are being overstrained.

After centuries of marginal existence thep eople of these countries are learning that poverty is not a necessary attribute of life. Increasingly the eradication of want and the fulfilment of individual capacity are claimed as rightful goals for all, not only for a fortunate minority. And to a large extent the Western World is responsible for this view, both by example and by sermon. The burning question is, how best to ensure its attainment.

The problems involved vary considerably because there is great unbalance in the abundance of fuel and mineral resources, in the availability of water, in dependence on agriculture and ·degree of industrialisation, in capital resources, in education, and in the restraints imposed by the traditions that have been inherited. Each country must therefore evolve its own characteristic approach.

Unfortunately, in spite of considerable aid over the past two decades, the discrepancies, at any rate at the extremes, are becoming worse. The data of Table 2 are averages over very large areas and therefore obscure the extremes, but they nevertheless establish the point I have made with disturbing clarity.

Thus for all developing regions the growth rate of GNP was only two-thirds that of Europe but the rate per capita less than half. This, in absolute terms, means that, on average, Europe added more than 10 times as much in income per capita as did the developing countries, which is a simple reminder of the low income base on which the latter have to build their development.

A crucial factor is, of course, the population increase, and in this respect the situation is unpromising. The United Nations data in Table 3 predict that, if recent trends continue, the 1960 world population of around 3000m. will have increased to around 7400m. by AD 2000, and within the less developed areas close on twice as fast as elsewhere. This increasing unbalance is not due to a rising birth rate but a rapidly decreasing death rate due to medicine. It remains to be seen whether other aspects of progress will correct this. Lord Blackett has estimated that it would be economically advantageous for India to spend £250m. a year on birth control to this end. Unfortunately, as pointed out by Barbara Ward, whereas in countries like ours, wealth came first and health followed after, in the developing countries the order was reversed.

The vital question is—how can the rising tide of expectation within the developing countries be made compatible with the consequences of such a population explosion?

International aid

The major contributors of financial and technical assistance to the developing countries are the members of the OECD Development Assistance Committee and the data included in this section are limited to them. The total annual net flows in period 1960-67 are shown in Fig. 1. In 1967 they amounted to 11·3 billion US dollars.

The official contributions include grants and loans made bilaterally to individual countries, as well as to multilateral organisations such as UNO and the World Bank. The private sector provides investment and credits, and is largely bilateral. There are also substantial grants, amounting to about 600m. dollars in 1967, made by private voluntary agencies which are not included in Fig. 1.

Official aid has increased only modestly since 1961—a good deal less, proportionally, than the private sector. In 1961 the total aid figure was 0.95 per cent of the GNP of the donor countries, but it had fallen to 0·75 per cent in 1967 when it represented less than 3s. per capita per week of the donor countries. Their official contribution was less than 2s. per week, of which almost one-half was in the form of repayable, interest-bearing loans. It can hardly be claimed that this amounts to generosity: nor would the internationally endorsed objective of a net disbursement of 1 per cent of the GNP of the developed countries by 1975. The problem of loan debt servicing, as illustrated by the UK data of Table 4, becomes increasingly acute.

The £207·9m. gross official aid amounted to 1·5 per cent of the total public expenditure within the UK in 1967, compared with defence 15·1 per cent, housing 6·5 per cent, education 12·0 per cent, and health and welfare 9·5 per cent.

These official aid figures cover financial aid for agreed special projects and for purposes such as budgetary assistance, compensation and pensions, land settlement, disaster relief, etc, and also disbursements for experts seconded for special projects, research programmes, educational developments, etc, and as grants to volunteers.

On the 30th June 1968 some 15 445 British nationals working overseas, including 1551 volunteers, were financed from official sources. Their fields of assignment are shown in Table 5. To these must be added an unknown, but large, number of men engaged within the developing countries on projects falling within the privately financed sector, and also almost 1300 who were serving similarly under the UN.

The official aid figures also cover support in the UK of selected students and trainees from the developing countries, as shown in Table 6, again for 30th June 1968. These

Lord Jackson was born in Burnley in 1904. After receiving his BSc and DSc degrees from Manchester University in 1925 and 1936, respectively, and a DPhil from Oxford in 1936, he was appointed Professor of Electronics at Manchester in 1938. During the 1939-45 war Lord Jackson served on many government committees and worked at the signals and radar research establishments of the Ministry of Supply. In 1946 he became Professor and Head of the Electrical Engineering Department at Imperial College and seven years later he joined Metropolitan-Vickers as Director of Research and Education. In the same year, 1953, he was elected FRS. He returned to his Professorship at Imperial College in 1961 and was appointed Pro-Rector in 1967. Knighted in 1958 and made a Life Peer in 1967, Lord Jackson was President of the IEE in 1959. He has been President of the British Association for Commercial and Industrial Education since 1962; is President of the Association of Technical Institutions and of the Electrical Research Association for 1969/70, and Chairman of the Television Advisory Committee and of the Engineering Advisory Committee to the British Broadcasting Corporation. He was President of the British Association for the Advancement of Science for 1966/67. He was a member of the University Grants Committee from 1954 to 1964; Chairman of the Committee on Manpower Resources for Science and Technology from 1963 to 1968; and a member of the Council for Scientific Policy from 1963 to 1968. (Deceased 17.2.70)

Table 3—UN Population Estimates and Predictions

Year	1960	1980	2000	2000/1960
Continued recent trends				
Total world	2990	4487	7410	2·5
More developed areas	854	1085	1393	1·6
Less developed areas	2136	3402	6017	2·8
Low projection, assuming decreased fertility				
Total world	2990	4071	5296	1·8
More developed areas	854	1006	1129	1·3
Less developed areas	2136	3065	4167	2·0

constitute about 10 per cent of the total number of young people from the developing countries being educated and trained in this country.

The problems of providing aid

The problems associated with helping the developing countries were summarised by the late Sir Andrew Cohen, Permanent Secretary of the Min. of Overseas Development. The widespread demand for independence during the past 25 years has brought the number of separate developing countries to over 80. All have many needs and, in some degree, look to the outside world to help satisfy them. But they, individually, decide what development is desired, and they, not any outside authority, determine what help is to be asked for, and from whom.

A prospective donor country must be prepared for much frustration and for prolonged and patient collaboration. No donor country is in a privileged position in any developing area, even in what were previously its dependent territories. The need for close coordination between donors can be complicated by national concern for influence, particularly perhaps where help has been sought from East as well as the West. Independence has unleashed aspirations of rapid progress at the very time when it has often removed much of the experienced expatriate staff required to produce it. Many nations want to bring about the kind of dramatic transformation achieved by Japan and by Russia, while lacking the informed and experienced leadership necessary for its planning and achievement. They may, for prestige reasons, seek to pursue projects which are technologically too sophisticated for their current manpower, are inconsistent with more basic needs of the community, but which a particular donor country may feel obliged, indeed may wish, to support.)

The donor's role

It will be evident from these considerations that there is no single formula for achieving effective aid. Just as developed countries differ widely among themselves, so a poor country seeking to develop rapidly can aim towards any one of a number of possible 'states of development' and can select from a variety of paths leading to the chosen one.

Essentially development is a process of exploration and experimentation—a process of tentative changes. Foreign aid is a major external stimulus to which the recipient country has to react. It brings its government, institutions and individuals into contact with a range of donor organisations and individuals. The donor government's personnel can, ideally, perform the role that a loyal opposition (often, of course, non-existent) would be expected to perform—the role of a partner in a necessary exchange of views, whose outlook may be sharply different and, therefore, inherently critical. More usually the role falls somewhere between giving advice and exerting pressure, and must be played with considerable care and sympathy. It should encourage the recipient authorities to re-examine their policies if necessary, but with no implication of sanctions if no action follows.

This raises an issue of fundamental importance, namely the relevance to the developing countries of a European or North American concept of technological development, of economics, sociology, etc. How far are we yet able to analyse the problems of underdevelopment without resorting to the methods devised and proved in our own very different circumstances? How far are our concepts geographically and historically limited to European experience? For instance, the application of modern production techniques, without regard to the local conditions, result all too often in a misuse of native resources. Building offers one of the most striking examples. In the new cities of the developing countries, buildings are going up which, though architecturally very modern, are functionally often quite unsuitable for the climate. Instead of local materials being used, these steel and concrete structures often require heavy imports, while the abundant local manpower has been replaced by imported machinery.

It was Sir Andrew Cohen's view that there was need for a more rapid movement towards integrated, and less self-interested, international effort, towards which, fortunately, considerable progress has been made in recent years. This could apply, of course, only to official aid. Private aid is usually more closely related to economic considerations and this may well explain the new American administration's decision to reduce its official aid and to replace it by strong stimulation of private investment. This it intends to do by establishing an Overseas Private Development Corporation which will have a much greater authority to underwrite risks, such as expropriation, currency changes and wars, than those previously possessed by Government agency and will not need

Table 4—The UK contribution

			£m.
Official sources	Gross grants 56% } loans 44% }		207·9
	Less loan repayments and interest*		57·5
Net			150·4
Private sources (estimated) net			124·9
Investment £82·4m.			
Export credits £42·5m.			
Total net			275·3
Percentage of Gross National Product			0·77

* Since late 1965 about 90 per cent of new loan commitments have been made free of interest, but on 31st December 1967 there remained outstanding for repayment some £988·3m.

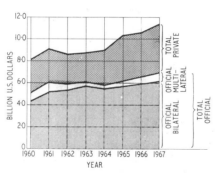

Fig. 1. The major contributors are the members of the OECD Development Assistance Committee: the greater part of the $1·3 billion has come from official sources

Table 5—Workers Overseas Financed Officially

Education	Development planning	Public Administration	Social Services	Works and Communications	Industry and Commerce	Agriculture	Health	Others	Total
5989	277	2085	223	2959	180	1571	1328	843	15 455

Table 6—Overseas Students and Trainees in Britain, Financed Officially

STUDENTS

Humanities	Education	Fine Arts	Law	Social Sciences	Economics	Natural Sciences	Engineering	Medical Science	Agriculture	Others	Total
82	839	84	74	271	321	208	516	545	285	26	3251

TRAINEES

Education	Development planning	Public Administration	Social Services	Works and Communications	Industry and Commerce	Agriculture	Health	Others	Total
54	17	223	50	183	111	45	177	39	899

Congressional approval on a year-to-year basis. This may well have the merit of maintaining, and perhaps increasing, the aggregate amount of American aid, but seems to me to carry with it the danger of giving less scope for exercise of the altruism which has been a prominent feature of official aid. Too much concern for the economic benefit to the private sponsor may mean too little concern for the basic needs of the recipient community.

The problems of progress

In order to make my discussion more realistic, I propose to refer specifically to India since, of all the poor countries of the non-communist world, India has had the best worked out plan and, moreover, is much the largest country, with a population of over 450m. For many years past her Planning Commission composed of men distinguished by world standards and assisted by experts drawn from both Western and Soviet countries — has been studying how India can most quickly and painlessly emerge from the essentially pre-industrial state of 1947.

The economic nature and magnitude of her problems may be judged from the summaries given in Tables 7 and 8 of her exports and imports during the April–September period of 1967. Exports reveal the dominantly agricultural character of the economy. Imports, on the other hand, show how the country is striving to improve agricultural productivity and to launch its own industrial development. Taken together, the tables reveal a discrepancy between exports and imports for the half year of some $600m., an embarrassingly large figure, in view of the international debt of over $5000m.

The significance of this indebtedness is evident from Table 9: of a gross aid disbursement to India during April–September of 1967 of $760m. almost 25 per cent was absorbed in debt servicing; a further 32 per cent was in the form of food aid after bad harvests of the preceding three years. The net aid for other purposes was only $326m., a substantial decrease on earlier years.

Few problems have caused more international concern during the past few years than that of the world's future food needs. The precarious situation is well illustrated by India's index of food production (100 for 1957–59) which fell from 108 in 1961 to 93 in 1965 and to 85 in 1966. Fortunately, there has been a considerable improvement during the past year, due largely to more favourable weather, but also to more productive methods and changes in rural societies.

These are encouraging signs and they reflect a trend to devote an increased proportion of aid to agricultural development. Two further contributory factors have been that the growing urban populations, which have relied heavily on imports, now buy more from the innumerable small urban producers; that scientists, teachers and students—perhaps reluctantly—have begun to accept that the promotion of agriculture does not necessarily mean commitment to backwardness and stagnation.

Nevertheless, the urbanisation trend is no less strong in the developing countries than in the developed ones, and it carries with it more severe changes in the demands made on education, in the character and conditions of employment and in the pattern of life for a rapidly increasing number.

Table 7—India's Principal Exports

Item	Value ($m.)
Jute manufacturers	163·5
Tea	96·6
Cashew kernals, coffee, fish, oils, etc.	78·9
Cotton and other fabrics, footwear, etc.	57·1
Engineering goods, iron and steel, minerals, fuels	56·0
Iron and other ores, minerals	54·9
Leather manufactures	36·9
Tobacco	31·3
Oil cables	26·7
Others	161·3
	763·2

Table 9—Inflow of Foreign Assistance

Item		Value ($m.)
Gross disbursement		760
Less Food	248	
Debt servicing	186	
Net aid flow (1–2 and 3)		326

Table 8—India's Principal Imports

Item		Value ($m.)
Consumer goods		
Cereals and cereal preparations		363·9
Intermediate goods		322·9
Fertilisers	72·1	
Mineral fuels	42·7	
Industrial (eg, cotton, jute, paper)	208·1	
Capital goods		505·5
Iron and steel	81·6	
Non-ferrous metals	74·2	
Metal products	10·1	
Machinery	278·9	
Transport equipment	47·5	
Other manufactures	13·2	
Unclassified		164·7
		1357·0

The road ahead

Clearly, for a long time ahead, progress will be largely dependent on the import of information, know-how and equipment, plus the assistance of experts, from the developed countries. The initial contribution of the developing countries' own technical manpower will be to help in the adaptation of the information, etc, to the needs of the local situation. (The spearhead of this manpower comprises those sent for special training overseas in respect of specific projects.) In my experience, arrangements which have involved a careful selection of the trainee, a clear definition of the further education and training to be received and of the function he is to perform on his return to his own country, have been extremely satisfactory. This does not apply to the much larger number who, more independently, enter postgraduate research schools and who, all too often, insist as a condition of return that they shall be allowed to continue their research, whether it has any relevance to the needs of their society or not. In consequence many of them do not return but contribute to the serious brain drain, notably to the United States.

The availability of young men sufficiently prepared to participate satisfactorily in postgraduate study or research, or in industrial training overseas is dependent, of course, on the educational system of their respective countries. In this context, one of the most gratifying features of the post-war period has been the growth of university facilities in the developing countries. Unfortunately this has not been matched by the provision of facilities for the education of technicians—a much more urgent need. But what cannot yet be provided are adequate facilities for good industrial and other training. I know of no better way in which we could give effective help than by making such facilities available on a much enhanced scale within our small and medium-sized firms and within our nationalised industries and public utilities. These would afford environments and types of experience much more relevant to the present state of the developing countries than the large manufacturing concerns, which for many years past have made an invaluable contribution to training of overseas students.

In conclusion, our attempt to strengthen our own technology and economy will, in my opinion, take us nowhere worth going unless at the same time we decide to reduce the discrepancy between the standard of living of the developing countries and the one we ourselves already enjoy.

Whether they go abroad to teach, practise or give advice, British engineers have a unique opportunity to continue our long-established relations with the developing countries on a new basis. Replacing the colonial administrator as a guide on the path to prosperity, today's engineer, to be effective, must show consideration of the newly independent nations' susceptibilities and adapt his methods to local conditions. In return he will broaden his outlook and render a real service to mankind. This article was published in 1964.

Fig. 1. East and West at Durgapur

The Challenge of the Developing World

by Frank Dunnill

Fifty years ago the role of a British engineer in Africa or Asia could be simply stated. Economies were undeveloped, local professional people more or less non-existent. Roads, railways, docks, harbours had to be built and maintained, buildings erected, power and water provided. But there was no effective pressure for social change, little danger of brusque outside criticism. International bodies capable of interference in this field had not been thought of. Politics scarcely entered into the sphere of interest of Public Works Departments, of Docks and Harbour Boards, of Railway Authorities. The expatriate engineer was, contrary to Donne's familiar assertion, "an island unto himself."

Today the position is, almost everywhere, very different—and the challenge, though as urgent as before, subtler and more complex. Asian and African national leaders are seeking not only 'the political kingdom' but extremely rapid economic growth. Technical and technological education and training, though difficult and expensive to provide, have a prominent place in every development plan. Local engineers are being brought as quickly as possible into positions of influence and importance; and simple patterns of government and administration have generally been replaced by arrangements which include a multiplicity of public corporations and large-scale technical and economic assistance from other countries.

What is the present and probable future role of British professional engineers in these rapidly changing countries? How, in practical terms, can they help these societies which so urgently desire to be on the move? What qualities of experience, character and temperament are likely to serve them best in this exacting situation?

Public works

If one looks, to begin with, at the public sector, it is at once clear that, however urgently developing countries aspire to have their public services staffed with their own people, there is still an interim need for people from this country.

The British Government is pledged to do what it can to help meet this need and the Department of Technical Co-operation, which is the Government Department responsible in this field, recruits about 130 professional engineers a year for periods of service overseas which form part of careers based in Britain. Normally engineers recruited in this way work as part of a team in the public works or water development or irrigation departments in dependent or newly independent Commonwealth countries. The scope of the work is usually far greater than would be the case in Britain and the areas for which an officer is made responsible are far wider. Considerable initiative and power of improvisation are needed to deal with local conditions and willingness and ability to accept responsibility are important.

A typical public works organization overseas will be responsible for the design, construction and maintenance of all government roads, buildings, bridges, airfields and urban water supplies.

The mechanical and electrical section maintains the transport and earth-moving equipment and all government electrical installations. It may also act as manufacturing unit, both for works and other departments.

Departments of water development and irrigation are responsible for the development of all sources of water, including preliminary trials, planning and construction. They also carry out irrigation works. Maintenance work is usually done by direct labour, construction by contract.

It is normal to have a headquarters unit in the capital which includes engineers in administrative posts or on specialist duties, and a number of works teams in the provinces, made up of junior professional men and technicians, led by a provincial or district engineer. A young engineer going out from Britain may find himself personally responsible, under the general guidance of his superior officer, for all stages of a scheme, from the initial survey, through soil investigation, design and estimates, to award of contract, supervision of construction and settlement of payments. Responsibility of this kind encourages a degree of self reliance that one would never get by watching others.

In most developing countries agriculture, and particularly the increased use of water, are of first importance. Many schemes of irrigation, land reclamation, soil and water conservancy are either planned or already under way and there will be more to come.

Advice and training

In addition to helping overseas countries to fill individual posts, the Department of Technical Co-operation is frequently concerned with the provision of consultancy services, sometimes in the context of large projects financed with assistance from Britain. Help has been given in a number of ways, for instance, in connection with the steelworks at Durgapur (Figs 1 and 2), which is one of three very large steelworks set up by the Government of India in the 1950's.

The decision to build the plant was based on a report submitted in 1955 by a Colombo Plan Technical Co-operation Scheme mission, led by Sir Eric Coates, which visited India in April of that year. In 1956 the Government of India contracted with a consortium of thirteen British companies to construct a works with a capacity of one million tons a year; this is now being expanded to a capacity of 1·6 m. tons. The British Government has lent £37 m. towards the foreign exchange costs involved and has provided various forms of technical assistance. British engineers have helped to run the plant and junior staff has also been recruited in Britain. At the end of 1963 there were 53 British employees at the plant. The present General Manager and two Assistant General Superintendents were recruited in 1962 by the Department and are paid by the British Government under the Colombo Plan Technical Co-operation Scheme.

From the beginning of the scheme there has been considerable emphasis on training. Not only have experienced British engineers gone out to do design work for the plant, but some 370 Indians have been trained in British steel firms.

The training element, present at Durgapur and in other projects receiving British technical assistance, is made explicit in schemes financed by the international agencies. When UNESCO, for instance, using funds made available by the United Nations Special Fund, sends engineers to Nairobi or Lagos and provides equipment for them, it arranges fellowships for local people, known in the jargon of technical assistance as 'counterparts', to come to more developed countries to be trained to replace the experts it supplies. British engineers of appropriate seniority and experience are needed to fill short-term assignments under the auspices of

[Courtesy: A.E.I. Ltd]
Fig. 2. Another view of the Durgapur steelworks which were built with British assistance

various international agencies. Most of these will be over thirty-five and possess specialist experience at a responsible level.

The education and training of local engineers at all levels is of obvious importance for the economies of underdeveloped countries as well as for their political self-esteem. It follows that a British engineer serving in a developing country ought to be capable of seeing himself, at least in part, as a teacher—a transmitter of knowledge and techniques eagerly desired by local people. If he can see himself in this light, the warmth—and duration—of his welcome will, generally speaking, increase considerably. And if he is prepared to work in a teaching institution—a university, say, a polytechnic or a technical college—he may well have before him a tolerably well-defined career and a task of considerable interest and importance.

Educational establishments

Examples abound of universities and other teaching institutions in developing countries employing British professional engineers and anxious to employ more. In many of these there is a reassuring continuity, so far as staffing and curricula are concerned, deriving from links of various kinds with British institutions.

The Delhi Institute of Technology, for instance, receives help, by way of staff and equipment, from the British Government and British industry and the standards are guaranteed by a relationship with Imperial College and by the presence of seven British professors. In Nairobi the engineering faculty of the University College which forms part of the University of East Africa has a character which owes a great deal to its contacts with Britain—formerly with the Regent Street Polytechnic and more recently with the University of London. Fig. 3 shows one of the laboratories and the students who use it.

The engineering faculty of the University of the West Indies has also gained a great deal from its special relationship with the University of London, which has helped the younger institution by sending external examiners and advisers and in many other ways.

Table 1—Who Recruits Engineers for Developing Countries

Agency	Overseas clients
Department of Technical Co-operation	
(a) Natural Resources Dept	Overseas governments (for direct service or under British Technical assistance programmes.
(b) International Recruitment Unit	United Nations specialized agencies, and other international organizations.
Inter-University Council for Higher Education Overseas	University of East Africa (University College, Nairobi)
	Ahmadu Bello University, Zaria, Northern Nigeria.
	University College of Sierra Leone.
	University of the West Indies.
British Council	Universities in India, Pakistan, Ceylon, South East Asia and Latin America.
Association of Commonwealth Universities	University of Malaya and University of Hong Kong.
Council for Technical Education and Training for Overseas Countries	Institutions other than universities providing post-secondary technical education.
Lt-Col E. C. Alderton, C.M.G., 33 Craven Terrace, London, W.2	University of Nigeria, Nsukka.
Mr J. A. Ata, London Office, 15 Gordon Square, London, W.C.1	Kwame Nkrumah University of Science and Technology, Kumasi.
Magdoub El Shoush, c/o Sudan Embassy	University of Khartoum, Sudan.

Moving the camera a little closer, let us consider what this means for an individual. What will a young man from a British university, who has taken a higher degree and wants to try his hand at teaching for a year or two, find on arrival? What problems—and what opportunities—will confront an older man with a background in a technical college or C.A.T.?

Different students and colleagues

The newcomer will find a student body very different in many ways from any he has known in Britain; most will come from simple rural backgrounds where few machines are seen; they are over-respectful of the printed word and apt to regard education as a means to a very specific personal end. A degree and a professional qualification can mean for them advance at a single bound from an existence scratched from the soil to a life of comparative affluence. Little wonder, then, that they are prepared to work hard for their academic and professional tickets, or that their view of what education is about is narrower than has been fashionable in 20th century Britain.

Among such students, where there is no family tradition of tinkering with things mechanical, practical work is neither familiar nor popular. The Vice-Chancellor of one African university told me recently that, poor as most of them are, those of his students who own bicycles will pay to have their punctures mended rather than do this themselves and lose face.

There is, on the other hand, a quite different tradition—one of mutual help within an extended family—which ensures that many African students carry with them throughout their lives family responsibilities much wider than any now considered normal in Britain. As regards their interest in politics, opinions vary. Only a few months ago two

In West Africa, too, there are engineering faculties receiving continuing help from Britain. The one at Zaria, in Northern Nigeria, is housed in buildings erected with substantial assistance from the British Government and staffed in the main with engineers from this country.‡ At Kumasi, the faculty forms the core of the splendidly housed technological university. It has a large group of British staff and is looking forward to a gift of £100,000's worth of equipment promised by Great Britain. Some of the facilities are shown in Fig. 4.

All these institutions are expanding, and new ones are being developed. It will be many years before their professional faculties can be adequately staffed by local people. Here is a great opportunity to help in the meantime. The Department of Technical Co-operation exists to maximize help of this kind and will meet anyone half-way who thinks he has a contribution to make. Table 1 shows a number of agencies concerned with promoting such recruitment for overseas work.

It would be quite wrong to think of these institutions as islands of British influence. They are, it is true, media through which whatever we may have of relevance and value can be transmitted.

But their main purpose is—and must be—to serve the countries in which they are situated. In Sir Eric Ashby's phrase, they must adapt themselves to their environment and contrive to put down roots in local soil. British staff serving in them must understand this, and be seen to understand it; and they must be capable of making their own imprint on a situation which is new and rapidly changing.

Fig. 3. The Engineering Faculty at University College, Nairobi, has strong links with London University

members of the same engineering faculty expressed conflicting views on this subject.

"Our students", said one, "are like students of technology all over the world—not much interested in politics."

"Of course they have political views," said another, "sometimes very strong ones. But their first object is to stay here and get their degrees; and they have the sense to keep their opinions to themselves in the meantime."

A British engineer engaged in teaching in an underdeveloped country will usually find a strangely mixed bag of academic and professional colleagues. Locally-born teachers are still few and far between in African universities, though numbers are larger in the Indian sub-continent and other parts of Asia. But in all these countries engineers from Britain will find themselves working alongside people from Poland, Yugoslavia, Israel, the Netherlands and many other countries. They will also find themselves working under leadership which is emphatically African, and within a framework which, with each year that passes, owes less to the presence of British academic administrators. Of five university institutions with engineering faculties in West Africa three have African Vice-Chancellors and all have African Registrars.

That these circumstances need not be crippling has been demonstrated by an engineer seconded from the University of London to help get engineering on its feet in the University of Nigeria at Nsukka. A single Briton, working from scratch with Dutch colleagues, an Afro-American administration and students ill-prepared by British standards, he has won golden opinions from all concerned with the development of this completely new university. The best proof of this success in unpromising circumstances is that other British engineers are being sought to replace him when he returns home.

The engineer's contribution

If teachers must wrestle with strange environments in overseas services, so, obviously, must practical engineers. The social background, the state of the economy, the administrative and political framework and, above all, the power situation, will all clearly have their effect. It would be wrong to generalize about most of these, but a glance at our own history and a brief comparison with that of less developed areas will help to highlight some of the more obvious considerations.

Effective communications, adequate power, large-scale industry, social services, reasonably satisfactory housing and sanitation—all the assets we have built up over a long period—scarcely exist in many of the overseas countries concerned; nor does the trained and educated manpower which is a pre-requisite of many of them. All these deficiencies shape the lives of local people and have a marked effect on temporary residents. So does the climate.

Some of these characteristics—the shortage of certain goods and services, differences of diet and methods of marketing, the great distances between inhabited places, and many others—affect all newcomers equally. But there are some of peculiar importance for engineers. For them the unfamiliarity of most local people with any aspect of engineering provides both an opportunity and a source of considerable frustration.

In these largely agricultural countries technology has tended to be a packaged import, somewhat outside the main stream of their affairs. Educationalists have, perhaps too often, thought in terms of producing a local elite, capable of taking over the reins of government and administration. For sons of peasants the office desk and the white collar are natural status symbols. They have gone on from secondary schools, modelled on our own grammar schools and public schools, to read predominently arts subjects in universities which have, until recently, perhaps too closely resembled some of our own, and very few have emerged who understand engineering.

There is thus, for any engineer concerned with development—as well as for any overseas government wishing to modernize its agriculture and expand its industry—the daunting first need to train local teachers.

There is at present no quick way out of this dilemma. The missing teachers must be found and trained and employed at rates which will retain them. They must set about the task of producing the missing engineers—and, especially, technicians—without whom no great advance seems possible.

What, in all this, is the potential role of a British professional engineer? Some aspects of it have already been mentioned. He may, by serving, even for a very short time, in some appropriate post overseas, help to redress the manpower balance. He may do jobs which local people are not yet trained to do, or he may take his place in the ranks of those engaged in the training and educational processes. He may, even while remaining in Britain, help to groom local engineers sent here for that purpose.

There is an additional, specifically engineering contribu-

Fig. 4. Engineering Laboratories at Kumasi in Ghana: the Faculty forms the core of a technological university with a large group of British staff

Mr F. Dunnill *is a Principal in the Department of Technical Co-operation and, as Joint Secretary of the Committee for University Secondments, concerned with the staffing and development of universities in developing Commonwealth countries. Educated at Queen Elizabeth's Grammar School Barnet, Mr Dunnill was in the Royal Navy during the 1939–45 war and has served in several Government Departments. He has written a number of books and essays about various aspects of public administration and was for some years a member of the Council of the Royal Institute of Public Administration.*

tion to this and similar overseas problems. Engineers can do more than fill gaps in an existing pattern: it is possible that they can help to do something fairly drastic about the pattern as a whole, certainly as regards education and training. In a developing country education is almost certainly too important to be left to teachers. New methods such as programmed learning, TV and radio, even the best means of transporting and accommodating pupils and the optimum size of schools, not to mention the problems of integrating the efforts of teachers, training officers and employers, are all subjects to which professional engineers have an important contribution to make.

It would be wrong to suggest that nothing is being done to explore these problems. A great deal of emphasis has been placed, in recent years, on the organizational and economic aspects of education in developing countries. But there is room for a great deal more and for the analytical and practical approach typical of a good engineer to be applied to some of these problems.

There is no reason why schools in Africa or Asia, where there is an even more acute shortage of competent science teachers than here, should not take short cuts into the nineteen-sixties by using television, language laboratories, learning machines and other technical aids. But without the influence, authority and skill of professional engineers, applied at the right point and the right time, it is unlikely that these and similar possibilities will be taken up with the seriousness they deserve.

Unlimited possibilities

This is only one example of an area in which experienced engineers can usefully co-operate with others to promote overseas development. Wherever a dam is being built, a harbour constructed, water brought to a new area or a wide river bridged, important social and economic consequences are seen on a scale which is not now common in Britain.

Engineers are at present engaged, to take only one example, in co-operation with planners, agriculturalists and sociologists, in the resettlement of farmers displaced by the Volta River project. Large areas are being cleared at great speed, new villages planned and built, crops sown mechanically and a new pattern of life prepared for the displaced villagers. And there are endless further possibilities.

Who can see Africans transporting fish from Lake Nyasa to Blantyre by bicycle without wondering what a slightly more sophisticated form of transport could do for the economy of Malawi? Fig. 5 indicates what has already been done in Malaya. Who can fly over tropical lakes and estuaries without wondering what the hovercraft and similar devices may have to contribute to the way of life of these countries?

This brings one to the most exciting aspect of engineering in an under-developed country—the greater possibility of being in on something completely new. There are few certainties in such countries: it is seldom possible to predict

Fig. 5. Fabrication of wagons for the Malayan Railways: Africa too needs modern transport and facilities for its maintenance

precisely how things will go. But, by the same token, it is seldom possible to set precise limits on the potential usefulness of one man—or the potential success of one project. Anything may lead virtually anywhere.

Bumping along a dusty road in Northern Nigeria with a distinguished Indian engineer I mentioned the shortcomings of even the best-planned manpower surveys and the difficulty of measuring precisely how many local engineers are likely to be needed in a given country over the next decade.

"It isn't a real problem," said my companion. "Even if you produce too many to start with and some are forced into technician's jobs while they are waiting for industrialization to catch up, this will do them and the country no harm in the long run. And by no means all of them will be forced down. Some will go up. In these open societies some will climb very quickly to posts of quite unexpected importance."

I thought, as he spoke, of one distinguished African Vice-Chancellor who once worked as an apprentice engineer on a banana boat and still makes a point of maintaining his own motor car.

"It will be a long time before saturation is reached," my Indian friend went on, "for two reasons. First, these new countries cannot afford to maintain artificial distinctions between disciplines. However conservative they may seem now, they will soon catch the wind and move towards the use of all their highly trained people, engineers and others, in a flexible way, so that an engineer's training and work will not distinguish him at all sharply from other applied scientists. This will make it easier to move sideways in response to the needs of the economy. The second point is even more important. It is when trained engineers are not immediately absorbed in industrial activities that they go off and invent things."

It is a nice thought that Africa has still to reach the point where optimum conditions exist for the emergence of a second Henry Ford and that Asia is still awaiting the birth of true prophets of engineering. Lurching along that red laterite road, I thought these events at least possible and the traditional capacity of Africa to produce some new phenomenon certainly no illusion.

Whatever one may think of the probabilities at this level, it is clear that new problems will require solution—and new and deep thinking will be needed. There is no reason why British engineers should not contribute, perhaps dramatically, to this process.

Organising Science in the Modern World*

by Lord Bowden, MA, PhD, MScTech, FIEE

The vast American expenditure on defence and space, science and education draws some of the best graduates from the rest of the world. Since national wealth depends on educated men and women, this makes the rich countries richer and impoverishes the poor ones. Such a distortion of the world's development may yet make the Marxist prophesies come true—if allowed to continue indefinitely.

Universities have suddenly become important all over the world. They are changing and growing everywhere and they are getting more expensive every year. The public used to ignore them—today every citizen is interested in them. In some countries they have become the most important sources of new ideas and new industries and many people think that they will determine the future of mankind.

The American universities are growing as never before. Every year they spend about £500m. on new university buildings and add a quarter of a million students to their university population. It is salutary for an Englishman to remind himself that all our English university buildings put together and their equipment are worth about £500m. and that they have about 200 000 students between them. Every year the Americans create another university system as big as ours. They plan to have 7m. students by the early 1970s; we shall be lucky to have 300 000.

The wealth of America makes it possible for almost everyone to go to college these days and universities have helped to make America rich.

I propose to discuss some problems of engineering and of education. We may have to conclude that education is not a panacea and that even engineering is not enough.

Custodians of wisdom

For centuries all the European universities had regarded themselves as custodians of wisdom. It was their task to preserve knowledge and to pass it on to the next generation. They studied the eternal verities and they ignored the changing world. A hundred and twenty years ago, a graduate said that in Cambridge:

> intelligent persons could not fail to observe that the subjects to which their attention was directed had no relation to any profession or employment whatever, and that the discussions connected with them had no analogy to those trains of thinking that prevailed in the ordinary intercourse of society. Our youth are not taught the sciences which it is the business of these bodies to teach.

Richard Watson became the first full-time Professor of Chemistry in Cambridge in 1760. He had wanted to be Professor of Divinity and he boasted that he knew no chemistry at all. His laboratory was one room which no-one wanted, and his greatest scientific achievement was to persuade the university to pay him a salary of £100 a year.

So much for science in my own university. No wonder that the first industrial revolution was made in England by self-taught men who knew nothing of our universities and

cared less; in fact the universities of England ignored and resented the whole of the industrial revolution and did their best to thwart it. In those days industrial secrets were in the hands of craftsmen and not in the heads of scientists. One of Arkwright's apprentices, named Samuel Slater, memorised details of the machinery and fled to America.

"—they ignored the changing world"

He built carding machines, drawing frames and 72 spindles which were driven by a water wheel, on Rhode Island, and in 1790 he started the American cotton industry. The British would have hanged him if they had caught him. They guarded their industrial monopoly as jealously, as ferociously and as unsuccessfully as the Americans guarded the atom bomb nearly two hundred years later.

The oldest centre of research in England is the Royal Observatory at Greenwich, but although it has been so important to mariners and to international trade, it has never studied anything but astronomy. The first real research laboratory we ever had was the Royal Institution, which is only 170 years old.

The idea of undertaking research in a university was first suggested by a Prussian civil servant named von Humboldt during the Napoleonic wars. Subsequently German university laboratories created organic chemistry and made the German dyestuff industry supreme for a century.

The Duke of Devonshire gave £8450 to build the Cavendish laboratory in 1871—less than 100 years ago. This was the first ever to be built for teaching and research in physics.

There were many who opposed the whole idea of teaching

* A condensed version of the thirteenth Graham Clark Lecture delivered to the three Institutions on 24th April.

Lord Bowden of Chesterfield *(a life peer since 1963) has combined a distinguished academic career with service to government and industry. Principal of the University of Manchester Institute of Science and Technology, he was Minister of State in the Department of Education and Science while on leave of absence from his university post in 1964–65. Born in 1910, Lord Bowden was educated at Chesterfield Grammar School and Emmanuel College, Cambridge. After working with the late Lord Rutherford 1931–34, he spent a year at the University of Amsterdam. Physics master at the Liverpool Collegiate School and chief Physics master at Oundle School in the pre-war years, he was engaged in radar research in England and America 1940–46. Employment with Sir Robert Watson Watt and Partners, and with Ferranti Ltd, Manchester (Digital Computers), was followed by his appointment as Dean of the Faculty of Technology, Manchester University, in 1953. He returned to his present post in 1964. He was chairman of the Electronics Research Council of the Ministry of Aviation 1960–64 and of the Science Masters Association in 1962.*

experimental science in a university. It was suggested that since university lecturers in Cambridge were all men of the highest moral character and moreover clergy of the Church of England, it would be impious to subject their conclusions to the test of experiment. This was less than a hundred years ago in the most enlightened university in England.

The subject was unpopular and the resources small. In 1895 the Cavendish spent about £1000 plus the salary of Prof. Sir J. J. Thomson. In 1912, when the electron was discovered in his laboratory, the budget rose to £3092 10s 5d for everything—teaching, research and salaries. Lord Rutherford became Cavendish Professor in 1920 and began the study of nuclear structure; he never had more than £2500 a year to spend, but he financed a dozen Nobel Prize winners out of it. Chadwick discovered the neutron there; Cockcroft and Walton built the first accelerator and disintegrated lithium and berylium; they split the atom and our modern world began.

The Cavendish Radio Section never had more than £50 a year to spend on apparatus before 1939. But it was there that Sir Edward Appleton won the Nobel Prize for elucidating the mechanism by which short waves are transmitted beyond the horizon.

Thirty years ago the Cavendish was the most famous laboratory in the world. I do not think that any academic institution anywhere will ever again maintain such a concentration of talent or make so many fundamental discoveries so extraordinarily cheaply.

The university laboratories in English universities prided themselves on the fact that their researches had no direct industrial application. To the day of his death Rutherford poured scorn on the idea that his great atomic discoveries would ever be of any commercial use.

The idea that universities should study problems of applied science which were important to industry first took hold in the United States. Prompted by J. S Morrill, President Lincoln established a group of universities endowed with grants of land from the Federal Government. From the

first these new universities were determined to study all the significant problems of society.

Their story must be well known to you. The Great American West was conquered in the laboratories of the Land Grant Colleges and their graduates tamed the continent. We owe two-thirds of all the food which is grown today in the USA to new crops and new techniques they developed and to the students whom they taught. They transformed American industry, they transformed the very nature of a university, and they helped to create the world as we know it.

If I had to choose, I would say that this was the moment when our modern world began. For the first time in history, industry and agriculture were studied in universities by scholars who believed that they should not ignore any of the problems of society. The effect of these colleges was dramatic.

Today there are enormous, magnificent, expensive university laboratories in England, in America and in Canada. Before the war, America concentrated almost entirely on applied research and was content to rely on the tiny laboratories of Europe to provide her with all the fundamental science upon which her industries depended. Vannevar Bush sounded the alarm in 1946 and demanded that America must become self reliant.

American universities bought up the best scientists they could find in the world, and today American laboratories are the best equipped in the world and they are doing most of the best research in the world. America is spending more on research and development than President Roosevelt had at his disposal for all purposes before Pearl Harbour.

European laboratories have grown too. The Cavendish laboratory spends £300 000 a year nowadays—a hundred times as much as it spent when I was a student.

Russian laboratories have grown even faster. They were smaller than American laboratories in 1950 but there are more research workers in Russia now than there are in America.

Not only is science moulding society—it has become a very important component of society itself. Of all the scientists who have ever lived, three-quarters are alive today! If the same growth continued for another 200 years, every man, woman and child and every dog, horse and cow in the world would be a scientist.

And several scientifically-based professions are growing in the same way. For example, if electrical engineering goes on expanding as it has done for 200 years, every Englishman will be an electrical engineer by the end of this century. It is clear that we are heading for trouble. Something has to give.

Something has to give

Ours is the first generation to realise how much our world owes to science. Yet, by a supreme irony, it will fall to us to limit its growth and perhaps to stifle its final flowering. We shall have to make drastic changes in policies which we and our ancestors before us accepted without question for more than 200 years. We may deprive our own children of fundamental human rights which have scarcely been questioned since recorded history began. It is all very alarming.

Perhaps we shall first be buried under a mountain of paper—I can scarcely get into my office for books I shall probably never read. But I shall ignore this possibility and concentrate on the mounting cost of science, which is beginning to worry both universities and governments all over the world.

The number of scientists doubles every 12 years but the cost of research doubles every five or six years. Unfortunately, this is five times as fast as the Gross National Product is growing. How much longer can this continue?

What is happening in America? The Pake Committee of the National Academy of Science suggested last year that the cost of pure research in the US universities should increase at 16 per cent *pa*.

But not even the US could contemplate a further 30-fold increase by the year 2000.

If the world stopped spending so much on science, would its wealth stop growing altogether? There is no doubt that the new, scientifically based industries are growing much faster than traditional trades.

But how much science must a country do itself in order to support its own industries? Science is world wide, its results soon become known everywhere and many prosperous industries depend today on the local application of scientific discoveries which have been made somewhere else. Neither Germany nor Japan is spending as much on science as we are, but the wealth of both countries is growing faster than ours—why?

It seems inevitable that science will have to stop growing quite soon. When it does slow down it will subject the whole scientific world to unprecedented strains. I think that there are already signs of strain, the Brain Drain is one of them.

Scholars and engineers have always wandered about the world in search of better facilities and better teachers. Students from the Caucasus went to Athens to study with Socrates. Throughout history, able intelligent refugees have brought wealth to countries which sheltered them. England profited enormously from the Huguenots who were driven from France by Louis XIV. A hundred years ago, when Lancashire was booming, it attracted skilled men and engineers from all over Europe. Modern British industries would never have developed without men who had been educated in Germany and in Switzerland. In turn, Englishmen provided the impetus to build the American railways and Swiss engineers designed many of their most important bridges. The atomic bomb would never have been built without the scientists who fled from Europe after Hitler came to power.

The growth of Australia and of America and Canada produced the most remarkable and sustained migration of which history has record. Millions of people crossed the Atlantic from Europe and despite their departure most of Europe became more populous and wealthier throughout the whole period of emigration.

The Brain Drain

But immigration to the New World has changed quite fundamentally since the 1939–45 war. In 1951 only four per cent of Canadian immigrants had professional qualifications. By 1960 the proportion had risen to 21 per cent. America now admits almost any well-qualified immigrant, regardless of race—but she does not want any more unskilled labourers. And the supply of scientists and engineers is running out all over the world.

"—*every dog . . . in the world would be a scientist . . . (p. 51)*"

Every country in Europe has begun to notice how many of her most valuable and well-educated young people are crossing the Atlantic.

The loss of graduates from England became serious only a few years ago. In 1961 we lost 198 PhDs to the US; in 1965 we lost 415—which is one PhD out of every six who graduated. That is bad enough, but America wants 10 000 more Physicists at this moment and we lost one PhD physicist out of every three who graduated last year.

A new American organisation has been established to recruit English engineers for America. They are, so I learn from the pamphlets, better than most American graduates. The time has come, so the advertisement says, to "hit Britain and to hit Britain hard". In the first three months of 1967, American employers interviewed twice as many students as they did in the whole of 1966. This ought to worry every university which depends on British industry for its income.

We are short of engineers and scientists in England; nevertheless we are glad to send our own men out to the under-developed countries and we have 16 000 overseas students in our universities. But can we afford to educate so many men for America?

Few people realise the value of these migrants to the community. It probably costs us about £20 000 to educate a PhD and his value to the community during his life-time as a source of new ideas may be as much as a quarter of a million pounds.

If we capitalise the value of those who have left England for America since the war, we have very much more than paid back the whole of the Marshall Aid. The men who went last year will be worth something like a couple of hundred million pounds to American industry during their lifetime.

This is as much as the interest on all the money which we have borrowed from the World Bank and on the total foreign exchange which the country possesses. It is one of the largest items in our balance of payments. It is as much as the budget of the Ministry of Overseas Development through which we finance all our overseas aid. It seems that we are contributing as much to American industry, in spite of ourselves, as we are to the development of all the under-developed countries put together. And so, to their dismay, are the Swiss: the total cost of educating their graduates who go to America every year is 30m. Swiss francs.

"—*engineers have always wandered about the world . . .*"

The underdeveloped world

But if the problems of the wealthy countries like Switzerland, England and Canada are difficult and unfamiliar, the problems which confront the poor under-developed emerging countries are much worse. Newly independent countries want to do everything at once—they want to establish themselves in the world, they want to get rid of their former masters—most of all they want to be rich. They know what education has done for America and for other wealthy countries; and they believe it will help to make them rich too—quickly, inevitably and painlessly. They are very impatient. They begin their march into the future by getting themselves some status symbols such as an airline, some impressive parliamentary and administrative buildings, and a university which must be like an English university; they imagine that anything else would be second best. They think that if they have status symbols the status itself will follow. The whole world sympathises with them.

I have seen a few new universities in Africa. Some of them are extraordinarily impressive. They will educate the poorest people in the world, and yet they have been the most expensive in the world to build. The University of Legon, for example, cost Ghana several times as much per student as the UGC allowed us to spend on universities in England. Legon bought an atomic pile from Russia to study nuclear physics.

That great English scholar and keen student of Africa, Miss Margery Perham, recorded the emotion and delight with which she heard boys on the Gold Coast acting Euripides. The new foundations were not to be African universities as such, but universities of the Western World and inheritors of the best that Europe could offer. It was a wonderful vision. Africans and their well-wishers were sure that African scholars would be able to administer their native lands as well as English Arts graduates had done; but no-one seems to have heard of J. S. Morrill and his 'Cow Colleges'—whose graduates tamed the continent of America.

A country whose poverty-stricken inhabitants suffer so much from disease due to inadequate diet and intestinal parasites, a country which has untapped natural resources but few roads and little industry, should have a different order of priorities. I have seen children leading their sightless parents to work in the fields because they suffered from 'river blindness', a disease due almost entirely to their insanitary way of life. I felt that they needed better drains, better water and more protein. I found it hard to understand what the graduates would do with some of their scholarly achievements.

Is their educational system really helping these unfortunate countries to modernise themselves?

All too often their graduates can find little outlet for their hardly won knowledge at home—their great expectations are rudely disappointed. Their education cuts them off from their own people and not all can find suitable work at home. They are literate—and they are unemployed. Many of them become dissatisfied and emigrate.

Thus it is always the poor and under-developed countries which lose their precious graduates although they are so desperately short of educated men. The new medical school in the French State of Dahomey has produced a total of 70 doctors of medicine. Only 16 of them remain in their native land.

America has been most generous in giving scholarships and grants to students from all over the world so that they can study in America. What happens to them? More than half the overseas students who went to the University of Iowa between 1950 and 1960 stayed in America. The proportion is increasing. Only 10 per cent of the students from Taiwan, Hong Kong, Korea and Greece return home.

The unemployed intelligensia

America has given more than $2 000 m. of aid to Iran since the war, but more than 85 per cent of the population are still illiterate: sixty per cent of the students who go from Iran to American universities stay in the States. No amount of aid will help a country to help itself if its educated men all leave it as fast as they can.

Everyone rejoices when young men come to England or America to learn what they can from us and then go home to help their kinsfolk, but all too frequently Western universities select the ablest men of their generation from primitive societies and take them away from their people for ever. They are depriving struggling countries of their natural leaders. Of course foreign students prefer to settle in America. A graduate from Taiwan can earn $8000 a year in North America but only $1000 at home. Why should he go back? Other students stay because they dislike the policies of the Government at home. Some students have been overeducated for jobs in their native lands. Some seem deliberately to choose subjects useful in America but without application at home.

Indian doctors are desperately needed at home where there is a dreadful shortage of medical men; a few of them might transform the lives of thousands of their fellow citizens, but we want them and we make it worth their while to stay here. Engineers from India and Africa help to man our factories, university teachers from Turkey, from Egypt and from the Argentine educate our young men. They replace our own people who have gone to America. This migration has begun to dominate academic and industrial policy in every country in the world.

India needs engineers and the work they can do, but half the Indian graduate engineers cannot get a job for a year after they graduate. The University of the Deccan has produced exactly 100 graduates in Anthropology in the past decade. Ninety-five of them are out of work and their professor has finally abandoned hope and emigrated himself. Men who cannot find a post in a university are cut off from their families by their education, and cannot go back to their native towns and villages; they drift aimlessly about in the cities—they are the dissatisfied unemployed intelligentsia, and of course they try desperately hard to find a job abroad. Who is to blame them?

For many years India has spent 40 per cent of her research budget on nuclear physics and only 8 per cent on agriculture. She has proved to everyone that her nuclear physicists are among the best in the world. They leave India for other countries where there are resources to match their education. Meanwhile the population grows, agriculture stagnates, and the country starves.

The dilemma which confronts these countries is appalling; even if they maintain enormous differences between the standards of living of the educated and uneducated parts of the population, most of their best people still leave. Even in Europe we have to face this very same problem. Poor countries can never hope to match American salaries.

The bystander may wonder if some universities have become the 20th century equivalent of the medieval monasteries.

The parallel is disquieting at times. One finds in

both a belief that their occupants must be free to occupy themselves as they please, that they belong to a world-wide society.

They demand magnificent buildings, they must live well, but they claim to owe nothing to society in return except what they themselves determine. They are engaged, so they say, in the study of the ultimate truths—of theology or of physics; they believe that the world is lucky to have them in it, they are sure that they are its finest flowers. It is true that the work some of these men do will lead to new industries which may transform our lives. But it is quite impractical to assume that all would-be research workers are potential Faradays. Society cannot give all of them everything they want. Someone has to choose between claimants.

How is the choice to be made?

In some ways one can compare the huge machines in Geneva and Brookhaven with St Peter's, Rome. Each represents the greatest technical triumph of its age. Each was built as a flamboyant gesture—for purposes which one can fairly describe as spiritual rather than material. Neither is of any practical use, but society would be infinitely poorer without them.

Will future generations see fit to compare John Adams who built the CERN machine, with Michaelangelo? Stranger things have happened.

Physicists should recall that St Peter's was built at the end of the age of faith, that it cost so much that it nearly bankrupted the Church, and that it provoked the wrath of Luther and inspired the rise of the reformed churches. The cost of CERN may yet be felt in physics laboratories in every English university.

There is tremendous pressure in every country in the world to spend vast sums of public money on equipment whose only justification is that the universities have educated men to use it—who demand it as the price for staying at home. Will the time come when Europe will deliberately export the more extravagant scholars to America in the hope that they will bankrupt the United States economy and leave less enterprising countries to survive in peace?

Stick to local problems

A scholar in an under-developed country finds it impossible to compete with colleagues who are studying the same problems in the West. His only hope of achieving real distinction must be to work in some unfashionable subject of great local interest. Nigerians might make fundamental contributions to African literature. Physicists in Ghana can analyse the geology of Africa. The Physics Department in Hong Kong has become famous by studying the ocean currents and the bed of the China Sea. Australian scientists discovered the effect of traces of metals on the health of sheep. They made their country prosperous and themselves famous.

The migration of scientists from one country to another is not unlike the movement of people within individual countries.

People move from the empty Highlands to the over-crowded South-East of England. The Americans are worried because their scientists leave less-developed States and flock to the east and west coasts. That is why they have decided to put the enormous new synchrocyclotron near Chicago rather than in California, and also why Congress insists that new medical research centres must be sited away from Washington.

The Russians have had to face the same problem. Until recently almost all their scientists and engineers lived in Leningrad or Moscow. About ten years ago, the Soviets decided to establish an entirely new scientific centre in the heart of Siberia, and last year I went to see the magnificent Science City of Novosibirsk. It has 16 research institutes, a new

"—impractical to assume that all would-be research workers are potential Faradays"

university, a special school, and about 35 000 inhabitants. It is very beautifully situated by the side of the Ob Sea. Young scientists flock there.

The Russians told me that Novosibirsk has cost about £200 m. in the last ten years and it has probably paid for itself twice over already. The Science City has organised the geological survey of the whole of Siberia. More than half the coal in the world is there; and unlimited quantities of iron and aluminium and rare metals; half the continent seems to be floating on oil, and they have found diamond fields in the north.

Novosibirsk has taken men of the cities into the wilderness and it will help to bring a whole continent into the 20th century.

Distorted pattern

The demand for scientists and engineers in the USA is insatiable. Three-quarters of the American research and development programme is now in the defence and space industries. These two giants are still growing rapidly and inexorably. Whole cities and States depend upon them and would go bankrupt without them, but does Washington control them any more? Was Mr Eisenhower right to warn his countrymen about their potential for harm? They had, he said, a momentum of their own, wholly independent of the need of the community—and of the world. The American defence and space budget next year is about $80 000 m.— more than the Gross National Product of Great Britain. It depends to an extraordinary extent on scientists and engineers.

No wonder these men are in short supply in America— no wonder they neglect less rewarding tasks at home— however important they may be—and flock there from all over the world.

The programme dominates the biggest industrial firms in America, and is beginning to play an ever-increasing role in the universities. Nearly 80 per cent of the research in MIT is paid for by the Department of Defence. Cal. Tech. engineered the Mars probe and was responsible for developing the propulsive mechanism, the control mechanism, the measuring equipment, the cameras, and for transmitting the data back to earth—a perfectly fantastic achievement upon which everyone concerned must be congratulated. But it may be a preliminary to a second bigger Mars probe which, so we are informed, might cost $100 000 m.

The whole pattern of education in the USA has been distorted, and it is becoming apparent that the educational system of the world is being distorted too. Our universities and the universities of poor countries in Africa and Asia may find themselves chained to the chariot wheels of the American Moloch which is to put a man on the moon by 1970; and this machine, let us remind ourselves, was created after the Americans had already produced the nuclear equivalent of about 100 tons of dynamite for every living human being and had established a method of delivering it to any point on the face of the earth.

The economy of the United States is in an extraordinary condition.

The real needs of the country's population, by which I mean their food, their houses, their clothes, their motor cars and so on, can be met by the efforts of about half the working force, so that the government must find work for the rest.

It seems to me that in her defence programme and space programme America has created the most extravagant, most sophisticated, and most dangerous system of outdoor relief ever devised by a great nation in peacetime.

To him that hath . . .

Has there been anything like it since the Government of Imperial Rome ruined itself—and the world around it—by spending 40 per cent of its revenue on the circus? But, you may say, America is a wealthy country and if she wishes to do so, she could afford to build Moon probes, Mars probes, probes to Jupiter and bigger and better rockets. I do not think this is true.

The riots which have taken place in San Francisco, in Watts County and in New York, are only the most obvious evidence of the discontent of the submerged part of the American population. It is preposterous that America should try to put a man on the moon when many of her citizens are living in conditions of squalor, devoid even of hope. The forests are being felled, essential ores are being used up, the Great Lakes themselves, the biggest areas of fresh water on the face of the earth, are rapidly becoming contaminated, and the St. Lawrence river which drains half the North American continent is reported to be unsafe for swimming below Montreal. Millions of citizens, unable to find work which suits their talents, abilities and education, view the future with anxiety and, meanwhile, the vast Moloch is moving slowly onwards, crushing beneath it everything in its way.

It is time for an observer to protest when he finds that fields in Africa, India and all over the world may remain uncultivated because of an attempt to put a man on the moon and to send a couple of probes to Mars.

We must keep our sense of proportion. We must

"—equipment whose only justification is that the universities have educated men to use it . ."

not forget that science has made our modern world, that it is transforming it before our eyes, that we could not survive without it and its products. We must remember, furthermore, that modern science is itself one of the supreme achievements of the human intellect.

The citizens of a civilised country today very properly want to have distinguished scientists among their numbers as much as they want painters and singers and musicians and architects and actors.

We used to think that these things were simple. When I was young there were not so very many scientists in the world and the country could afford to indulge their eccentricities.

But we have eaten of the Tree of Knowledge. Science has become too important to be left to scientists.

The whole pattern of development throughout the world is being disturbed by the toll that America is levying on countries less wealthy than herself. How much longer will scholars be free to move about the world? They have done so for thousands of years, but if they do so in future, will it be to fulfil the Biblical adage:

> To him that hath shall be given and from him that hath not shall be taken even that which he hath.

This text has perplexed many theologians, but Karl Marx based his whole political creed upon it. He was sure that in a capitalist society, the rich must become richer while the poor become poorer.

In every industrialised country in the world the enormous increase in productivity, the pressure of organised labour and the social conscience of the electors have disproved this fundamental theory of Marxism. Wealth in an industrial country does not depend on poverty in the same country; machines do the work which once was given to the labouring poor.

Poor countries took the advice of well-intentioned liberal scholars who hoped to help them by building an educational system, but America is taking their natural leaders and their educated men.

Last year the GNP of America increased by more than the total GNP of the African continent, which is hardly changing at all.

It is beginning to seem all too probable that Marx will be proved right after all in a way he did not foresee. The rich countries are becoming richer at the expense of the poor countries because educated men go to America because America is spending so lavishly to combat the menace of the Communism which Marx inspired in Russia and China. And it is this chain of events which seems to me to be the supreme irony of history.

The American universities impressed themselves upon the world because they first saw how to study the problem of society in an academic environment. Will their unlimited wealth and their preoccupation with the problem of defence and space ruin them—and deprive half the universities of the whole world of their staff?

Shall we be forced to restrict the movement of men about the world as we have long restricted the movement of goods—and do so in the name of progress? That indeed would be a tragedy.

The only great countries whose graduates do not flock to America are China and Russia. Will mutual suspicion between East and West keep these Communist States free from the thraldom of Marxism as it was originally expounded by the sage himself?

Time alone will tell.

In a comprehensive review of the engineer's position in 1964, the author stressed the importance of thinking ahead and ranged from education to management and the structure of industry. His naval experience illustrated some of the points he made.

The Shape of Things to Come— A Challenge to Engineers

by Vice-Admiral Sir Frank Mason, KCB, FIMechE*

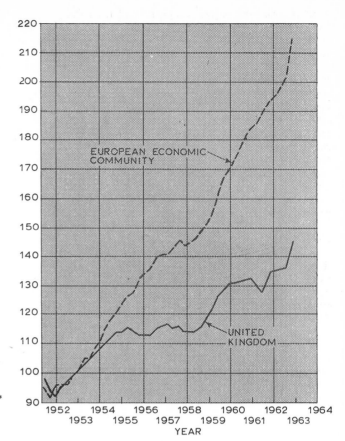

Fig. 1. Industrial production 1952–1963 (1953 = 100)

One might be permitted perhaps to wonder why it is necessary to talk about the role of the engineer in the twentieth century; it might be argued that it should be self-evident. I would agree that it should be but all the indications are that, to a large proportion of the community, it is not. Why this should be so is primarily a matter of history, and the proper understanding of history is of immense importance.

What I am concerned that we should do is to examine the present situation in the United Kingdom, consider whether it is satisfactory from the national point of view and, if it is not, suggest ways by which it might be improved.

Engineering and our economy

Why is it important that there should be a proper understanding of the role of the engineer in this country? Perhaps Table 1 will make it clear. The engineering industries contributed 46 per cent of direct exports in 1963. In addition there is their indirect contribution, for they provide many of the tools required by the other exporting industries.

Although our industrial production is increasing, it has not increased, in the long term, at the same rate as world trade, so we are not maintaining our share of the world markets as shown in Fig. 1. At the same time, our population is growing and with it our need for food and raw materials; for our only native raw material available for manufacturing industry in any quantity is coal. We have, however, one other very important native asset: our brains, and it seems to me to be quite vital that our exports should reflect this undeniable fact. So, if we are to live, let alone increase our

standard of living, our products must contain a high proportion of brainwork; put more simply, they must be very advanced, which means in practice that they must be well developed before being placed on the market. They must also be what people want, at a time when they want them and at a price they are willing to pay.

There is no future for us in making products which everyone else is making or can make, for the emergent countries will soon make for themselves those standard items of equipment which the technically more advanced countries have perfected over the years, and for which the techniques are widely known. Scientific instruments are one example of where we should be in the van. Advanced products, in the main, do not require a vast home market but they do require imagination, enterprise and courage on the part of both manufacturers and users who must be prepared to share the risks between them. Switzerland is an example of a country with small resources and a small home market, living on advanced products as well as tourists. Japan is perhaps an even more direct comparison; she has taken the need for quality and reliability very much to heart.

There is another form of export; the engineers themselves. As Mr Geoffrey Bone and Dr D. G. Christopherson said to a conference of headmasters held in the Institution of Mech. Engineers overseas countries, formerly part of the British

* *Formerly Engineer-in-Chief of the Fleet. This was his Presidential Address to the Institution of Mechanical Engineers.*

Empire, no longer require administrators from this country but they do need, and need very badly, our engineers.

The satisfaction of this need requires young men, and indeed young women, who are prepared to go where they are needed, whether on some specific assignment or on more permanent service: in so doing they will help to apply the technical knowledge and experience of Great Britain to the urgent problem of raising the standard of living in places where this falls far short of that in their homeland. Looked at from the point of view of our own economy, if we do not meet this need, other countries will, to the detriment of our trade and the advantage of their own. From what I know of the younger generation, I have no fear that this need will go unfulfilled; but they must be given imaginative leadership.

So these are some of the reasons why engineering is important to the nation.

Applying the results of research

It is often said that we are good at research and even development, but not so good at converting the results of these two essential activities into saleable products. Another variant of this is that we often initiate an idea, only to allow it to be exploited abroad. I fear that there is quite a lot of truth in these beliefs, although the reverse is also sometimes true.

My own experience agrees with the belief that we are not behind in research and even development; it is in the final stage of exploitation where we tend to lag behind others. Often, of course, the idea and the work done to show developments to be feasible are ahead of their time; no use can be seen for them and they go into the limbo of forgotten things.

Table 1—The Engineer's Contribution

Income generated in 1962	(£ million)
(a) Engineering and allied industries	3 600
(b) All manufacturing industries	8670
(c) Gross national product	25200
(a) as per cent of (b)	42
(a) as per cent of (c)	14

Percentage of total exports in 1963	
Machinery	28·8
Road motor vehicles	12·2
Aircraft, aircraft engines	2·7
Ships and boats	1·0
Professional scientific and control instruments	1·8
Total engineering products	46·1

Figures by courtesy of the Central Statistical Office

Table 2—Firms in England, Scotland and Wales Manufacturing Mechanical and Electrical Products

Number of employees	Number of firms	Total number of employees	Average number of employees
11–24	2285	41 000	18
25–99	5166	268 000	52
100–499	2427	518 000	213
500–999	426	298 000	700
1000–1999	249	351 000	1410
2000 or more	131	509 000	3885
Totals	10 684	1 985 000	186

Figures extracted from the Ministry of Labour Gazette, April 1962

But papers are published and placed in libraries the world over and are there for reference when, perhaps years later, someone realizes that the work can now be applied, for example, because the knowledge of materials has caught up.

If we ourselves put more effort into the rapid application of the knowledge gained by research, this might not happen so frequently and, as a result, we should be able to exploit to our material benefit the fruits of our own research. There is a somewhat pressing need to do this because similar ideas seem to arise in different countries at about the same time; the country which gets in first reaps that particular harvest. Moreover, the application of today's research is the only way of paying for the research of tomorrow.

A forward-looking unit

A corollary of this is that any live concern needs what I would call a forward-looking unit (it could be one man), whose business it is to look not just at the next generation of products but at the next generation but one. It is no good saying that we cannot afford it; we just cannot afford to be without it if we are to survive. And yet, one of our greatest problems is how to get the results of research into industry. We are apt to think of industry as consisting of a few giant firms but this is not so. Table 2 shows the latest available figures relating to the sizes of firms manufacturing mechanical and electrical goods in Britain. The greater part of British industry consists of medium and small sized firms, many without a single graduate on their staff, and it is these that need to be fed with the results of research and development, a great deal of which is to be had for the asking, or even just the reading. But so many do not seem to know that they have a need, and all too often proffered help falls on stony ground.

The reason for this is a failure of communication; there is no one at the receiving end; hence my plea for a forward-looking unit. Before this can come about, there will have to be a better understanding of the part which can be played by young men and women who have graduated from universities and other institutions of higher education. These same people would also help to further what can properly be called the scientific approach to the problems which face their firms; in other words, arriving at solutions by reasoning from first principles instead of a too widespread tendency to proceed by trial and error, relying on experience alone.

I speak from some knowledge of the difficulties which face establishments like the National Engineering Laboratory in getting the results of their work applied in industry. I should not be so concerned if time was of little consequence but I feel almost vehemently that it matters a great deal and, in my view, time is no longer on our side; if it ever was.

Hitherto a great deal of the stimulus for research and development has come from the requirements of defence; this is now becoming less so and we must look elsewhere for some, at least, of our inspiration. The military requirements are apt to be very exacting and this is their chief merit, as far as the economy as a whole is concerned. I welcome, therefore, the policy of the Department of Scientific and Industrial Research in placing civil development contracts with firms who have promising projects which, however, contain speculative features involving too big an element of technical risk.

At present the scale of this joint operation falls far short of what is needed if the policy is to be really effective; moreover it calls for a lightness of touch on the financial controls, not usually associated with Government. The Federation of British Industries, in a report on the needs of research and development, have suggested working up to £50m a year.

This is a large sum, perhaps too large, when it is remembered that Government money, however lavish, will not achieve the desired ends without the support of Industry.

Our industrial structure

I am concerned about a tendency to regard the purchase of licences to build to foreign designs as a substitute for doing our own research and development. I have often heard this suggested as being a cheap way of buying the results of other people's work. It is one thing to buy a foreign licence in order to gain time by cutting a particular corner; it is quite another to try to live by it. This can only lead to disaster, to atrophy, whereby we cease to be creative and drive all our best and brightest young men and women to other lands where their talents will have the scope denied to them here.

To a limited extent this is already happening, although we need to keep a sense of proportion and remember that we have attracted gifted people from abroad, particularly during the period immediately following the last war. Throughout our history, persecution on the European Continent has enriched this land.

I have referred to the large firms which are really groups of firms. It has to be remembered that they carry heavy overheads because of their very considerable effort in research and development. This is a grave handicap, to which must be added apprentice training for which these firms carry a disproportionate share of the load. I hope that the Industrial Training Act will result in ways being found to recompense them for this added burden.

Indeed, I suggest that an even more fundamental problem needs to be investigated and solved: no less than the structure of the British engineering industry, for it is all too evident that in some sectors there are too many firms competing for the same market. The industry with which I have been associated during my career—shipbuilding and marine engineering—is a case in point. While foreign countries are not by any means to be emulated in all that they do, some of them show a very different picture from the situation in this country.

For example, in the United States there are only two firms in the field of heavy power plant and these two also supply the marine requirements. In this country there are four, and these make only a small contribution to marine engineering.

If we do not put our house in better order, we risk being undersold, simply because our present advantages of lower costs will have been more than cancelled out by the concentration of production in fewer hands in our competitors' countries.

There is a great deal to do and I should like to turn your attention to the role which engineers must play in maintaining the greatness of our country. May I first try to dispel some misunderstandings about this role.

The engineer and the scientist

Confusion exists, and it is quite general, about the difference between a scientist and an engineer. Certainly since the 1939–45 war, science has come, quite properly, into the forefront of the national consciousness, but all too frequently it is not understood that the engineer gives expression to the findings of science; if, indeed, it is realized that such an activity exists. If the role of the engineer is thought of at all, it is associated in the popular mind with craftsmanship and this idea persists in places where more informed views might be expected.

So it may not be out of place to describe briefly what each does, bearing in mind that both the scientist and the engineer receive comparable fundamental academic training.

The scientist is primarily concerned with discovering the facts of nature and the laws which govern physical phenomena. He has to take things apart, so to speak, and by observation, analysis and deduction formulate the laws by which natural processes work. This frequently means evolving theories to fit observed facts, and testing them. The process is often a long one and may span several generations: Above all, the scientist is concerned with the elucidation of facts; compromise, as the term is ordinarily understood, forms no part of his stock-in-trade. From time to time, however, he must pause to evolve temporary laws from facts which he knows to be incomplete, laws which a later generation may even disprove.

On the other hand, the engineer takes the findings of the scientist and uses them to build something which will be of direct benefit to mankind. Engineering consists in applying power and materials of all sorts to the service of man; to lighten the labour of earning his daily bread, to enable him to move more rapidly from place to place, and generally to lift mankind into a higher plane of existence where he can enjoy cultural pursuits with more leisure.

I am not saying that this is what actually happens, for we have not yet arrived at such a state, but in time we will. What is now called automation is simply the acceleration of a process which has been going on for a long time and is rapidly bringing to the fore problems of retraining and the use of leisure with their social consequences. In order to use the findings of science, the engineer has to satisfy many conflicting requirements. These will not be only technical, but also commercial and financial and, most important of all, personal ones.

The resolution of the diverse elements in the situation demands good judgement and brings in the art of compromise. It is indeed an intellectual challenge of a high order to the engineer and calls not only for imagination but also for courage.

So, although the scientist and the engineer start their academic life in much the same way, their roles are complementary, but different.

Engineering is a science-based art, and the engineer should be able to indicate to the scientist areas in which knowledge is scarce. It is axiomatic that the scientist and the engineer must work in close understanding and harmony if the results of their work are to come to full fruition. In this respect, the present state of affairs leaves much to be desired and falls far short of the need.

In trying to reconcile all the factors involved in producing a new piece of equipment, the engineer may well have to go beyond existing experiences in the application of physical law. He will then be involved in experiment; an activity which, though not strictly research, uses the techniques of careful measurement, observation and deduction which are the normal routines of the scientist. So it comes about that scientists are often employed by large firms and organizations for this type of experimental work and the distinction between scientist and engineer becomes nominal. It is undoubtedly the engineer, however, who must take the decisions which depend on the results of such experiments, and shoulder the responsibility involved.

Some confusion has arisen in the minds of administrators between the role of the scientist in pushing forward the frontiers of what is theoretically possible, and that of the engineer in judging what is practicable, within the limits of time and finance.

What, then, is the full scope of the engineer's activity if he is to be able to make such judgements? In the earlier stages of the industrial revolution, about the time when this Institution was founded and earlier, the great masters did everything themselves. I mean by this that they personally worked out what was needed to meet the requirements of any given situation; they made their own calculations from what knowledge existed at the time. They did their own experimental work, reaching out into the realms of the unknown. They made their own drawings and had a hand in the actual manufacture. As the demand for the products of engineering increased, and it did very rapidly, this situation changed and the master had to look to others to undertake some of the detailed activities. This involved him in training, and even educating, the young men he employed. In addition he was frequently his own salesman, since no others were equipped to persuade those responsible for finance of the potentialities of the new field of engineering. Inevitably also he had to turn more and more attention to the management of his firm as it increased in size.

Today the position is vastly more complicated and although I deem it imperative that the overall responsibility should come back to one engineer, many other engineers who have become specialists in the various branches, are required to play their part. Whatever their specializations may be, their common need is to be well grounded in the principles underlying all engineering and to possess professional competence of a high order.

I should like, if I may, to quote from the introduction of the Parsons Memorial Lecture which I had the honour to deliver before the North-East Coast Institution of Engineers and Shipbuilders in 1955. In referring to Sir Charles Parsons and the Royal Navy, I said:

> Today we are doing our utmost to be worthy of this heritage. On each side of us giants in the field of technology, as in other fields, have arisen to overshadow our past pride of place. With far smaller resources we have nevertheless set ourselves as a country to match them. Our chief asset is still our technical genius, but it is no longer vested in the tremendous stature of one man's inventions and pertinacity. We have to turn to the great industrial groups and associations where the wisdom of what is still a comparatively small number of men acts as the focus of teamwork of a large body of technical experts in particular fields.

In other words, teamwork is involved. Of course teams require leaders who also need to be good engineers, and the two things are not always synonymous; I should add that the teams will not necessarily be entities in themselves, although I should like to see this happening more and more. I suggest that such teams, even if they do not always include scientists and financiers, will certainly need to have ready access to the physicist, the chemist, the metallurgist, the mathematician, and the accountant. Put in another way, engineering, like peace is said to be, is indivisible.

The engineer as a manager

If this indeed is the role of the engineer in the twentieth century, why is it not better understood and recognized for what it is, challenging, exciting and satisfying? It is easier to pose this question than to answer it. The origin of the misunderstanding lies back at the beginning of mechanical engineering as we know it and may well be found in the name chosen by the early masters wherewith to describe themselves and others. They used the term 'engineer' (derived from the Latin *ingenium*) to describe both the man who designed and built the machine and the man who operated it. On the continent of Europe the term 'engineer' was reserved for the former and the word 'machinist' or its equivalent for the latter: and in most other countries there is a proper understanding of the role of the engineer. In some countries the title is safeguarded by law.

As has so often happened in my own experience, the engineer has been his own worst enemy by concentrating too assiduously on the demands of his chosen vocation without considering sufficiently its impact on others, or theirs on him. If the engineer is to influence policy, he must be prepared to shoulder wider responsibilities than those of engineering alone; those prime considerations of time and cost within which all engineering responsibilities have to be discharged. This will certainly mean stepping outside a relatively secure and familiar world into a more exposed one, where the cold winds of finance and competition blow keenly.

Having said this, I should like to add that it is easy to suggest, as is often done, that the engineer is necessarily a narrow specialist, incapable of understanding the world of affairs and commerce, while at the same time assuming that the accountant, the lawyer and above all the brand known as an 'administrator' are immune from such restricting influences. A belief seems to be current that knowledge of what one is administering is unimportant.

That one needs to have an educated mind to manage an activity properly is indisputable. But perhaps one may be permitted to question whether a classical education is the only means of training the mind to deal with the scientific and engineering problems of the twentieth century. There are other disciplines which are equally effective and more apposite. But whatever the discipline, some postgraduate study in management is an urgent requirement and I am very glad, therefore, to see the proposal made by Lord Franks for two new Colleges of Management, one to be associated with the Imperial College of Science and Technology and the London School of Economics, the other with the University of Manchester.

All the disciplines have this in common: they require a thorough grounding in the English language; not least science and engineering, in which clarity of expression seems to grow less as the need for it increases.

Having thus touched briefly upon the engineer's place in management, I should like to mention the other functions which the original engineers performed in building up the industry of our country. We must continue to see that properly educated engineers are available to fulfil them all.

Research, design and manufacture

If one were to ask a layman whether he thought that an engineer was likely to be concerned with research and development, he might think that, possibly, the engineer had something to do with development but I very much doubt whether he would associate him with research. And yet, applied research is as much the province of the engineer as it is of the scientist.

In any very advanced project, such as a large plant of new conception and novel features, one would expect to find the influence of the scientist predominant in the early stages but the engineer would be responsible for the ultimate result. In the earliest stages, there would have to be a survey of existing knowledge and research might be needed to fill any gaps which were identified. The results of pure research would be surveyed and any applied research needed put in hand. One might call this a combined operation of scientists and engineers. As the knowledge is gathered in and the gaps filled, development work starts on specific features of the project and this stage is almost entirely in the hands of the engineer.

Concurrently, design studies will be undertaken; initially

as part of the combined operation; but as the emphasis moves on to detailed design, the engineer will be in full control. The team which he builds up will, however, need to consult other people whose knowledge and advice is necessary to a successful outcome.

The field of design is easily recognized as the activity, above all others, through which the engineer is able to give expression to his knowledge, his wisdom and his professional ability. I believe British industry has suffered from delegating too much responsibility in this field without adequate professional engineering supervision. However brilliant the concept, a design is only as good as its details and too many of the engineer's first class design concepts are apt to fall into disrepute because of lack of attention to detail in design as well as manufacture.

This may arise from a growing tendency to divorce manufacture from design by interposing another stage, thus weakening, if not entirely removing, the responsibility of the designer for ensuring that his concept can be made as effectively and cheaply as possible. I think that designing and planning for economic production must go hand in hand and, indeed, be the overall responsibility of one man. The Feilden Report, produced for the Council for Scientific and Industrial Research, makes interesting reading on this subject.

There has been much discussion of late about the teaching of design, and quite properly too, for opinions differ sharply; but it seems likely to me that this is not a subject for teaching by rule. An introduction to this vital activity, however, could be given even in the senior forms of secondary schools and could be followed up in the universities by an exposition of the philosophy of design.

The creative activity called design is, I believe, one which can only be truly absorbed by precept and practice or, if you like, by sitting at the feet of a master. This could take the form of a postgraduate activity after the graduate has had some enlightenment of what actually goes on in the drawing office and in the workshop; an important object being to acquire an understanding of the people who work in these places, as well as a knowledge of what machines can, and cannot, do. The present virtual divorce of art from technical design did not exist in earlier centuries.

The influence of the customer

A most important, indeed vital, sphere in which the engineer is perhaps not as numerous as the circumstances require is what is usually called 'sales'.

In mentioning sales, one ought perhaps to differentiate between the relatively simple activity of selling standard articles, including spare parts of complicated equipment, and selling the complicated equipment itself. The latter activity requires both a thorough understanding of what one is selling and of what the customer really needs. The proper salesman in this case is the representative of the design department.

Market research may well be needed to support both kinds of activity. Here is the direct contact with the customer whose needs are the wellspring of design and production and, through these, influence research and development. One cannot design in a vacuum but must meet a known or potential need: discovering what are the customer's wants and helping him to formulate it are certainly within the role of the engineer. So is the complementary activity of service after sale from which comes the feedback to the designer of information about performance and accessibility. Analysis of failures in service is an indispensable part of this process.

Perhaps I could illustrate this by describing briefly how it works in the Royal Navy, as far as the propelling machinery is concerned. For this purpose the Navy can be thought of as a very large firm, with a strong scientific and engineering side of its own, which is the customer for the propelling machinery of a warship. It is in fact an informed customer with large resources.

The actual customer is the Fleet, represented by the Naval Staff who, by various means, not the least of which is first-hand operational knowledge, evolves a need for a new sort of warship. This is discussed with the technical people who, of course, have to consider the ship as a whole.

From these discussions the requirements for the propelling machinery emerge. It is a long and somewhat complicated process, facilitated by the fact that all the parties (including the Naval Staff divisions themselves) contain seamen and engineers who have been at sea and will go to sea again, probably in the very ships they are evolving. So there is a common thread running through their work.

When the 'staff requirement', as it is known, has been settled, it is translated into a technical specification for the guidance of the contractor who carries out the detail design and manufacture under the oversight of the technical side of the Navy. In recent years, the evaluation of various con-

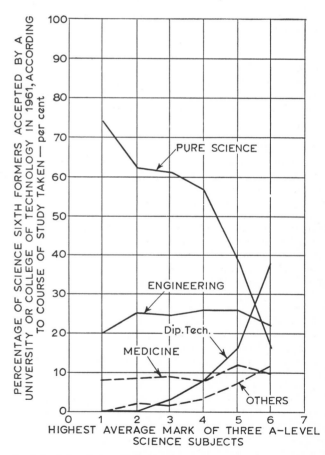

Fig. 2. Sixth form boys in England and Wales according to G.C.E. A-level marks and choice of study at university level (sciences and technology). The grades range from over 75 per cent (1) to between 30 and 40 per cent (7) as the average mark for three A-level subjects

(Reproduced from Technology and the Sixth Form Boy, published by the Department of Education, Oxford University)

cepts for new types of propulsion machinery as well as the preparation of the detailed technical specifications, have been much facilitated by the creation of a unique body called the Yarrow-Admiralty Research Department which, under development contracts, looks ahead to what types of machinery can be considered for future warships within five to twenty years. It produces data on which the Navy Department of the Ministry of Defence can judge the way in which the navy staff requirements for a new class of ship can best be met. Finally, it takes a major part in producing or arranging the production, not only of the necessary specifications, but also of models and mock-ups, sometimes to full scale.

These specifications are the result of research in which the products of industry best suited to meet the particular requirements are identified, evaluated and woven into an optimum machinery installation. This process, which takes place in close association with the Navy Department, also reveals where further development and research are needed.

Thus the naval engineer officer in charge of a machinery project in the Navy Department is carrying out the roles of: manager of the various design contracts under which firms are asked to contribute to the project; designer of the overall installation; salesman (showing that the naval staff requirement is fulfilled); and, to some extent, overseer of manufacture. He may well be involved in development of machinery at some time in his life and his training is based on experience of operating and maintaining naval machinery at sea. The versatility which is demanded of its engineers by the Royal Navy is typical of the versatility demanded of the whole engineering profession.

Much attention has been paid to the education of these officers by the Royal Navy and it is equally necessary that our nation should take stock of the education of its engineers for industry.

Too few engineering students

I think the reasons why there are too few engineering students are social and historical. Consider the way in which our educational system has developed.

Full-time general education is relatively new and full-time technical education has really only just started. Previously, except for the universities and the Royal Dockyard Schools, technical knowledge was acquired in industry and by evening study at technical colleges and schools. Out of this grew the National Certificate Scheme, the very success of which possibly delayed the introduction of full-time technical education.

Aforetime, therefore, full-time general education was only obtainable in schools possessing varying degrees of independence, and these schools were, and still are, staffed in the main by university graduates. On the science side are graduates in pure science: masters so educated naturally want their brightest and best pupils to follow the way they themselves know.

There are, as far as I know, few engineers teaching science subjects in the secondary schools of this country, so there is little knowledge to be had in these schools of the nature of engineering. As things stand, most of the bright boys opt for science and are later attracted to research, which is thought to be glamorous, leading to fame, whereas engineering, if it is thought of at all, is associated with crafts, notably metal work, and is adjudged dull. It is little realized that all activities, including research, have their chores and call for much tedious work, the high spots being few and far between.

Moreover, I would presume to wonder if those who do research, particularly in universities, do not tend to get caught up in a closed circuit which sometimes has little real contact with the world of men and affairs outside.

It is not always realized that good engineering demands good human relationships and one is not dealing merely with ironmongery but primarily with people. For the professional engineer, as he progresses in his career, the control of people will be the major preoccupation. This point transcends all others in importance.

All this may seem to be naive in the extreme but such information as has come my way attests the need for it to be said. Remedies which have been proposed are varied and probably slow. Television, sound radio and the Press can each play a part but the case must be put over properly, otherwise it can do more harm than good.

Probably the best way is for engineers of standing to use all the personal contacts they can; particularly those who have access to governing bodies of schools and people of eminence in the world of education, especially the universities. Headmasters have mentioned to me the great value of works visits, but they have been emphatic that such visits must be organized and carried through with great care and in close co-operation with the school, otherwise, quite literally, they do more harm than good. The scientist can help greatly by making it plain that engineering and other technologies present as big an intellectual challenge as science. I welcome, therefore, the close and active interest which is being taken by The Royal Society in this matter.

I greatly fear that the entry requirements of the universities and colleges have tended to turn the secondary schools into cramming establishments whose teaching is aimed at the sole object of passing the prescribed examinations. I deplore specialization before the foundation of learning has been laid, but this seems to be the way in which we have moved. Secondary schools are ceasing to be places of education.

Perhaps the universities will be far-sighted enough to rescue us from this ominous situation; it is within their power to do so.

I am certainly not advocating the teaching of engineering as such in schools, but I am asking that the application of what is taught should be brought out in ways which would make plain the relevance of physics, chemistry and mathematics to the everyday things of life; for example, the internal-combustion engine, in which most boys and some girls are interested. This seems to call for an applied science laboratory, such as exists in a few schools. The true significance of a workshop would then become apparent, whereas, by itself, it is liable to give a false impression of the nature of a professional engineer's calling.

The trends in secondary schools shown in Fig. 2 were elucidated in 1962 by the Education Department of the University of Oxford in their excellent report on *Technology and the Sixth Form Boy* and it is not possible to indicate on the graph a direct comparison with the situation on the continent of Europe but there is no doubt that the engineering profession there holds a much greater attraction for the school leavers of higher ability.

Table 3 reproduces, from the same booklet, the average ranking of occupations, taken from a sample of over 1000 individuals who were over 16 years of age. It will be seen that the professional engineer comes very low on the list.

It is evident that the schoolboy, his masters and his

parents do not think that engineering is a worthwhile occupation and, until they can be made to see that it is, we shall continue to drift on to a lee shore which, as you know, is a hazardous position in which to be. The drift takes the disturbing form of producing twice as many scientists as engineers, whereas the exact opposite is needed and, equally worrying, the quality of the engineering entrants falls short of that of the science students, and of the need.

While on the subject of schools, I suggest that we need another 'Industrial Fund'. The last one, raised to help independent schools to enlarge their science laboratories, was a most successful and generous venture. Could there not be another one, this time for equipment, the requirements for which are apt to be both numerous and expensive and often beyond schools' unaided purses? Perhaps it could take the form of a joint venture with Government backing.

Dangers from university expansion

I am rather troubled about some aspects of the very large expansion which is now taking place in the university world. A danger seems to me to lie in the direction of failing, chiefly during what is called the 'bulge', to get enough teachers of the necessary calibre, simply because there may not be enough such people in the country to meet all the demands which continue to grow at an increasing rate. Flowing from this is the inherent risk of lowering the standards of learning associated with universities.

One is tempted to wonder whether enough attention has been paid to the possibilities of reducing the failure rate by such a step as critically examining afresh both the curricula and the methods of selection. It is the candidate's potential contribution to society which has to be assessed, not his ability to store and reproduce information.

Another likely result of university expansion I find even less attractive: that the possession of a degree will become an essential passport to every kind of job which has any prospects. We are a fair way on to this situation now, when it seems to me that a degree is laid down as the minimum requirement, often quite unnecessarily.

If this is carried too far—and this is always the tendency during expansion—we shall close the door on some talents we can ill afford to lose, for not everyone is good at booklore and passing examinations. Those who fail may well be endowed with natural gifts which no amount of learning will implant. We must have enough flexibility to find a place for them in our nice tidy plans, for we are dealing with people and people are our most precious asset.

I think that we are faced with something of a contradiction. On the one hand, we have the desire for a status symbol, such as a degree and, on the other, this age seems to be preoccupied with equality to the point of obsession. Certainly we must have equality of opportunity and in considerable and increasing measure we already have it, but if we pursue the quest for equality to its logical conclusion, we shall end by denying opportunity to the gifted and embracing instead the mediocre and the second rate.

Too much maths?

While speaking of curricula, I should like to interject a perfectly wicked thought which has long haunted me and which I have voiced on occasions. Are we losing a lot of good potential engineers by pitching our mathematics requirements too high? I am no mathematician myself and I believe that the subject in its higher flights comes easily to relatively few. I cannot rid myself of the feeling that we

Table 3—Social prestige accorded to various occupations (1962)

Occupation	Average ranking
Doctor	2·84 (high)
Solicitor	4·47
University lecturer	4·52
Research physicist	4·80
Company director	4·86
Dentist	5·80
Chartered accountant	6·28
Professional engineer	6·65
Primary school teacher	6·93
Works manager	7·63 (low)

Reprinted from Technology and the Sixth Form Boy, *published by the Department of Education, Oxford University*

lose a lot of good people by our insistence on too high an academic standard for the run-of-the-mill engineer.

I would go as far as to say that only in exceptional cases would the essential characteristics required of the engineer, as I have been trying to portray him, be combined with outstanding mathematical ability.

Another fear I have is that the new technological universities may be tempted eventually to go for the full-time course. Emulation and expediency may well tempt them this way, for in many respects it is easier than the sandwich course. But if they do fall into this temptation, I think they will be doing engineering a disservice because the sandwich principle has its own distinctive contribution to make, particularly in the spheres of design and production where, if the shortage of designers and production engineers is anything to go by, we are especially weak.

Another influence which is working against the sandwich principle is the increasing difficulty of finding firms prepared to accept students for industrial training. I feel sure that there are many firms which could provide excellent training if the benefits to industry of the sandwich principle were more widely appreciated. The burden of industrial training is at present unevenly distributed and I am hoping that the problem may receive early consideration by the newly appointed Industrial Training Boards which, in my view, must embrace the professional engineering student, as well as the technician and the craftsman. Until this happens, I appeal to firms, great and small, to re-examine what they are able to offer towards the satisfaction of this most important need.

Before leaving the subject of technical education, I think we ought to heed what has happened recently in the U.S.A. We seem to be in danger of becoming too theoretical in our own approach; of forgetting that engineering is an art as well as a science.

This is the way the Americans have gone since the war; with the result that industrial employers are now complaining about the kind of graduate produced by the universities. They say that the graduates know too little about the practice of engineering and the factors which govern engineering decisions.

The Ford Foundation, therefore, has financed a number of Engineering Residencies, as they are called. Individuals who have held the degree of Ph.D. for two years are seconded to industry for a further two. There they are placed with senior executives who are faced daily with making important engineering decisions.

This is a radical way of dealing with what must be a serious situation and I hope that we shall not be so misguided as to put ourselves in a similar position. We certainly shall do so if we allow technical education to become too academic and theoretical.

Reaction to the challenge—the C.E.I.

How is the engineering profession reacting to the challenge of the times? The observer is faced with a proliferation, not only of learned societies but also of qualifying bodies. Many are incorporated bodies with Royal Charters. It is a bewildering situation even to the informed observer. To the uninformed, who are legion, it must seem chaotic. In some important respects it is.

This situation has exercised the minds of many people for a long time. In 1962, a determined attempt was made to bring some order to the scene; this resulted in the formation of the Engineering Institutions Joint Council, now known as the Council of Engineering Institutions. A start had to be made somewhere, so, initially, it has been based on those institutions named in Appendix 1 to the White Paper *Scientific and Engineering Manpower in Great Britain* 1959.

The success of this body is vital to the well-being of the engineering profession but, as with all such co-operative ventures, it will be achieved only by patience, goodwill, give-and-take, and by vision. "Where there is no vision the people perish" is a great truth from the Book of Proverbs, and this applies to engineering institutions as to others set up by mankind.

So far, the C.E.I. has been given all these essentials but there is no room for complacency. Much remains to be done before the profession will be able to speak with one voice through the Council; but until this happens, it cannot expect to be accorded that attention in the counsels of the nation which the contribution of engineering to the national well-being would seem to merit.

To whom can the Government turn for authoritative advice if not to a body which should be able to claim that it represents the profession? I am not referring here to the day-to-day work of the Administration. There is an urgent need for something more; for a ready source of high-level engineering advice to which the Government can turn when faced with policy decisions involving, as they often do, judgements on engineering matters.

Therefore, the Council of Engineering Institutions must not fail to meet this national need. It will only succeed if its constituent members keep steadily before them the true significance of their task.

The role of engineering is so important that it is our bounden duty to put our house in order. Moreover, the importance of this role is beginning to be realized outside the world of engineering and with this realization comes the comprehension of its fragmentation.

Standards there must be in all things and the Council is no exception. It must have its own standards both to establish a minimum requirement for its membership and to act as a measuring rod. It is aspiring to create a series of examinations which will contain as large a common element as possible, while providing sufficient flexibility to enable the individual requirements of the constituent institutions to be met.

The basis is the existing Common Examination to which a number of the institutions already conform. I hope this will result in a combined qualification whereby the Council establishes a common academic standard and each member institution sets its own professional requirements. The compliance with both would be necessary before anyone could call himself a Chartered Engineer.

Keeping up to date

I have said that the Council must have its own standards. A standard is something that stands upright and nothing will stand upright without good foundations. So it is important that the foundations should be as well laid as possible. But knowledge tends to advance so rapidly these days that what was good enough for the last generation will not do for this one, so how does one keep abreast of advancing knowledge?

As one advances in years one becomes less of an expert, although one hopes one advances in experience, and possibly in wisdom. In the middle ranks, however, it is necessary to keep abreast: how does one do it?

An eminent friend of mine has suggested that everyone should have to requalify at intervals but I feel this is asking too much. I suspect a lot of the older generation would cease to practise as engineers if examination hurdles, of which there are already too many, had to be faced in middle life. Nevertheless, I agree that there is a need and I would have thought that it could have been met by postgraduate or refresher courses, if necessary at separate institutions set up for the purpose.

Mr G. S. Bosworth has referred to this aspect in a series of articles published in May and June 1964 in *The Engineer*. Of particular value is his clear statement of the education and training which industry requires of its engineers. He has suggested a pattern of postgraduate courses in engineering science which is markedly different from that generally evident today. He advocates advanced courses, particularly in product technologies, which would present students with the problem of designing and making actual equipment needed by a laboratory or works and which would be undertaken within realistic limitations of time and cost. These proposals merit the most careful joint study by industrialists and educationalists. Independently the Institution has been giving much thought to these matters and on much the same lines.

It must be recognized that the specialized classifications of these courses would differ from the formal ones with which we are familiar, and which are embodied in the professional institutions today. Increasingly, the products of engineering do not fit neatly into the formal engineering classifications. We must consider how best to educate and train our students to fit them for the task of engineering the needs of today and tomorrow. I suggest that this presents a great challenge to the statesmanship of the Council of Engineering Institutions.

Where do the technicians fit in?

By technicians is meant the very large number of people whose qualifications are such that they are ineligible for membership of any of the constituent bodies of the Council of Engineering Institutions. They are diverse, both in attainment and responsibility, and include those who are well qualified, perhaps to professional level, in some limited aspect of technology. Their contribution is vital. Without them the technologist could not function.

The number required to support him varies with the technology and the nature of the particular job but technicians certainly outnumber technologists by many times. I have deliberately used the word 'technologist' in this section of my Address because in a real sense all those normally lumped together as 'technicians' are engineers.

As in everything else in this Address, I am voicing my personal views, so I will go on to say that they should be known as engineers and not technicians. This being so, it is necessary to differentiate between the technologist and the technician, and the answer is ready to hand in the terms 'Chartered Engineer' and 'Engineer'.

I think that the Council of Engineering Institutions must help these engineers to put their house in better order, simply because the Chartered Engineer has a moral and practical responsibility for the training and experience which they receive. Unfortunately the proliferation of their organizations is even greater than among the Chartered Engineers.

I would like to see some positive link forged; it could take the form of affiliation of one or more of these organizations with the appropriate constituent member of the Council of Engineering Institutions, which would interest itself in the activities of its affiliates by advising, for example, on their standards. In order to do this, it would need to be represented on their governing bodies.

The corollary of this is that these engineering associations could set up an organization parallel to the Council and affiliated to it in the form of reciprocal representation. The object of all this is to bring coherence into the 'technician' world of today and assist it to raise its standards to meet the ever-increasing demand. We neglect these duties at our peril.

What happens in the Royal Navy

Perhaps I could simplify this rather involved picture by describing what happens in the Royal Navy which, for a long time, has had very clear distinctions between the technologist, the technician, the craftsman and the semi-skilled man. The first is the Engineer Officer, trained at the Royal Naval Engineering College and elsewhere to standards required by the Navy. Fig. 3. shows a typical class at the R.N. Engineering College at Manadon. These are broadly comparable to a university degree and give exemption from the qualifying examinations of this Institution. In addition, a growing proportion of officers are reading for, and obtaining, an external degree of the University of London.

I should like to emphasize the word 'officer' or if you prefer it, 'manager'. For, from the very early days of his service, he has to assume responsibility for the activities and well-being of others. He can be employed in any capacity, both ashore and afloat, and much thought is applied to giving him as wide an experience as possible, not only in technology but also in the administration and training of personnel; both officers and ratings.

The second category is the officer who is promoted from the Artificer Branch and, more rarely, from the Engineering Mechanic Branch. He is now known as a Special Duties Officer and is a specialist in the maintenance and operation of ships' machinery, but plays little part in design and shore administration, except in so far as he forms part of the staffs of Artificer and other rating training establishments. In Her Majesty's ships he is interchangeable with the junior commissioned officers and in small ships is often the only engineer officer. Much attention has been given to his training of late, which now includes a period at the Royal Naval Engineering College, with a view to increasing his versatility.

The third is the Artificer, trained by the Navy in its own establishments to a high degree of craft skill and operating ability and to the academic standard of the Ordinary National Certificate. He does the skilled maintenance work on board ship and comprises the skilled element of the steaming watches. The more senior rating can take charge of machinery units in a large ship and the whole machinery of a small one, where in effect he is *the* marine 'Engineer Officer'. He has

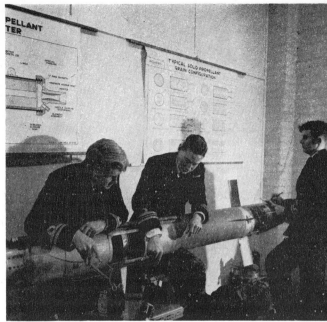

Fig. 3. Engineer officers at Manadon receiving instruction on a rocket test vehicle

always been a highly skilled man but latterly much attention has been given to his engineering training, as distinct from his craft training, for more and more, in these days of highly rated machinery, he must be a diagnostician. This is the source from which the majority of Special Duties List Officers are taken.

The fourth is the Engineering Mechanic, semi-skilled in maintenance and skilled in operation. He is the modern descendant of the former Stoker, and even more than those of the Artificer, his training and education have been revolutionized. The effect is to take some of the load off the Artificer, particularly some of the skilled watchkeeping at sea and the more routine portion of maintenance afloat.

The Engineer Officer does not merely influence the training of all the other three; he actually lays down the policy and carries it out by commanding the establishments at which the training is undertaken and by forming the core of the instructional staffs which in the rating establishments also contain Special Duties Officers and Artificers.

Correct balance

The Navy is always concerned to ensure that it has a correct balance between the technologist and the technician. It does this by careful planning of its personnel needs and the detailed complementing of its ships and establishments.

I wonder whether there is an object lesson here. In our planned expansion of professional engineering training in universities and colleges, have we given due weight to the numbers and training of those now called technicians? Put simply, are we not paying too much attention to providing the professional element while remaining unable to train enough technicians? And shall we be faced in a few years' time with graduates doing work which is really in the field of the technician, merely because there are enough of the first and not enough of the second?

If this is a real risk, and I believe it may be, we are creating conditions which will give rise to a disgruntled generation. I understand that the matter is being looked into, but it seems to me that the professional engineer and technician requirements need to be taken together. It is, of course, an immense task, compared to that of the Navy, but it is nevertheless vital to find out where the correct balance lies in each industry and in each technology.

I would say, in passing, that the Chief and Petty Officers constitute the backbone of the Service and their opposite numbers in industry are the foremen. Throughout the period of my own career, and subsequently, the Navy has paid increasing attention to the education and training of the Petty Officer and Chief Petty Officer. In particular, the engineering side has developed the training of the Artificer as a Chief Petty Officer who has to take charge of others and is not merely a skilled craftsman. The Navy knows from experience that when the Petty Officer is neglected, the officers tend to lose touch with their men and in the ultimate case the result is mutiny. After all, the Petty Officer is the one who is in closest touch with the rank and file and should know what they are thinking.

Is it too fanciful to wonder whether one cause of industrial unrest is the failure of managements to educate, train and support their foremen and thus to establish a firm line of communication between management and men? It is certainly not fanciful, because it is within my own experience, that the spirit of any organization comes from the top. The spirit of a ship stems from its Captain whose personality and principles penetrate his command and affect even its humblest member. Napoleon once said "There are no bad soldiers, only bad officers"; my own experience testifies to the truth of this observation.

Practical training

May I at this point make another interjection, this time about what is called 'practical training'. I submit that there is no essential difference between what is required in this respect of a naval engineer officer and a professional engineer ashore. The officer does not need to be a skilled craftsman; therefore, although he is and must be given training in both machine and hand tools, he does not spend the time on them which the Artificer Apprentice requires to do. With so much to learn, it is not possible for the professional engineer to spend too much time in making himself a proficient craftsman.

I should like to illustrate this by what happened to me. I am not a natural craftsman, nor have I spent undue time in being taught to be one. I was, however, given the opportunity to acquire experience which proved to be quite invaluable.

As a Midshipman at the age of 19, I volunteered for engineering. Being a Midshipman in a Gun Room afloat was in itself a salutary experience, but an even greater one awaited me. The Engineer Officer of the ship (it was, in fact, the Queen Elizabeth, then flagship of the Grand Fleet) put me as mate to the senior Chief Engine Room Artificer.

With him I struggled with inaccessible nuts in impossible places, sometimes in the cold bilge and sometimes in some ill-ventilated hot spot. With him, I learned to hate the draughtsman who had never had to maintain his own creations and the installer of machinery who never went to sea. He could say anything he liked to me as long as he called me 'sir' occasionally and he taught me many things, including nautical language. I am eternally grateful to him for all he taught me, not least the Artificer's way of looking at people and things. I shall always remember his withering scorn for the semi-skilled man and his mixed views about the officers.

Can we not do something like this for our young men by putting them to constructive work such as erection and repair, when they have received instruction in the essentials of manufacture?

The essence of the matter

Man is a trinity; body, mind and spirit. We spend much time and money on the body; we are beginning to spend a little on the mind, but the things of the spirit are largely left to look after themselves. And yet the spirit is the mainspring of the other two: a mainspring which many of us do not wind up and which, when a sudden demand is made upon it, is unready to respond to the emergency; for it is in emergency that the power is most needed.

As engineers we ought to know this and as engineers we need this source of unseen strength more than most, for while, certainly, we are dealing with our own inanimate creations, these same creations demand great integrity of thought and action because they cannot think or forgive and are utterly dependent upon our own qualities, which they will reflect.

It is on these creations that we depend for most of our essential everyday requirements and they must be given meticulous care in design and manufacture; and attention to what I have called elsewhere the 'unconsidered trifles'.

A split pin or pipe joint badly fitted, possibly because of poor design or layout, can put a large machinery installation out of action with far-reaching consequences to individuals and the community for whom it is providing some essential service. In my experience, it is the unconsidered trifles which let one down, simply because they may seem insignificant and do not, therefore, receive the attention they merit. Moreover, with the great increase in complexity, and therefore in the number and variety of people engaged in a project, be it design, manufacture, operation, or maintenance, individual responsibility for final functioning is much less direct than it once was. This carries with it the risk of lessening the feelings of personal accountability which should be ever present in the minds of people dealing with machinery.

All-round integrity is not simply a gift with which some people are endowed while others are not. It can be cultivated by anyone who is prepared to recognize the need for it and the humility which goes with it. As an instance of humility we would do well to remember with Sir Charles Parsons that, when we speak glibly of creating, we, in fact, create nothing but merely rearrange what has already been created.

These are things of the spirit and although spiritual things are often thought to be irrelevant to the busy world of today, I am convinced that they are more necessary now than they have ever been, if man is to work with purpose and vision. The wear and tear inseparable from living in this world applies to the spiritual, as to the physical and mental.

I should like to end by quoting one of my former chiefs; a wise and lovable superior and a first class engineer. He said to me one night at sea: "I always add an eleventh commandment to the existing ten—'Thou shalt not be found wanting'." If we are to survive as a source of influence in a world to which we have so much to give from our rich store of political and social experience, acquired over many centuries of more or less orderly evolution; a world in which many so desperately need all the help we can give them, we must heed this additional commandment.

Let us see to it that, as far as in us lies, we are not found wanting in meeting the challenge of our day and generation.

Technological Forecasting

by W. H. G. Armytage, MA*

The Author of 'Yesterday's Tomorrows' looks at the roots and present efflorescence of technological forecasting and offers a few observations on its relevances rather than its excesses. He sees it as part of the rising anticipatory temper of modern society which is reflected in most fields of social policy where forecasts depend on sophisticated estimates as to how technology might develop.

In the early 1920s the Institute of Patentees kept a book for engineers and government departments to suggest what inventions were wanted. As might be expected, the Institute got a number now familiar to us; a speaking cinema, a method of utilising atomic energy, a process for instantaneous colour photography, etc. When an engineer from Sheffield was asked for his suggestions he listed four: a ferrous or non-ferrous alloy with double the tenacity of, and less brittle than any then known; an anti-fog illuminant; an unerodable refractory for furnaces; and an anti-roll device for ships. But, like a cautious Yorkshireman, Hadfield held that

> there was little or no possibility of providing results by analogy, interpolation or extrapolation. Intelligent anticipation may be of considerable assistance, but certain knowledge can only be derived from a complete scheme of correlated research.[1]

Such a complete scheme of correlated research was set in motion in 1929 when the first engineer to become President of the United States appointed a group of scientists to survey American society. They detected a need to anticipate the impact of technological change on society and suggested that a national advisory council of scientists, educationists and economists should be established to do so.[2] And so it was the National Resources Committee which organised a study of technological trends and their social consequences. This study was described by those who made it as "the first major attempt to show the kinds of new inventions which may affect living and working conditions in America in the next 10 to 25 years". It also indicated some of the social consequences of such inventions and the national policy needed to adjust to them.[3]

This report owed much to S. Colum Gilfillan who, from the history of 19 major inventions,[4] showed that an average interval of 174 years elapsed between the proposal of an idea and the first patent and yet more time before its first important use.

In other words, many inventions likely to be important in the future were already in existence.

But more important still was Gilfillan's conclusion in 1938 that there seemed to be a clear case for a committee of technical men, social scientists and students of production.[5] Twenty years later the Director of Research at the General Electric Co. warned that the passion for prophecy could be overdone, since constant repetition sometimes made prophecy seem to become fact. To him the crystal ball really had become clouded ten years ahead.[6] Today, Gilfillan's own technology

of forecasting[7] has been supplemented by others and a large number of study groups have taken shape[8], the subject is being studied at graduate level,[9] agitation has begun for a British think tank,[10] and last year President Nixon ordered the establishment of a National Goals Research Staff. Under the direction of Leonard Garment, this is to forecast future developments and assess the longer-range consequences of present social trends. In doing so it will also estimate what alternative goals might be attainable given available resources, and integrate the results of similar research being carried on by government and private groups.

Its first assignment is to illuminate goals for the two hundredth anniversary of the USA in 1976. Starting on 4th July 1970, an annual report will be prepared for public discussion of some key choices and their implications, so that Americans can enjoy the luxury and responsibility of making decisions.

Engineering intuition

Before climbing the decision tree or taking a critical path, can such matters prospective be put into perspective? For, whatever its fruits, a tree is only as strong as its roots, and the roots of technological forecasting lie deep.

Anticipatory techniques are as old as civilization and, by the time of the Greeks, had become an integral part of religion. Like most human activities it has been improved by specialists: weather-forecasting, economic and business forecasting, social and political forecasting are now part of daily life.

Today, technological forecasting is becoming respectable. I don't mean the inspired doodling of Leonardo, the Marquis of Worcester and even Newton. Nor do I mean the intuitive forecasts that enabled Defoe to predict Boolean algebra, H. G. Wells the tank, or Hugo Gernsback radar.

Nor do I mean the sort of prediction that is built into science which led Clerk Maxwell to predict that the viscosity of a gas would be independent of the pressure or Rutherford and Chadwick to predict the neutron. For, in general, scientific theories tend to be judged by the categorical forecasts to which they lead.[11]

But I do mean the interaction of the second with the first, ie, the application of the scientific method to the analysis, and forecasting of technological change. Consider, for instance, the steam engine. Its parturition was long, but as early as 1773 it was cast for the role of propelling a carriage

This is a greatly condensed version of his Thomas Hawksley Lecture.

50

"—population tends to increase faster than the means of maintaining it"

with twin cylinders and rear wheel drive. The 'caster' (or should I say forecaster) nearly persuaded Matthew Boulton to build one.[12] He was Erasmus Darwin, a Lichfield doctor, who envisaged the steam engine powering "flying chariots through fields of air"[13] 18 years later.

Like a true technological forecaster, he added that within half a century some other forces might be discovered which would do the job equally effectively. We know too that his forecast influenced Sir George Cayley to work on aircraft, who in turn influenced Wilbur and Orville Wright.

Yet another of Darwin's anticipatory artefacts was a rotating four-bladed propeller on a vertical shaft. Indeed, amongst the 75 subjects in which he advanced new and correct ideas or proposed, and in many cases developed, new inventions and techniques, were speaking machines, canal lifts, a hydrogen and oxygen motor and the theory of evolution, later developed by his grandson.[14]

Darwin's technological fantasy, the *Botanic Garden*,[13] showed how what might be called the anticipatory imperative was emerging.

Statistical inference

The anticipatory imperative is perhaps better illustrated by the activities of the Rev. Thomas Bayes, a Presbyterian minister at Tunbridge Wells. His theorem is concerned with finding the revised probability of an event after a trial, given prior probabilities.[15] Bayes' discoverer, Richard Price, worked on mortality tables and his work led Thomas Malthus to state that population tends to increase faster than the means of maintaining it.

The labour of computing such tables led Charles Babbage to build his famous computers and suggest mechanisms to accomplish another mental process, namely—prediction.[16]

Such ideas attracted economists like W. S. Jevons, who tried to build a logic machine and was also interested in large-scale forecasts. He entered the realm of energy forecasting and in 1865 prophesied that British coal reserves would last for another 110 years.[17] This was in striking contrast to a forecast four years earlier that they would last for another 1000 years[18] and to another two years earlier that they would last for another 212.[19] Subsequent Royal Commissions, made yet different forecasts.[20]

That these energy forecasts could have been made at all was partly due to the establishment of a department of mining records in 1845 under Robert Hunt and to an algebra of anticipation, known as probability theory, which had its early roots in gambling, insurance and weather forecasts.

Following Bernoulli, Bayes and Boole, Jevons preached statistical method to economists.

He embarked on his vast *Statistical Atlas*, plotting trends since 1731 with a view to analysing them. Some of his results were interesting: the best indication of panic, he found, were the number of patents, the number of Acts of Parliament and the number of bricks manufactured! Later he developed ideas of Sir William Herschel and R. C. Carrington that commercial crises were predictable in terms of the sun's heat and harvest fluctuations. He was (said Lord Keynes to the Royal Statistical Society) the first theoretical economist to survey his material with the prying eyes and fertile controlled imagination of the natural scientist . . . But (and this was in 1936), it was remarkable how few followers and imitators he had had in the black arts of inductive economics.[21] One of those fellows was that sober British economist, Alfred Marshall, who warned that

> Greater risks are taken when no attempt is made to forecast the future while considering methods of action or inaction that will largely affect the future than by straining inadequate eyes in reading such faint indications of the future as can be discerned by them.[22]

Keynes himself, in his own words, "took wings into the future" during the depression in 1930. To combat the pessimism of the revolutionaries and the pessimism of the reactionaries he called attention to the vast acceleration of technical change which would enable food to be produced as efficiently as coal or cars. To him the depression was due to a new disease—technological unemployment. And he went on confidently to predict that the standard of life in progressive countries 100 years hence would be four to eight times as high.[23]

Between Jevons' own logical machine[24] and the computerised Social Accounting Matrix now operating at Keynes' own university of Cambridge, the 'black art' was getting under way amongst economists.[25] Perhaps the best known of these is C. W. Churchman of the University of California who has given us the concept of relevance for decision making.[26] His ally is the analogue computer for which he is developing a game entitled 'Technological Innovation'. Such techniques have to be sharpened in Britain too, where national planning has grown on the skills developed through corporate planning in the USA. Listen, for instance, to the chairman of the NEDC.

> A plan is partly a prediction, partly a series of decisions aimed to achieve the prediction and partly a benchmark against which both the predictions and the decisions can be measured to see whether they need to be changed.[27]

The stimulus of war

"Depend upon it, Sir", said Dr Johnson "when a man knows he is going to be hanged in a fortnight, it concentrates his mind wonderfully". Such concentration has produced war games, simulation and teaching machines. Imagination has been screwed to the obsessional level needed for technological change.

We need go back no farther than the war against Napoleon when visions of floating castles, balloons and tunnels were triggered off by the threatened invasion.

The real pioneer of this new form was an engineer officer, Sir George Chesney in his *Battle of Dorking* (1871). This fantasy showed how modern technology made possible a surprise attack by the Germans whose 'fatal engines' disposed of the protecting British fleet. Its influence lasted till the outbreak of the first World War'.[28]

English engineering journals certainly followed the fashion, and a young man, who was to make his name by his famous annuals of fighting ships and aircraft, Frederick T. Jane, wrote a story of torpedo ships and added:

> may our warfare of the future long be confined to the pages of books; as, indeed, it will be, so long as foreign nations know that we are ready to tackle the lot of them if need be.[29]

When the first world war did come, engineers found themselves rethinking techniques of mass production. At the American Ordnance Bureau, H. L. Gantt successfully argued that planning should be based on time rather than quantities. Hence his famous chart. Walter Rathenau, managing director of the German AEG, visualised factories "so completely mechanised that one man would suffice to keep the whole clockwork of production going"[30].

"—visions of . . . balloons . . . were triggered off by the threatened invasion"

Apprehensive strategies were intensified by the development of intuitive forecasting (the term 'science fiction' was first used in 1929).[31] But war was the greatest stimulus.

In 1940 an evaluation committee of the US National Academy of Sciences began to study the technical options of the gas turbine. One member, Theodore von Karman, submitted a report on the subject in 1944, in which he forecast supersonic aircraft and intercontinental ballistic missiles. By 1959 he was supervising the study by some 100 Air Force personnel of various problems (including space). He later initiated long-term studies for NATO of 15 or 20 military and scientific topics, each updated in rotation at five-year intervals. Member countries have an opportunity of preparing working papers which are then submitted to 'brainstorm' groups which arrive at an agreed conclusion.

The US Army's Chief of R and D began to call for reports from each of the technical services which, in 1955, resulted in a *Technical Capabilities Forecast*. Today the Army Material Command has its own Technological Forecasting Branch. The US Air Force prefers *ad hoc* study groups representative of universities, private corporations, federal agencies and so forth. These have produced some notable reports. The Office of Aerospace Research has also sponsored two symposia on the subject.

The Marine Corps has Syracuse University on contract for this purpose. In 1964 it produced a four volume study of the United States and the world as anticipated in 1985.

Such Service prospicience vastly impressed the academics and several conferences and publications have resulted.

The American nutrient

It was the Americans who provided the extra nutrient for such technological forecasting. According to de Tocqueville:

> It can hardly be believed how many facts naturally flow from the philosophical theory of the indefinite perfectibility of man or how strong an influence it exercises even on men who, living entirely for the purpose of action and not of thought, seem to conform their actions to it, without knowing anything about it.[32]

His words, though written in 1835, apply to American speculation about tomorrow. Jules Verne's idea of a multivaned helicopter (in *Robur the Conqueror or The Clipper of the Clouds*) was taken from the young American writer, Luis Senarens who, from the age of 14, had been selling science wonder stories for boys.[33] Nor should we forget America

of the year 2000 as projected by Edward Bellamy in *Looking Backwards* (1888); the best and most influential of many such fantasies.

But after the 1914-18 war English social moralists like Huxley, Orwell and C. S. Lewis unashamedly used their projective techniques[34] not to sing the praise of a mechanical millennium but to excoriate it.

Some of these myriad rivulets of speculation, though some of them were disorderly and shallow, were channelled by operational research and systems analysis during the last war.

Systems analysis was, after 1948, the special concern of the Research and Development (RAND) Corporation. Working for the US Air Force under contract, it developed cost-effectiveness techniques and, working for the Department of Defense, the Planning Program Budgeting System. It rediscovered the educational role of games. In doing all this it inevitably and deliberately became involved in technological forecasting, for which it developed a system of iterative opinion sampling, known as Delphi.[35]

Delphi dates back to 1964. In the original experiment some 82 respondents were subjected to a questionnaire asking them to name inventions or scientific breakthroughs, both necessary and possible, during the next 50 years. Responses were graded into quartiles and a median date. The 82 respondents were subjected to three more questionnaires designed to elicit a significant consensus for all the items.

Other attempts to improve intuition are described as brainstorms, buzz groups and so forth. It is even claimed that a harvest of ideas can be gleaned by gazing at the human boss—or navel—a technique known as Omphaloskepsis, or by simply allowing researchers a long rein, so that they can emulate the three princes of Serendip and discover something while looking for something else.[36]

But omphaloskepsis and serendipity are not enough in a world of cost effectiveness, so the Ford Foundation established Resources for the Future Inc. in 1953 to consider social problems like land use, water, energy, minerals, pollution, and regional development. Such application has prompted further internal efforts at forecasting up to the year 2000.[37]

RAND has fathered several other forecasting institutes. In 1956 it hived off the System Development Corp which specialises in technological forecasting in a management information system and is building up an electronic data bank from which system models can be built. This method according to

"—a technique known as Omphaloskepsis . . ."

Hasan Ozbekhan, enables 'possible futures' to be conceived, compared and cased.[38]

Also in 1956 the General Electric Co established a group for technical military planning. Today 20 per cent of its work is directed to management. This group applied to business the lessons learned from missile systems. Its Centre for Advanced Studies stresses the interaction of many trends, rather than simple extrapolation.[39] Their work seems to indicate that trend curves can be smoothed by implicit factors and variables derived from the behavioural sciences.

Recognition of technological forecasting was virtually assured by 1958 when the Stanford Research Institute developed its Long Range Planning Service. Financed by some 400 organisations, each paying $3000 to 4000 a year, it makes forecasts from 5 to 15 years ahead.[40]

More involved in the applied physical sciences is the Samson Science Corp and its affiliate the Quantum Science Corporation.[41] It forecasts output figures for electronics firms, based on forecasting of inventions and on an appraisal of the individual qualities and competitive chances of the companies concerned. In 1967 it started a simulation programme.

Abt Associates Inc. have prepared a report for the National Commission on Technology, Automation and Economic Progress and the reports of this commission contain numerous other papers on the methodology of technological forecasting. Abt reported that just under 100 full scale models of "feasible conformations of the future world" had been generated and expressed the hope that more should be systematically produced and translated into technological, economic, social, political and military capabilities and requirements. Abt used network techniques like PERT and CPM.

Networks

These networks stemmed from the ever increasing magnitude of the enterprises on which the Department of Defense was engaged. In Britain the CEGB had been working along similar lines.[42]

Historically Du Pont's seem to have anticipated them with the Critical Path Method, devised and applied by James E. Kelly and Morgan R. Walker in 1957. PERT is an analysis in terms of events, CPM in terms of activities.

These are acronymic ancestors of over a hundred variant systems that fall into four categories: time-based only; or time-based with costs; or cost-based to optimise time and cost; or resource-based. Among the latter is Critical Path Scheduling (or Critical Path Planning and Scheduling), whose deviser, Mauchly, had been a member of the Du Pont project and left it in 1959 to set up his own firm.[43]

Forecasting processes have become correlated and multi-dimensional as computing facilities have improved. In Russia statistical forecasting has been successfully applied to geophysical problems like rain, earthquakes, discharges of rivers and age levels of ground water.

In the biological world, forecasting is used to predict the appearance of pests like the Colorado Beetle, potato canker, caterpillars, etc.[44]

Technological forecasting has four aspects: intuitive, exploratory, normative and feedback.

Intuitive is best exampled by the Delphi methods. Exploratory forecasts extrapolate present trends. Normative forecasts, by going back from a desired future to the present, hope to open up a technological path that connects them. Feedback forecasts represent the interaction of exploratory and normative in computer simulation.

Herman Kahn and Anthony Wiener, with the co-operation of the American Academy of Arts and Sciences, produced for the 'Commission on the year 2000', not only some new views but new concepts.[45] The most interesting of these is the post-industrial society in which "the organisation of theoretical knowledge becomes paramount for innovation in society and in which intellectual institutions become central in the social structure".[46]

A similar need is felt by business; witness the Detroit Edison Co commissioning Constantine Doxeiadis to look to the year 2000 with a view to obtaining some contour lines of the city which they would be called upon to illuminate.[47]

Detroit Edison were only one of 500 to 600 US companies engaged in technological forecasting in 1965,[48] when they were estimated to be spending 1 per cent of the total R and D expenditures.[49,50] Among the UK firms commended by Jantsch for their forecasting activities were Unilever[51] and Esso U.K.,[52] Royal Dutch Shell (1955) and Vickers (1960).

As the inevitable concomitant of planning, the imperative need for forecasts is now well established. When mathematical economics returned to favour in 1954 in Soviet universities after being prohibited by Stalin as 'playing with funny figures', it appeared as 'planometrics'.[53]

Bertrand de Jouvenal suggested an institute to isolate the probably attainable from the logically possible futures. The former should then be amplified in the light of their impact on individuals and presented for democratic choice.[54]

"—engineers do worry about the consequences of their works . . ."

The same idea prompted Olaf Helmer of RAND Corp to enlist the support of the National Industrial Conference Board. This independent research and reporting body, on which most of the industrial and academic interests of America are represented, began in 1966 to move towards an Institute for the Future, which is now engaged not only in orienting various groups in society towards the future but providing some of the services required for such study.[55]

Cautionary noises

Of course a massive brief can be assembled against such practices. By 1961 the sheer number of forecasters led Professor P. A. Samuelson to claim that they were too much in touch with each other, so much so that he compared them to many Eskimos crowded into the same bed who all turn over together.[56] Then there was Lee du Bridge in the following year expostulating that "Scientists and engineers do worry about the consequences of their works. But neither they nor anyone else has discovered how to avoid or even to predict these consequences".[57]

Historians, the prophets in retrospect, point out that the novel technological event always remains unnoticed by the predictor's elbow. For example, H. G. Wells ruled out air

transport as a serious means of communication in 1902, J. W.
Campbell in 1941 rules out a moon rocket because it would
have to weigh a million tons to carry a pound of payload,
Lindeman and Vanneyar Bush deprecated the competitiveness of ICBMs and the Astronomer Royal made his celebrated
'utter bilge' speech a year before *Sputnik*. Even Gilfillan, the
noted exponent of the principle of equivalent invention, failed
to notice the jet or the helicopter, whilst Palmer C. Putnam,
the long-term forecaster for the American Atomic Energy
Commission, took a very superficial view of the impact of
these inventions.[58]

In business life, technological forecasting must, as Igor
Ansoff remarked in 1965, allow for the '$2+2=5$ effect' or
'synergy'.[59] So, too, Donald Schon disparages the potential
of technological forecasting, even though he is the director
of OSTI (Organisation for Social and Technological Innovation) established at Cambridge, Massachusetts.[60]

Forecasting is still only 25 per cent science and 75 per cent
art as Dr Harold Linstone reminds us, adding that the
approach has to be interdisciplinary, the planners being mixed
with the doers.

Only a few months ago Sir Solly Zuckerman questioned
the assumptions made from the trends and rate of technological progress in the past, and the intellectual capacity and
authority of the normative forecaster.

> The more the technologist—and scientist—understands how events
> move in the political sphere, the more he is able to participate, either
> directly or indirectly, in the decisions of politics, the more society will
> benefit and the less there will be to fear.[61]

But such participation involves information, and opportunities to check forecasts which may in fact be Government
propaganda. One of the major motivations of management is
to latch on to an industry which has a future and where
management communications are good,[62] for example an
American electronics company had a policy of having 'as
many general managers as possible'. These at all levels had
access to literally all data about the company and were
informed of all plans and policies. But this company's
attitude is by no means general.[63]

Conclusions

Enthusiasts may claim that technological forecasting
enhances central co-ordination and intensifies the search for
diversification. In collaborating on a simulation model, the
specialist is exposed to aspects of his work that he might otherwise ignore, ie, he takes part in a systems analysis.[64]

Other virtues of such heuristic homework are that it
improves insight into a problem. A wise old British economist
once wrote, "Explanation is simply prediction written
backwards: and when fully achieved, it helps towards
prediction".[65]

Throughout this article I have referred to the anticipatory
imperative in modern life. It is reflected in all sorts of ways,
eg, in the rise of a predictive dialogue or debate.[66] Gloomy
predictions too are useful. For instance, in 1862 Hofmann,
a pupil of Liebig, said that beyond a question England would
become the greatest colour-producing country in the world,
sending her coal-derived blues to indigo-growing India, her
tar-distilled crimson to cochineal-producing Mexico, etc.
He was right, about the wrong country, for a decade and a
half later that role was to be Germany's.[67]

He could not foresee the establishment, by the German dye
industries, of great research teams concentrating on the
costing, testing and evaluation of ideas thrown up by their
own researchers. A. N. Whitehead realised that the greatest

invention of the 19th century was 'the invention of the method
of invention'.[68] This was to be brilliantly exemplified by

"—as many general managers as possible"

Haber and Bosch
who nullified the
prophecy of Sir
William Crookes
in   1898   that
nitrate   shortages
would   lead   to
starvation.   Their
synthetic   process
for obtaining nit-
rogen from the air
shows how necess-
ity can father as
well as mother in-
vention.[69]

In   conclusion
let me quote from
a   report   by   the
US   National
Academy of Sciences in 1967:

> The most important invention in the pursuit of modern (as opposed
> to older) applied science is the big mission-oriented industrial or
> government laboratory. In fact modern science can hardly be discussed
> without reference to these homes of applicable science. These institu-
> tions derive their power from three sources: (1) their interdisciplinarity
> and the close interaction between basic research and application; (2)
> their methodology for precipitating and organising coherent effort
> around big problems; (3) their ability to adapt their goals to the re-
> quirements of their sponsors'.[70]

One could almost say of them, as Robert Frost said of the
dreamers of better worlds, "they do not believe in the
future they believe it in".

REFERENCES

1. HADFIELD, SIR ROBERT A. *Metallurgy and its Influence on Modern
   Progress with a Survey of Education and Research*, 1925, p. 301,
   Chapman and Hall, London.
2. PRESIDENT'S RESEARCH COMMITTEE ON SOCIAL TRENDS. *Recent Social
   Trends in the United States*, 1933 (2 vols.), Government Printing
   Office, Washington.
3. U.S. NATIONAL RESOURCES COMMITTEE. *Technological Trends and
   National Policy*, 1938. Government Printing Office, Washington.
4. Actually selected by readers of the *Scientific American* as the most
   important of the years 1888 to 1913. *Scient. Am.* 1913. Vol. 109,
   pp. 63, 243, 352.
5. GILFILLAN, S. C. In *Technological Trends and National Policy*
   (*op. cit.*), 1938, p. 18.
6. SUITS, G. *Suits: Speaking of Research*, 1965. John Wiley & Sons,
   London.
7. GILFILLAN, S. C. 'The Prediction of Technical Change', *Rev. Econ.
   Statists.* November 1952. Vol. 34, No. 4, p. 3688.
8. JANTSCH, E. *Technological Forecasting in Perspective. A Framework
   for Technological Forecasting, its Techniques and Organization: a
   Description of Activities and Annotated Bibliography*, 1967, p. 96.
   Organization of Economic Co-operation and Development, Paris.
   This is by far the most comprehensive study of the art.
9. BRIGHT, J. R. *Research, Development and Technological Innovation*,
   1964. Richard D. Irwin, Homewood, Illinois.
10. FLETCHER, R. 'Poised for the Great Leap', *Guardian*, 1969, 28th July,
    p. 9; *The Times*, 1969, August 27th, p. 3.
11. TOULMIN, S. *Foresight and Understanding. An Enquiry into the Aims
    of Science*, 1961, p. 18. Hutchinson, London. See also MATHIESON, T.
    'The Unanticipated Event and Astonishment', *Inquiry* 1960. Vol. 3,
    p. 1.
12. KING-HELE, D. *Erasmus Darwin*, 1963, p. 157. Macmillan and Co.,
    London.
13. DARWIN, C. *The Botanic Garden: the Economy of Vegetation*, 1789.
    Vol. 1, p. 292.
14. KING-HELE, D. *Op. cit.*, 156, and *The Essential Writings of Erasmus
    Darwin*, 1968, p. 199. MacGibbon & Kee, London.
15. See BARNARD, G. A. *Biometrika*, 1958. Vol. 65, p. 393.
16. BABBAGE, C. *Passages from 'The Life of a Philosopher'*, 1864, p. 62.
    Longman, Green, Longman, Rogers & Green, London.
17. JEVONS, W. S. *The Coal Question*, 1865. Macmillan and Co., London.
18. HULL, E. *The Coal-fields of Great Britain. Their History, Structure
    and Duration*, 1861. E. Stanford, London. This was later revised up
    to 1905.
19. ARMSTRONG, SIR WILLIAM. 'Address to the British Association 1863',

British Association for the Advancement of Science, Newcastle Meeting, 1864. John Murray, London, p. 52.

20. ROYAL COMMISSION ON COAL. H.M.S.O., London 1903, Cd. 1724, Cd. 1725; 1904, Cd. 1990, Cd. 1991; 1905, Cd. 2353, Cd. 2362, Cd. 2363, Cd. 2364.

21. KEYNES, J. M. *Essays in Biography* (edited by Geoffrey Keynes), 1961, p. 268. Mercury Books, London.

22. MAYS, W. 'Jevons's Conception of Scientific Method', *The Manchester School*, 1962. Vol. 30, p. 233.

23. MARSHALL, A. *Industry and Trade*, 1923 (4th edn), p. 506. Macmillan, London.

24. KEYNES, J. M. *Essays in Persuasion*, 1952, pp. 359, 364. Rupert Hart-Davis, London.

25. GRUNBERG, E. and MODIGLIANI, F. 'Predictability of Social Events', *J. Polit. Econ.*, 1954. Vol. 62, p. 465. See also TRESS, R. C. 'The Contribution of Economic Theory to Economic Prognostication', *Economics*, 1959. Vol. 26, p. 194.

26. CHURCHMAN, C. W. *Prediction and Optimal Decision*, 1961. Prentice-Hall, Englewood Cliffs, New Jersey.

27. CATHERWOOD, H. F. R. *Britain with the Brakes Off*, 1966, p. 117. Hodder & Stoughton, London.

28. For a good account of this genre see CLARKE, I. F. *Voices Prophesying War 1763-1984*, 1966, p. 42. Oxford University Press, London.

29. *Ibid.*, p. 85.

30. URWICK, L. and BRECH, E. F. L. *The Making of Scientific Management. Vol. 1 Thirteen Pioneers*, 1966, p. 89. Pitman, London.

31. MOSKOWITZ, S. *Explorers of the Infinite: Shapers of Science Fiction*, 1963, p. 240. The World Publishing Company, Cleveland and New York.

32. DE TOCQUEVILLE, A. *Democracy in America* 1835, Part 1, Chapter 2 (trans. Henry Reeve) 1835-40. Saunders and Othey.

33. MOSKOWITZ, S. *Op. cit.*, p. 116.

34. MOSKOWITZ, S. *Seekers of Tomorrow: Masters of Modern Science Fiction*, 1965. The World Publishing Company, Cleveland and New York.

35. HELMER, O. *Social Technology*, 1966. Basic Books, London.

36. GREEN, E. I. 'Creative Thinking in Scientific Work', in *Research, Development and Technological Innovation*. See above Vol. 9, p. 118.

37. LANSBERG, H. H., FISHMAN, L. L. and FISHER, J. L. *Resources in American Future-patterns of Requirements and Availabilities 1960-2000*, 1963. The Johns Hopkins Press, Baltimore, for Resources for the Future Inc..

38. OZBEKHAN, H. *The Idea of a 'Look Out' Institution*, 1965. System Development Corporation, Santa Monica, California.

39. PAINE, T. O. 'The City as an Information Network'.

40. LOVEWELL, P. J. and BRUCE, R. D. 'How We Predict Technological Change', *New Scient.* 1963. Vol. 12, p. 370.

41. JANTSCH, E. *Op. cit.*, pp. 251 and 96.

42. LOCKYER, K. G. *An Introduction to Critical Path Analysis*, 1964, p. 2. Pitman, London.

43. The best exposition of all this that has come to my notice is WOODGATE, H. S. *Planning by Network*, 1967 (2nd ed.), p. 23. Business Publications Ltd, London.

44. See for instance IVAKHNENKO, A. G. and LAPA, V. G. *Cybernetics and Forecasting Techniques*, 1967, p. 1. American Elsevier Publishing Co., New York, and LISICHKIN, V. A. and GUISHIANI, J. *Prognostika*, 1968.

45. BELL, D. *et al. Daedalus, Towards the Year 2000: Work in Progress*, 1967. American Academy of Arts and Sciences, Boston, Mass. KAHN, H. and WIERNER, A. J. *The Year 2000. A Framework for Speculation on the Next Thirty-three Years*, 1967. Collier-Macmillan, London.

46. KAHN, H. 'Notes on the Post-Industrial Society', *Publ. Interest*, 1967, Nos. 6 and 7.

47. DOXEIADIS, C. *Emergence and Growth of an Urban Region: The Developing Urban Detroit Area*, 1966, p. 1. Detroit Edison Co., Detroit.

48. JANTSCH, E. *Op cit.*, p. 256, citing the McGraw-Hill survey.

49. JANTSCH, E. See above. Vol. 8, p. 271.

50. *Ibid.*, p. 260.

51. BRECH, R. *Britain 1984: Unilever's Forecast—An Experiment in the Economic History of the Future*, 1963. Darton, Longman & Todd, London. He has described his method in *Planning Prosperity—A Synoptic Model for Growth*, 1964, p. 45. Darton, Longman & Todd, London.

52. WAGLE, B. 'A Review of Two Statistical Aids in Forecasting', *Statistician*, 1965. Vol. 15, No. 2.

53. ZAUBERMANN, A. 'New Phase Opens in Soviet Planning', *The Times*, 1965, 2nd February.

54. DE JOUVENAL, B. *The Art of Conjecture*, 1967. Basic Books, New York.

55. *Futurist*, 1968. Vol. 2, No. 4, p. 68.

56. SAMUELSON, P. A. *Problems of the American Economy*, London School of Economics, Stamp Memorial Lecture, 1961. 1962. L.S.E.

57. DU BRIDGE, L. 'The Shape of the Future', *Engng Sci.*, 1962, p. 13. California Institute of Technology, California.

58. PUTNAM, P. C. *Energy in the Future*, 1953. Van Nostrand, Princeton.

59. ANSOFF, H. I. *Corporate Strategy: An Analytic Approach to Business for Growth and Expansion*, 1968. McGraw-Hill, New York.

60. *Futures*, 1968. Vol. 1, No. 2, p. 271.

61. ZUCKERMAN, SIR SOLLY. 'Society and Technology', The Trueman Wood Lecture, *Jl R. Soc. Arts*, 1969. Vol. 117, No. 5157, pp. 624, 627.

62. JANTSCH, E. *A Study of Informational Problems in the Electrotechnical Sector*, DAS/RS/64.250 (Restricted), 1965. O.E.C.D., Paris. MYERS, M. S. 'Conditions for Management Motivation', *Harv. Busin. Rev.*, 1966. Vol. 44, No. 1, p. 58.

63. JANTSCH, E. *Technological Forecasting*. See above. Ref. 8, p. 269.

64. SHAPLEY, L. 'Simple Games: An Outline of the Descriptive Theory', *Behavl Sci.*, 1962. Vol. 7, p. 59.

65. MARSHALL, A. *Industry and Trade*, 1919, p. 7. Macmillan and Co., London.

66. KRETZMANN, E. M. J. 'German Technological Utopias of the Pre-War Period', *Ann. Sci.*, 1938. Vol. 3, p. 417; KRYSMANSKI, H.-J. *Die Utopische Methode Eine Literaturund Wissenssoziologische Untersuchung Deutscher Utopischer Romance des 20. Jahrhunderts*, 1963, p. 31. Cologne.

67. BEER, J. J. *The Emergence of the German Dye Industry*, Illinois: Studies in the Social Sciences, 1959. Vol. 44, p. 24. University of Illinois Press, Urbana:

68. WHITEHEAD, A. N. *Science and the Modern World*, 1933, p. 120. Cambridge University Press, Cambridge.

69. COATES, J. F. *Chemical Society Memorial Lectures*, 1951. Vol. 4. The Chemical Society, London. For the actual development see HABER, L. F. *Chemistry and Industry in the Nineteenth Century*. Oxford University Press, London.

70. HAMMOND, R. P. *Applied Science and Technological Progress. A Report to the Committee on Science and Astronautics by the National Academy of Science*, 1967. U.S. Government Printing Office, Washington.

Physicists Have Led the Way

by Sir John Cockcroft, OM, FRS (Hon. MIMechE)

As innovation in engineering becomes more and more science-based, the scientist—and in particular the physicist—must become a member of the research and design teams. Recent history shows many examples of technological advance in which physicists have led the way.

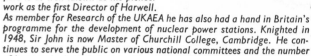

Sir John Cockcroft *studied electrical engineering at the Manchester College of Technology and obtained his practical experience with 'Metro-Vicks' before he went up to Cambridge in 1924. He has worked with many distinguished engineers and physicists, including Lord Rutherford, Kapitza and E. T. S. Walton. In 1934 he took charge of the Royal Society's Mond Laboratory and he was also a University Lecturer in Physics at that time. He became a Fellow of the Royal Society in 1936 and three years later he was appointed Professor of Natural Philosophy in the University of Cambridge. His war-time work in the atomic field is as well known as his pioneer work as the first Director of Harwell. As member for Research of the UKAEA he has also had a hand in Britain's programme for the development of nuclear power stations. Knighted in 1948, Sir John is now Master of Churchill College, Cambridge. He continues to serve the public on various national committees and the number of honours he holds is legion. (Deceased 18.9.67)*

The history of engineering development shows that in a large number of important cases the innovation has come from physics and physicists and this process is likely to continue.

Electronics

The development of electronics is a classical case. Maxwell and Hertz were the pioneers whose work led to Marconi's development, and so to radio engineering. This was followed by J. J. Thomson's work on the discovery of the electron, and the work of Fleming and de Forest on the thermionic valve.

During the pre-war years, the physicists, Watson-Watt, Wilkins, Bowen, were the principal innovators introducing long wave length radar in Britain. The development of microwave radar was due primarily to the initiative of the group of physicists who were brought into radar work on the first day of the war. After a few weeks of operational experience on chain radar stations they dispersed to their laboratories and to Government establishments to work on new developments of radar. The Oliphant group returned from Ventnor to Birmingham University. Their brief experience reinforced their appreciation of the need for high power at very short wave lengths—10 centimetres or thereabout—to provide much more directional radar beams.

At that time the short-wave magnetron, producing about 10 watts, was known, but a thousandfold increase in power was required. To physicists this obviously meant a cathode emitting amperes, a high anode voltage of the order of 10 000; and high Q-tuned circuits to define the wave length. The first innovation was the resonator Klystron, operating at a wave length of about 7 cms. We called it the VFO—Valve for Oliphant.

Almost immediately afterwards came the invention of the cavity magnetron by Randall and Boot, producing about 10 kW. I saw this first in the laboratory of Birmingham University, attached to a diffusion pump; shortly after this I saw a model operating on the Downs above Swanage. Physicists had in the meantime developed the antennae and also the microwave receiver, using a crystal mixer.

By this time the engineers were following up and developing the hardware but the initiative in the development of short wave radar continued to remain with the physicists. They developed the airborne radar sets; the H_2S sets for blind bombing; devices such as 'Gee' for accurate pinpointing of targets and for navigation; equipment to enable anti-aircraft guns to follow targets automatically. Other developments in which physicists played an innovating role were the countermeasures to magnetic mines; and proximity fuses for anti-aircraft and artillery shells.

The great development of solid state physics was initiated in the 1930s by the work of the theoretical physicists, such as A. H. Wilson and N. F. Mott in this country, and Slater and others overseas. This bore astonishing fruit in the post-war years. The most remarkable case was the development of the transistor by the physicists of the Bell Telephone Laboratories, Bardeen, Brattain, Shockley and others. This has revolutionized electronics and made possible the present generation of high-speed computers and extraordinarily compact circuits for space travel work. These developments would not have been possible without a very profound knowledge of solid state physics and quantum electronics.

More recently we have seen the development of the masers and lasers, based on the work of the physicists Shirlaw and Townes. The maser, an amplifier of microwaves, greatly helped the development of satellite communications. The laser, in both solid state and gaseous form, is finding many applications both in applied science and technology in the course of which physicists and engineers are collaborating in the development.

The development of the so-called 'hard superconductors'—alloys which are superconducting at appreciably higher temperatures and in the presence of strong magnetic fields, derives from the pioneering work of physicists in the last 50 years. The metallurgists have taken a hand by making the new alloys in forms which can give reliable performance, and engineering physicists have built superconducting coils to produce fields of up to 120 KG. We now expect that a great deal of physical apparatus which would normally require 10 MW of power will in future use superconductors and enormously reduce its power consumption. Perhaps, in time, the technology will creep into conventional engineering equipment, once the physical laboratories have shown the way.

Nuclear energy

The development of nuclear energy owes a great deal to the innovations of physicists and chemists. Otto Hahn discovered fission; Enrico Fermi built the first atomic pile and most of the theory of nuclear reactors was worked out by theoretical physicists. Thereafter the work became the joint responsibility of engineers, physicists, chemists and metallurgists. A breed of reactor physicists developed to build and operate 'zero energy reactors' for testing the nuclear properties of full-scale reactors before they are built.

The reactor engineers are, of course, responsible for design and construction and Goodlet and Moore were typical innovators in their design study of the Project PIPPA in 1950, a gas-cooled, graphite-moderated reactor power station. Thereafter project teams designing and constructing reactors consisted of many specialists who each contributed to the design but they always included the reactor physicist.

The main responsibility for the operation of nuclear power stations has been in the hands of engineers but senior physicists have made a considerable contribution to development through operational research. It is perhaps worth noting that the Members of the Atomic Energy Authority for Engineering and for Production are both physicists by initial training.

The interaction of physics and engineering has been seen at its best in the design and construction of large-scale nuclear physics accelerators such as the £10 m. proton synchroton at CERN; the $130 m. linear accelerator at Stanford; and the large-scale radio telescopes.

The CERN proton synchroton, which produces protons of 28 BeV energy, was designed on the new principle of strong focusing of the circulating proton beam, invented by physicists immediately after the war. The design of the 300 m diameter magnetic racetrack, which provides the guiding and focusing fields and the radio frequency accelerating system to speed up the particles, required elaborate calculations to ensure that the circulating protons would stay in stable orbits when travelling a few hundred thousand kilometres round and round the racetrack.

The leader of the project team was an engineering physicist; he was helped by nuclear physicists, radio engineers, magnet designers, civil engineers and theoretical physicists. An engineer with conventional training would have been lost in this work.

An even larger project is nearing completion at Stanford University in California. This is an electron linear accelerator, 3 Km long, designed to produce electrons of 18 BeV. This has required the development of new waveguide structures for the accelerator, sealed off megawatt klystrons operating on centimetric wave lengths; and enormous auxiliary systems, such as vacuum pumps, focusing and bending magnets, spark chambers. Here again physicists had the overall direction of the project, with major help, of course, from numerous engineers.

The design and construction of radio telescopes, such as Joddrell Bank, the Cambridge interferometer telescope, or the Parkes radio telescope in Australia, have followed the same pattern of development with excellent results.

Physicists also play a leading part in the development of new forms of instrumentation, which is now so important. They have been responsible for the development of mass spectrometers, electron microscopes, electron spin resonance spectrometers, modern high vacuum techniques and cryogenic equipment. They have shared with electrical engineers the development of digital computers.

Another group of physicists, the geophysicists, have played a leading part in the development of seismic exploration techniques for oil and minerals. They have to work closely with the engineers who take part in the surveys and carry out the drilling.

Engineering physicists

During the centenary celebration of the MIT there was a discussion on the training of engineers. Professor Gordon Brown, Dean of Engineering, divided engineers into three broad classes. The first group are the engineers who take part in enlarging the boundaries of science and technology; the second are the designers; the third are the operators. He pointed out that "the first group are composers rather than arrangers. Their work is predominantly intellectual and it depends on a profound knowledge of science. They may work as scientists but their knack of seeing the useful, rather than searching for the unknown, characterises them as engineers. Perhaps the best description of this group is engineering physicists'."

An undergraduate course is not sufficient for this group of engineers. They require postgraduate training, especially in advanced fields of applied mathematics, such as statistics, quantum theory, information theory and linear programming. These courses are available in the great engineering schools of the United States. How many of the engineering schools of this country provide them?

There is a strong tendency today for some electrical engineering firms to recruit almost as many physicists as engineers. The physicists tend to go into the research and development sections of the organisation. They should be interchangeable with engineers in many new projects.

The organisation of the Bell Telephone Research Laboratory, which has a long record of successful development, is of special interest. The research department has about 500 professional scientists, chemists, metallurgists, physicists and engineers. Their job is to select promising fields from current developments in pure science and to carry out research work to see whether the developments are likely to be of interest commercially.

This group has, within the last ten years, been responsible for transistors, hard superconductors, masers and lasers and satellite communication. The second major group are the system analysts, predominantly engineers, whose job it is to study the most economical ways in which new developments can be integrated into engineering systems. The third group are the development engineers whose job it is to turn the work of the first group into practical hardware. The second and third groups outnumber the first by about four to one. Finally there are the production engineers in the associated companies.

The examples I have given illustrate the modern trend towards an interdisciplinary attack on the problems of engineering, a trend which is inevitable as innovation becomes more and more based on science. One of our largest electrical engineering companies today recruits about one fifth of its new professional staff from physicists and, no doubt, recruits also a number of chemists and mathematicians.

The Institute of Physics and the Physical Society have informed me that about one third of their 10 000 members work in industry.

Collaboration

Is engineering then applied physics? To this question I return a decided negative.

The design engineer today has to synthesize the contributions of a broad spectrum of technologies into his design—many of the problems, such as heat transfer, vibrations, acoustics, are essentially physical; others are the concern of the materials scientist; others, such as combustion problems and chemical compatibility problems, concern the chemists.

The modern design engineer has to be able to appreciate the contribution of these disciplines and to take account of them in his synthesis. He does, therefore, require a science-based training, with at least the first year out of three devoted mainly to the basic sciences important in his profession.

There are no serious administrative problems in combining the efforts of engineers and other scientists in projects, provided that the head of the design team appreciates the importance of their contributions and sees that their views and work are fully considered.

When I was concerned with atomic energy problems, the engineering design group invariably formed a committee which contained representatives of all the groups contributing to the project.

The design committee met at approximately monthly intervals and the work of the contributing groups was fully discussed and taken into account.

One of the difficulties may be to foresee all the physical or chemical or materials problems which may arise in a project; we have had many examples of this in the past. No doubt second sight is required to foresee all possible troubles, but a great deal can be achieved, provided the engineering organisation has an adequate quota of scientists from the important disciplines looking out for such possible problems.

Co-operation between the professions of engineering and physics can be best facilitated by arranging joint technical meetings to discuss projects of mutual interest. The British Nuclear Energy Society has been very successful in organising such meetings in their field of interest.

The Relevance of the Social Sciences

by A. B. Cherns, MA

Why do people become engineers? How do they behave when they work together in groups? What are the effects of different training methods on their careers? These are some of the many questions of interest to engineers which the social sciences are concerned with. In a number of colleges, engineering students are now encouraged to think along these lines. Once the interdisciplinary chasm is bridged, engineers and social scientists find much to learn from each other.

Now that Departments of Social Science are being set up in some Colleges of Advanced Technology the general conceptions and methods of these sciences will not be entirely outside the experience of future engineers. Some universities also offer opportunities for engineering students to learn something of these subjects. There must be many engineers, however, who know little or nothing of the social sciences and some who regard themselves as no worse engineers in consequence. Although this is not my major objective, I shall be gratified if a few of them change their minds after reading this article.

The social sciences differ from the natural sciences in their topic of study. While the physical sciences are concerned with the nature and behaviour of matter, and the biological sciences are concerned with the nature and behaviour of living matter—in particular with living organisms, the social sciences take as their subject the study of man in society. Broadly speaking, psychology (which also has claims to be a biological science) is concerned with the behaviour of human individuals, their capacities and their development; social psychology deals with the effect on the individual of his membership of social groups. Sociology takes as its basic unit the social group and studies the interaction of such groups and the social roles of their members. Some social institutions are subjects of more specialized work: political institutions are studied by political scientists, financial and economic institutions by economists, legal institutions by lawyers.

Anthropology takes whole societies as its subject matter. It can do this, without indulging in meaningless generalities, largely because the cultures so far studied have usually been small, relatively isolated and complete in themselves. Tribal societies, while more complex than is generally believed, are nevertheless relatively simple in comparison with advanced societies.

Have the social sciences enough in common with the natural ones to justify the use of the term 'sciences'? In theory, what the natural sciences have in common is their use of observation, hypothesis, experiment, more observation aimed at confirming or refuting the hypothesis, further experiment, and so forth. In fact, very little scientific endeavour is conducted in this way; this account of scientific method is to some extent an elaboration of the philosophers. Nevertheless, science does progress by experiment and observation, and the essence of a scientific hypothesis is that it can, in principle, be confirmed or refuted by actual and repeatable observation of facts.

A complex subject

The subject studied by the social scientist is complex and variable and in many cases inaccessible to direct experiment. Mathematical treatment of the social scientists' operations has, therefore, had to depend on the development of statistical methods and techniques. Although experiment in the strict sense is so far impossible in astronomy, the techniques of systematic observation have, for centuries, entitled it to be considered a science. The biological sciences have only recently developed the refinements of experimental technique that are now enabling them to take giant strides. The social sciences are often compared, as to their stage of development, with the biology of fifty years ago. Although this comparison offers no guarantee of similar rapid development, social scientists generally work in the faith that the 'breakthrough' is just around the corner.

More than anyone else, the engineer has changed the face of society and social life. The immense material changes he has wrought have precipitated equally far-reaching but unplanned, and often unforeseen, social changes. The greater his material power, the more he is obliged to study the environment in which this power is applied. If you can only make a little explosion you need not worry about its climatic effects; if you can make explosions measured in megatons (or gaga-tons, or whatever the next fashionable multiple will be) then, if you are wise, you will start concerning yourself with the science of meteorology. Now that the engineer can do so much to alter society, he should be interested in the study of society. Furthermore, he himself, his motives and the pressures guiding and directing his behaviour, are part of the subject matter of social science.

In this article I shall not be concerned with these themes on a global scale but with their humbler counterparts in the more intimate sphere of the engineer's work in industry.

'— precipitated far-reaching but unplanned, and often unforeseen, social changes'

How relevant is social science to the engineer and to his job? It may tell him something about himself and his background, his abilities and his personality. It is certainly concerned in suiting his training to the actual job he performs. But, most important of all, the findings of social sciences may, in various aspects of his work, help to shed a little light on the numerous obstacles in his path.

Why they become engineers

Fortunately we have available preliminary results from a very recent study conducted by Professor Hutton of the Department of Mechanical Engineering, with Dr Gerstl of the Department of Industrial Relations, both at Cardiff University College, who have generously allowed me to quote them. The research was partly financed by a grant from the Human Sciences Committee of the Department of Scientific and Industrial Research.

One thousand Full and Associate Members, and Graduates, of the IMechE were interviewed during 1962. Even allowing for the rapid increase in the last generation of the engineering profession, it comes as somewhat of a surprise to learn that no more than 4 per cent were the sons of Chartered Engineers. Another 17 per cent had fathers who, though non-chartered, were described as 'engineers', a notoriously vague description. It is likely that some of the remainder had a connection with engineering as the fathers of 29 per cent were skilled manual workers.

When it comes to choice of occupation a third of the sample mentioned a parent or near relative engaged in engineering, a quarter giving this as a major reason. A practical bent was the reason most frequently mentioned (55 per cent) and given as a major influence by 30 per cent. Nearly two fifths mentioned that their academic aptitudes, mainly mathematics and science, were suitable for engineering training, 13 per cent gave this as a major reason.

While 23 per cent* thought that the good prospects offered by engineering had influenced them, none gave prominence to this aspect.

One's first impression from these figures is that entry into the engineering profession is about as haphazard as other choices of career. However, the data from this research are being analysed now and a full account of the findings will not be available for some months.

Creative intelligence

Another very recent research study has interesting implications concerning the abilities and personalities of engineering students. Dr Liam Hudson of Cambridge has been experimenting with a battery of tests, including a test of high-grade intelligence, on boys in the sixth forms of public and grammar schools. Some of the boys tested were interviewed as well.

His first finding of interest is that these tests readily discriminate between boys pursuing arts courses and those in science sixths. Speaking generally, the scientists have high scores on conventional tests of intelligence, are well able to manipulate spatial concepts (e.g., to visualize the appearance of a diagram when rotated or reflected in a mirror), with comparatively low verbal ability and restricted vocabulary. Their interests tend to be practical and outdoor, rather than cultural.

Of the scientists, those undertaking engineering courses, or planning to become engineers, tend to have general

* These percentages add up to more than 100 per cent because some respondents gave several reasons.

'— tend . . . to be philistine in their approach to the arts'

cultural interests and to be philistine in their approach to the arts. They are not, in general, the most academically able of the scientists and, though scoring quite high marks in intelligence tests, seem far less intelligent in general conversation.

The predictable differences between arts men and scientists induced Dr Hudson to explore further recent American concepts of 'creativity' or 'divergence'. Experiments have suggested that people who are creative dislike the discipline of directed and repetitive tasks like those involved in conventional intelligence tests, and are better at tasks which encourage tangential thinking (called divergence, range and quality of the answer are the chief criteria).

Dr Hudson applied an American test of divergence to his sample. The boys were asked to give as many meanings as they could to a list of ambiguous words and as many uses as they could for a list of everyday objects. They were also asked to write a brief autobiography, to comment on a number of controversial statements and to illustrate them with a drawing. The physical scientists, while coming highest in conventional tests, score lower on these tests of divergence.

It is impossible at this stage to say whether people with conventional minds but high ability choose to study science or whether the methods of teaching science in schools put a premium on conventional, 'convergent' ways of thinking. Convergence of this kind tends to be associated with limited emotional range, conventional political attitudes and a narrow range of interests. It also appears to be associated with a satisfactory integration into school life.

Even if everything that Dr Hudson has found proves to be reproducible with other samples from other schools and if his interpretations of his data stand the test of close examination, it only concerns future engineers. Nothing so far indicates that these findings must be true of earlier generations of schoolboys, although we may feel that many of them are.

In any case, concern is often expressed at the possibility that fewer creative minds may be finding their way into engineering science today, just when a bold, adventurous outlook is greatly needed. Mr Bone's suggestion that boys who would make good managers are deflected from the engineering profession would seem to receive support from these studies, provided we are justified in assuming that 'creativity' is desirable in a manager.

One other recent study should be mentioned here. D. W. Hutchings of the Department of Education at Oxford has shown that, as is widely believed, academically abler boys prefer pure to applied science as their subject of study at the University. On the whole, too, the Dip. Tech. gets a lower cut off the academic joint than engineering degree courses.

As far as his survey shows, this preference for pure over applied science is not a general feature of Continental experience.

For the preference shown in Britain there are several reasons. It appears that, even when boys plan to be engineers, they have little knowledge of careers in the technological field or of what is available in the way of further education. Furthermore, similar courses bear different names at different places. Schools prefer to teach science as a pure discipline with practical applications rather than as the interaction between the needs which evoke discoveries, the ideas that inspire them and the advancing techniques which make them possible.

Lastly, the choice of applied science commits a boy to a career while the choice of pure science delays the final decision. Engineering is thus a career held in comparatively low esteem among academically minded schoolboys and their teachers.

Training of engineers

A considerable number of studies are at present concerned with such aspects of this topic as: what an engineer needs to know; what is the effect on his subsequent career of the different routes of advancement; methods of teaching engineering subjects; the impact of liberal studies on students of technology; the effectiveness of sandwich courses; and social aspects of student life during training.

Dr Marie Jahoda, now head of the new Department of Psychology and Social Science at Brunel College of Technology, has studied the impact of the sandwich principle on the attitudes of students to work and to their professional careers. She has pointed to the crucial effect of the first industrial training period experienced by students on the formation of their professional values. Among the interesting findings of this study are that, in their first academic year, students who have come to the College through the O.N.C. route do as well academically as those with G.C.E. 'A'. The best industrial experience is given by project work, the worst by production work at labouring level. In between come training school, production work at technician level and a 'perambulatory' course (walking round with, rather than sitting by, Nellie). Students benefit from experience in go-ahead firms, where they are given increasing responsibility, where they seem to be of some use, are made to work reasonably hard and where discipline is good. The impact of good industrial experience is shown by their increased interest in their college course and by their greater certainty that they have chosen their future occupation well.

Out of the experiences gained in this study, Dr Jahoda has been led to conclude that different students have very

'— experience . . . by production work at the labouring level'

'. . . The impact of the appearance of an object can be studied by psychophysical methods'

different learning 'styles'; that methods which suit one student may not be the best for another. The Ministry of Education is now backing a study designed to explore these differences in learning behaviour to identify the preferred style of individual students and to seek to devise flexible methods which could match the conditions of learning to the student's preferred style. This study is of obvious importance to all forms of further education.

Another study is by Mr N. L. Day, a lecturer in engineering at Oxford Technical College, who has made a critical assessment of the education and training of professional engineering students in technical colleges. His first finding of interest is that half the students who pass H.N.C. ultimately acquire professional status and that courses are needed to turn the other half into high-grade technicians. He criticises the haphazard and materialistic principles guiding technical education and the unnecessary training wastage caused by the systems of teaching and examining. According to Mr Day 'Theoretical engineering studies are a medley of unco-ordinated and unrelated topics'. His major recommendation is that 'professional engineering courses should be regarded as instruments of a liberal education, designed to provoke the student into asking fundamental questions concerning truth, faith, God'.

If this looks bizarre at first reading, perhaps so much the better, from Mr Day's point of view. He has faced the question, 'what is engineering training for?' and offered an answer from which he goes on to derive a syllabus aimed 'to give a co-ordinated view of engineering activity and its social implications'. If his syllabus were adopted, the changes might well in the end prove more apparent than real for, as we all know, society has a dispiriting habit of absorbing reforms without being reformed. Nevertheless, these illustrations were chosen to demonstrate the extent to which the comfortable underlying suppositions of our training systems and methods are being scrutinized.

Ergonomics

In considering the relevance of social science to engineering as such, one immediately strikes that sensitive spot of today's engineers—the question of design. The appearance of products is, of course, in the domain of aesthetics. In so far as it can lay claim to the title of a science, it is an evaluative rather than a true social science and one whose study has been strangely neglected. The impact of the appearance of an object on individuals can be studied by the psychophysical methods, the standard tools of psychometrics. From studies of this kind and of the more familiar techniques of the social survey, more scientific predictions can be made than are possible by the normal guessing methods.

The contribution of social science is stronger when we turn to the fitness for use of a design, particularly the design of a machine. Another article in this Journal will be concerned more specifically with the topic of ergonomics but I shall briefly mention it here.

It is by now fairly well understood that a machine or tool must be designed with the characteristics of the user in mind. Yet it took a very long time for this simple concept to be formulated, understood and accepted by the only person who can do anything about it—the engineer.

This is not so surprising, if we take into account, first, the fact that the training and interests of the engineer dispose him to regard his machine or tool as part of a mechanical system, if not as an end in itself. The challenge is to master the material. Also, it is only comparatively recently that machines have become sufficiently fast or complex to make demands on the human operator for which he is inherently less fitted than for the merely manual dexterity which was formerly required. Again, only comparatively recently has the study of man's abilities and aptitudes advanced far enough to indicate what skills man is best fitted to acquire. We now know that he is comparatively poorly equipped to acquire and exercise skills requiring immediate memory or prolonged vigilance or judgements of the future state of systems which respond sluggishly to his actions. These are factors as important to the engineering designer as the qualities and characteristics of the inanimate materials with which he works.

These issues are rapidly becoming crucial in the design of automated plant. A cry frequently heard is the need somehow to select and to train people who will be efficient at process operation and process controlling tasks. A good deal of original research into these problems has been, and is being, conducted by Dr Crossman at Oxford and Mr Broadbent and his team at the Medical Research Council's Applied Psychology Unit in Cambridge. Essentially, these tasks consist of maximum watchfulness and minimum activity. The operator has to monitor a series of signals of which only a very few demand action; but these exceptional signals must not be overlooked. Corrective action may take the form of an adjustment whose results will not be fully apparent for many minutes, even hours, after the action.

Ingenious ways will doubtless be evolved for modifying these tasks and training people to perform them somehow. Relying on the amazing plasticity of human performance, we shall muddle through. Maliciously, one almost permits oneself to wish that, just once, a superb, highly instrumented plant will be built which no-one can learn to operate!

Yet much of this is unnecessary the difficulties are of our own creation. They occur only when we consider everything about the mechanical system except the man. It is now a commonplace that we are concerned not with mechanical systems but with man-machine systems in which the mechanical component must be designed to perform those tasks least well performed by the man.

Automation and the social unit

When he has fully dealt with this situation, the engineer will find that the social scientist has an even larger pill ready for him to swallow. Because, unfortunately, the man-machine system is only one element of a socio-technical system. This is a somewhat more complex and more sophisticated concept and one to which it is difficult to do justice briefly. In its present form we owe its formulation to studies at the Tavistock Institute of Human Relations. Essentially, we must consider the entire activities of our functional unit, which can hardly be less than the factory or plant. The activity of this technical sub-system is imposed by the nature of the materials with which it is supplied, the markets to which it sells, and the state of knowledge in the physical sciences upon which it is based. These factors interact so that an advance in scientific knowledge may precipitate a switch in the raw material preferred, or a change in the relative costs of different raw materials may make more profitable a fundamental change in engineering technique.

This technical system is concerned primarily with the flow of materials and the transmission of energy. Interlocking with this at all points is a social system concerned primarily with the flow of information. Its form is dependent upon the physical demands made by the technical system, upon its interactions with other social systems outside the factory and upon the inherent dynamics of all social systems. Clearly, a change in the technology will have an immediate impact on the social system, which will adjust to a new equilibrium. These adjustments will, in turn, affect the operation of the technical system and may operate in unexpected and unwelcome ways so as to diminish the desired effects of the technological changes.

The most dramatic illustration of this kind of effect is the Tavistock's study of the longwall system of coal-mining. Here the impact of the magnificent new coal-cutting and loading machinery was to break up the traditional small groups. Re-constitution into large functional shifts resulted in a situation where the social interactions between groups performing complementary functions were antagonistic instead of co-operative. Previously, each team would be responsible for all the jobs in its area of coal-face—cutting, loading and transporting the coal, securing the pit roof, and so on; the entire cycle of operations. Each member understood the significance of the work done by his fellows and could appreciate the reasons for slower work when this was necessitated by the local conditions of the seam or by the need for more careful propping. After the introduction of the new machinery, one shift would cut the coal, another load and transport it and the third prepare the pit for the next cycle. In these circumstances, each shift blamed the preceding one for delays and difficulties. Only a fraction of the anticipated improvement in productivity was achieved; and this was accompanied by the lowered morale resulting from social disintegration.

Automation brings other familiar problems which are studied by social scientists. Under the auspices of the former European Productivity Agency, groups in several countries have been looking into its impact on clerical workers. Clearly, the nature of clerical jobs is changed when computers are introduced and many traditional jobs are completely eliminated. Redundancy, which may not be severe, is not the only, nor necessarily the major, resulting problem. The man or girl whose job it is to operate or to 'mind' complex machinery is performing a task far closer to that of the factory operative than to his or her previous role. The special relationship between clerical workers and managers, expressed in the description of clerical workers as a staff, and the associated higher status (still preserved, despite the reversal of pay differentials), all this may be lost. Computers offer a few high-grade jobs and a great many more routine, low-grade jobs, with little expectation of promotion from the latter to the former.

Research, particularly the British work at Liverpool University, is showing that the impact on management of

'These specialists . . . may assume an elite role'

office automation is no less than its effects on the white collar worker. There is a small but increasing band of specialists who control the computer which, once installed, provides the data essential to management decisions. These specialists represent an important new element in the ranks of management and may assume an elite role within it.

As more and more specialists become necessary to management so does management itself tend to become professionalized. At the same time, as a manager, the specialist is required to adopt a more general outlook. This is indeed one of the central problems confronting engineers in industry. More managers should certainly be engineers, but more engineers will then stray from their own field of expertise.

The human factor

In Britain, the spur to the study of people at work came from the 1914–18 war. In response to the need for vastly increased output of munitions, the hours of work at armament factories were greatly extended. For a while output was raised but soon diminishing returns were experienced. A further increase in hours of work resulted in a lowering of output per hour and even in a smaller total output. At the same time, sickness absence increased. The Health of Munitions Workers Committee, then set up, sponsored the studies of fatigue which led to the work of the Industrial Fatigue Research Board, later retitled the Industrial Health Research Board. After the 1939–45 war, the work of the Board was taken over by the Medical Research Council. Fatigue, hours of work and rest pauses were only a few of the matters investigated by these boards.

This work provided the first demonstration to many engineers that, on the one hand, human beings did not obey mechanical laws relating input to output; but that, on the other hand, their behaviour, though complex, could be studied systematically. The techniques for the analysis of human behaviour involved indirect, rather than direct, methods of measurement (notably the psychophysical ones). Experiment and statistical techniques resembled those of the agricultural research worker, more than those of the physical scientist.

Engineers concerned with problems of heating, ventilation, lighting and noise have learned from the research of the psychologists and physiologists that quite sophisticated experiment is needed to determine what factors make for comfort and that these factors interact in complex ways. It has, however, proved possible to determine standards of

comfort to which engineers can work. The individual human factor is thus familiar to many specialist engineers; but what of the social factor?

One well-known experiment, conducted over thirty years ago in America, has had reverberations throughout the field of industrial psychology; it has shaken confidence in the results of some of the earlier studies. The experiments of Elton Mayo and his associates at the Hawthorne plant of the Western Electric Co. were intended to be a thorough study of the factors affecting efficiency and output. The outcome indicated that the awareness of the workers of being studied was itself a factor in the results; that, in isolation, a small group tends to adopt its own norms of behaviour and output. The lessons that have been drawn from these studies by social scientists have been innumerable and some concern the engineer.

First, he cannot hope for an easy or simple evaluation of the results of any changes of equipment, methods or techniques he may introduce; for a proper evaluation, the methods and statistical techniques of social science are needed. Secondly, changes in methods which alter the size of working groups, their skills or their supervision are likely to have unpredicted results. The structure and organization of industrial firms are problems not of mechanical engineering but of social engineering.

Causes of accidents

I should now like to turn for a few moments to the discussion, from the social scientists' point of view, of an aspect of industrial behaviour that is often puzzling and frustrating to the engineer who is, as a rule, very conscious of safety at work. His training emphasizes safety factors in design and, as we have seen, the typical personality structure of the engineer may dispose him to caution. Nevertheless, despite his best endeavours, accidents occur and it almost seems at times that people are bent on having them.

This is a topic on which every variety of human and social scientist has something useful to say. We have already described the poor performance of the human being at tasks involving vigilance or prolonged attention. It is obvious that, in tasks of this kind, accidents are very likely to occur. Another physiological limitation of the human being is the rate at which he can process information. With practice and experience, the individual can handle more and more information by taking advantage of redundancy in information received and, possibly, by reducing his monitoring of his own motor responses. But there is a point beyond which he cannot go; his capacity is then fully occupied. Any extra or emergency signal arriving at this stage will have to wait until a channel is clear and the operator may be unable to react to it in time. Accidents arising from this kind of failure can be avoided by better ergonomic design of the job.

There are also other causes of accidents, far less simply dealt with. Investigations extending back to the work of Farmer and Chambers for the Industrial Fatigue Research Board suggested that certain individuals are more prone to accidents than others, and a great deal of work has gone into attempts to define the accident-prone personality. It is still not certain that accident-proneness is more than a statistical artifact; certainly the removal of accident repeaters does little to reduce the accident rate of a group. It remains true, however, that anxious people are likely to have more accidents than others and that, in some cases, an accident, like a neurotic symptom, may serve to remove the victim from the situation provoking the anxiety. The isolated individual in a group is

'— require evidence of recklessness as a token of masculinity'

question presents itself. To what extent are differences in intelligence merely differences in the rate of learning, rather than in the complexity of any subject which can be absorbed? In other words, if the average man could take his own time about it, could he master subjects hitherto thought to be beyond his range? Although it seems likely that complexity will still discriminate among individuals, we may have to recast quite substantially our ideas about intelligence and human learning.

Conclusion

In this article I have ranged fairly widely over those topics which are of interest to both the engineer and the social scientist. In some of these fields, notably in ergonomics, their joint approach to problems has been most fruitful. In other fields which I have not mentioned, engineers and psychologists are giving each other new ideas to play with. This is particularly so in the case of computer engineering and cybernetics. The digital computer has provided insights for the psychophysiologist into the possible modes of operation of the human brain; and the study of what the brain can do and how it seems to do it suggests solutions to some of the problems of the computer engineer.

In most of the fields which interest both, a joint attack will probably have to wait until the concepts of the social scientist become more clearly defined; the study of accidents and their avoidance, however, provides an example of how, even now, the engineer and the social scientist can profitably work together.

What the engineer needs at the moment is the ability to spot a social science problem when he sees one. In order to do so and to avoid the disappointments consequent upon tackling a difficulty from the wrong end, he will have to learn much more about social science, what it has done and what it can do, than he knows at present. The introduction of the social sciences in engineering colleges is therefore a most encouraging development. For the next generation of engineers an article of this kind should be superfluous.

likely to have more accidents than his more popular companions.

The anthropologist and the sociologist take us further still. Groups differ in the extent to which accidents are regarded as inevitable or as the responsibility of the individual. They differ also in the amount of risk they regard as justifiable. Some groups, notably adolescent ones, place a positive value on risk-taking and require evidence of recklessness as a token of masculinity. There are, clearly, complex social factors governing the probability of accidents in a given group; therefore we need not be surprised to find that the removal of one accident repeater may leave the accident rate of the group unchanged. The reduction of accidents must, again, be regarded as a problem in social, as well as mechanical engineering.

Teaching machines

One last topic I wish to discuss is that of the teaching machine. The idea is naturally one to interest the engineer, particularly as, at first sight, it appears to offer possibilities of replacing the human being at the level of the higher mental functions. In fact the word 'machine' is a little unfortunate in this context and the emphasis on the replacement of the teacher may be wrong at the present time. Essentially, the point of the teaching machine is the development of programmed instruction.

Empirical studies of learning seem to show that, because one learns an error one has made as easily as a right response, it is more efficient to avoid the possibility of the pupil's making errors in the course of learning. Each pupil has a characteristic optimum rate of learning and the variations between pupils' rates makes any kind of class teaching a relatively inefficient compromise. Teaching programmes are usually illogical, both in order and in presentation and show logical discontinuities.

The virtue of the teaching machine is that it requires a programme which proceeds at an even rate from step to step with sufficient redundancy to ensure that each new step will present the pupil with the virtual certainty of a correct response. In these circumstances, each pupil can progress at his own speed and a new, very interesting, and disturbing

Mr A. B. Cherns *was educated at St. Paul's School and Trinity College, Cambridge, where he graduated in Mathematics and Psychology. He served with R.E.M.E. during the 1939-45 war, attaining the rank of Captain. In 1947 he undertook a year's work at Cambridge in the Nuffield Unit for Research on Ageing and, from 1949-54, he worked in the Air Ministry and was responsible for research and advice on officer relation and training and selection for ground trades. He was then appointed Chief Research Officer at the R.A.F. Technical Training Command where he worked for five years. In 1959 he became Head of the Human Sciences Section, D.S.I.R. and Secretary of the Human Sciences Committee. Mr Cherns is a Fellow of the British Psychological Society and a Member of the Sociological Association, a member of the Council of the Ergonomics Research Society and the Universities Industrial Relations Association. He has written a number of papers and contributed to the symposium SOCIETY published by Kegan Paul.*

A technician at the Department of Surgery, Hammersmith Hospital, monitoring physiological conditions during heart operations. The console serves two operating theatres and a recovery ward, reduces overcrowding, cuts infection risks, and gives surgeons a quick idea of progress during complex operations using heart-lung, hypothermia and other apparatus

[Courtesy, Honeywell Controls Ltd]

Biological Engineering

by Ronald Woolmer, VRD, FFARCS

The biological sciences are expanding in many directions. They are becoming more and more dependent, for this expansion, on the physical sciences. This means that biologists and engineers are having more to say to each other; but because they don't speak the same language there is often a defect in communication.

In the first year of their training the medical student and the budding biologist take a course in physics and chemistry, about equivalent to 'A' level of the G.C.E. Though in their later work they may gain some practical experience of the applications of physics and engineering to biological problems, their formal training in the physical sciences goes no further than this.

As the medical student leaves physics and chemistry behind for elementary biology, for pharmacology and bacteriology, for anatomy and physiology, for pathology and *materia medica*, and then goes on to clinical surgery and medicine, to obstetrics and public health and to the other branches of medical science which he is required to study in his six-year training course, it is not surprising that his early brush with the fringe of the physical sciences leaves little impression on him. This creates a situation in which a medical practitioner or a biologist may have a problem which, he is vaguely aware, has a physical or a mechanical basis, but which he is quite unable to formulate in terms intelligible to an engineer.

Equally, the engineer goes through his training without acquiring any knowledge of living systems. He knows nothing of the cell, the brick from which all living organisms are constructed. He does not realize the significance of biological variation, which permits apparently identical tissues or organisms to react differently to the same set of circumstances. He is unaware of the consequences of the introduction of foreign tissue into the animal body; or of the infection which invariably follows a wound or incision unless the implements used are first sterilized to remove the all-pervading bacteria. He is not brought up to the idea that in the animal body one cannot (or until recently could not) take a component out of service for repair.

He is not used to thinking in terms of a pump, such as the human heart, which is designed to run continuously, without

servicing, for up to a hundred years, which has an output rapidly and continuously adjustable between one and nine gallons a minute, and which weighs less than twelve ounces. He is unfamiliar with the structure of living organisms, and his mental approach is necessarily different from that of the doctor or biologist.

It is evident, then, that there is a defect in the biologist's knowledge of mechanics and the physical sciences, and in the engineer's knowledge of biology. This defect *could* be remedied by adding appropriately to the curriculum of the two disciplines.

Such a suggestion, however, would be vigorously opposed by the educational authorities who are constantly assailed with requests to cram more subjects into already crowded curricula. And, indeed, the average general practitioner can demonstrate little need for a deeper knowledge of the physical sciences; and the average engineer does not find himself involved in biological problems.

The need for interpreters

Since there is no hope of evolving a generation of biologists well versed in the principles of engineering; or of engineers for whom biology holds no mysteries, it is inevitable that professional biologists, and professional engineers, should talk different languages and think along different lines.

When communication is necessary between people who speak different languages, it is effected through interpreters, who are familiar with both languages. Who are, or who should be, the interpreters in this instance? They are the biological engineers. They do not hold a degree or diploma in Biological Engineering; such a thing does not exist. A few—in the whole of Britain they can be counted on the fingers of two hands—have degrees both in engineering and in medicine or biology. The others are either engineers or biologists who have drifted by interest or been forced by necessity into the area of overlap, and who have acquired their knowledge of the other discipline without formal training.

Biological Engineering is a wide subject and the work they do is very varied. A few examples must suffice to indicate its scope.

Certain diseases, of which poliomyelitis is the commonest, cause partial or complete paralysis of the breathing muscles. Survival can then be assured only by artificial respiration, which may have to be maintained for the rest of the patient's life. Its provision is a problem for the biological engineer. The early 'iron lung', in spite of its unfortunate shape (it became known abroad as 'the English coffin'), saved many lives, but it was a crude affair. Its modern successors, the various types of mechanical respirator, are a much more satisfactory blend of mechanical efficiency and biological acceptability. A good respirator must be absolutely reliable, as unobtrusive as possible, and must allow the patient to be mobile and accessible to nursing care. It may be required to accompany him in a wheel-chair or in a car or even in an aeroplane, so that it must be adaptable to different power supplies. It must generate the right form of pressure wave to force air into the patient's lungs, and the pressure pattern must be adaptable over a wide range. The fact that there are now a number of satisfactory machines available shows that medical men and engineers have co-operated successfully in their endeavours.

Artificial organs and limbs

Cardiac surgery has made spectacular advances in the past decade. No engineer would think of repairing a pump while it was running, but that is what the surgeon had to do, and still does, in the correction of mitral stenosis—the narrowing by disease of the channel from one chamber of the heart to another. The success of this operation led to the hope that other cardiac defects—in particular the presence of a gap in the wall which separates the left side of the heart from the right—might also prove amenable to surgery. To achieve this, it was mandatory to take the heart out of service for repair.

This entailed the design and construction of an artificial heart to replace the real one, and an artificial gas exchanger to replace the lungs, which for technical reasons had to be by-passed with the heart.

The difficulties are formidable. Whenever blood comes in contact with air, or with almost any material other than the lining of the blood vessels, it clots. The silicones were able to provide a surface which did not engender clotting. The cells of the blood would not withstand the battering and the accelerations imposed by any ordinary pump, and a special 'peristaltic' pump had to be devised.

The blood had to be kept from contact with the air and all bubbles had to be carefully removed from it. The entire mechanism of the pump and conducting tubes had to be made so that it could be taken to pieces and sterilized by boiling. No foreign matter, such as lubricating oil, could be tolerated within the system. The output had to be controllable over a wide range; and, of course, the pump had to be completely reliable. Still other limitations had to be imposed and, indeed, the specification for the artificial heart, if it had been written in engineering terms, would have made the boldest designer quail.

It is perhaps fortunate that it never *was* written out in precise terms. The medical men who wanted it wouldn't have known how to do so. But they co-operated, on an informal basis, with engineers of various sorts. Much effort

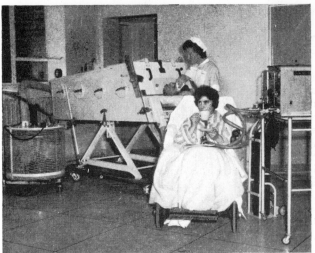

[Courtesy, W. G. Pye Ltd]

The old-style 'iron lung' compared with the far more flexible Barnet ventilator, shown on the right

was wasted through defects of communication, through the pursuit of unsound ideas and through the empiricism that had to be adopted; but by the vigorous application of the methods of trial and error, simultaneously used at a number of different centres, several satisfactory systems have been evolved.

There is still room for improvement, but it is now possible to maintain a patient on an artificial heart for two or three hours at a time: long enough to permit an elaborate repair of the defective organ, using machines designed jointly by surgeons and engineers.

A fruitful field for collaboration between the engineer and the medical man is the design of artificial limbs. A wooden stump for a leg and a metal spike for an arm, which had to suffice for Long John Silver and for Captain Hook, are no longer good enough. Modern below-knee amputees can pass as normal in a crowd, and the most successful of them can win dancing competitions. Upper limb prostheses are more complex and challenging. Much ingenuity is being devoted by biological engineers all over the world to designing artificial arms and hands with their own motive power to do duty for the natural muscles.

Provision of a satisfactory power source is a formidable problem: bringing it under voluntary control is an even stiffer one. One group is attempting to pick up the minute electrical potentials which accompany muscular contraction—actual or attempted—and use them to control a powered prosthesis for a paralysed limb. If this approach is successful, the subject will merely have to 'will' an action of his paralysed extremity, for the powered system automatically to carry it out. But the obstacles are formidable. It is difficult to choose electrode sites for the pick-off signal which are reproducible from day to day; the signal-to-noise ratio is low; and the co-ordinated movements of a natural limb are infinitely more complex than anything which can yet be handled by man-made systems.

Analytical and production techniques

Biologists and medical scientists are of necessity becoming aware of the developments of automatic control in engineering, and the more forward looking are eager to apply them to their own problems. One of the results is the *Auto-analyser* which is now appearing in the biochemical departments of our larger hospitals. This introduces into hospital practice the type of production engineering developed by the chemical industry.

In a large hospital several hundred biochemical estimations may be required every day. They will include the amount of sugar in successive samples of urine, the amount of urea or creatinine or phosphates in a specimen of blood, and so on. Each estimation follows a prescribed course, with the addition, in turn, of precise quantities of certain reagents, until the attainment of a definite end-point, which is often the development of a particular colour, or degree of turbidity, in a solution. Such estimations are normally done by skilled laboratory technicians but, in fact, the individual steps are a matter of mere routine, and there is no inherent difficulty in designing equipment to carry them out automatically. This is what the *Auto-analyser* does and, in a hospital large enough to justify it, great saving of time and glassware can be achieved.

Time and motion study of other hospital activities reveals many instances where better engineering, and the application of modern methods of information handling and control, could result in increased efficiency and economy. The 'electronic nurse' which has received some publicity lately, is a case in point. Much of a human nurse's time is spent in obtaining and recording information about the temperature, pulse rate, respiratory rate and blood pressure of a number of patients.

Equipment to obtain and register this information automatically is not difficult to design, and its use would save much nursing time. When dealing with human beings,

however, strictly engineering considerations may be overshadowed by psychological ones; and many patients would benefit more from having their temperature recorded wrongly by a pretty nurse than correctly by a machine.

Research techniques

Medical and biological research depends for much of its technical progress on engineering, and the 'methods' sections of nearly all papers describing research in biological and medical journals are descriptions of applied engineering. The type of engineering which is thus applied is often electrical, because that is the branch which lends itself most readily to measurement and control. The power available from biological systems under study is usually small and biologists are very dependent on electronic amplifiers. Indeed, the oscilloscope and the electronic black box are as commonplace today in a biological laboratory as they are in one devoted to the physical sciences.

In addition to this general dependence of biological research on engineering technology, there are some important special applications. An obvious example is the electron microscope. This has opened up a new vista to anatomists and physiologists of an importance which can scarcely be overstated.

Its development is the fruit of co-operation between physicists, mechanical engineers, specialists in electron optics and biologists.

From this random sampling of the field it is evident that collaboration does take place between biologists and engineers, and that the necessary interpreters do exist. What is their professional status, and in what departments or institutions does their collaborative work go on? The question is more easily asked than answered.

Engineering facilities

In the Medical Research Council's main laboratories at Mill Hill there is an Instrument Section whose members collaborate with biologists. Very few biological research teams in the smaller units include an engineer or physicist; in most cases the engineering development has become yet another function of the laboratory technician.

The biological or biophysics departments of universities and medical schools undertake similar work and have a comparable staff structure.

In a few Regions of the National Health Service, Regional Physics Laboratories have been set up. These are headed by able physicists, and staffed by qualified physicists and engineers, and by trained and semi-trained technicians. They have been set up to provide the medical staffs working in hospitals with the engineering facilities, consultative and practical, which they need. The work of a Regional Physics Laboratory is varied and interesting. It may include such things as the design and construction of surgical instruments for special purposes, making applicators for radiation therapy, designing a cryostat for rapid fixing of tissue removed at an operation, producing a more efficient dialyser for an artificial kidney, advising on the handling of radioisotopes, and collaborating with hospital architects on the design of new buildings. The director of the department confers on equal terms with the senior consultants and administrators of the hospital.

Such Regional Physics Laboratories provide a rare example of machinery specifically set up to provide hospital doctors with proper engineering facilities and consultation. Only two or three of them exist, but similar work is going on in many of

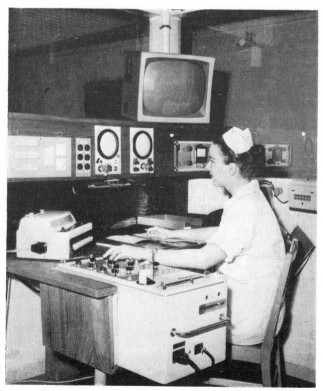

[Courtesy, Elemaschonander]

Monitoring temperatures and supervising wards from a central position at the Foraxkliniken, Sweden

the larger hospital physics laboratories, and in the physics departments of medical schools throughout Britain.

Certain aspects of biological engineering are important to the armed services. These concern such things as the habitability of fighting vehicles, high altitude physiology, the response of the human body to variations of atmospheric pressure and gravitational force, underwater vision, the effects of blast and radiation, chemical warfare and so on. Various institutes and laboratories, such as the Institute of Aviation Medicine at Farnborough, exist to study these problems. They are staffed by biologists, physicists, engineers and technicians of various grades in the armed forces or the civil service.

A number of commercial organizations are concerned with the life sciences. These include surgical instrument manufacturers, parts of the chemical and pharmaceutical industry, makers of laboratory equipment, and firms connected with electrical and electronic engineering. With developments in 'medical electronics' the last group have become more active recently.

Cobalt therapy units, linear accelerators, ordinary X-ray equipment, image intensifiers, television cameras and receivers are examples of the 'heavier' sides of these activities.

In light electronic engineering there are a host of applications, such as hearing aids, cardiac pacemakers and defibrillators, mechano-electric transducers of various types, electrical devices for the measurement of temperature and pressure at various points in the body, electrical methods for analysis and control, specialized illuminating systems and so on.

Though, on a commercial scale, the market is a small one, a large number of firms are interested in supplying medical and biological equipment, and they have on their staffs people concerned with development and marketing who have been trained in various branches of engineering, and a smaller number who are qualified in medicine or biology.

A heterogenous group

It will be seen, then, that the people actually applying engineering to biology and medicine are a heterogenous and unorganized group, some with and some without a formal training, scattered rather thinly through hospitals and medical schools, biological laboratories and commercial organizations, the armed services and research institutes.

There is no professional organization which caters specially for them. The nearest to it is the Hospital Physicists Association, but this caters, as its name implies, for physicists, and it is only in exceptional cases that non-physicists can join. Furthermore, in the past the H.P.A. has been concerned mainly with radiation physics in the diagnostic and therapeutic X-ray departments of hospitals, although the activities of the association are broadening rapidly at present. Those with the necessary background can of course belong to the appropriate engineering institution, and it is of interest to note that the I.E.E. has set up a Discussion Group for Medical Electronics. The British I.R.E. has established a Medical and Biological Electronics Group and joint meetings between the two are now being planned. There is also the International Federation for Medical Electronics, to which interested people may belong as individual members, but which is mainly intended to co-ordinate national societies where these exist.

Until recently there was in Britain no organization, professional or academic, which embraced the gamut of bio-

Professor Ronald Woolmer *was born in 1908 and educated at Rugby School and Oxford University. He did his medical training at St Thomas's Hospital and was Senior Resident Anaesthetist there from 1934–36. During the 1939–45 war he served as anaesthetist in the Royal Naval Medical Service. From 1946–56 he was Lecturer, Senior Lecturer and Reader in Anaesthesia at the University of Bristol and Honorary Consultant Anaesthetist to Bristol Royal Infirmary. In 1957 he became the first Director of the newly created Research Department of Anaesthetics in the Royal College of Surgeons of England and in 1959 he became the first Professor of Anaesthetics in the College when the British Oxygen Chair was endowed. Professor Woolmer's great ambition in recent years was to achieve the maximum co-operation between the professions of medicine and engineering. He was one of the founder members of the Biological Engineering Society and the first President of that Society. He died on 8th December 1962 and this was the last article he wrote.*

logical engineering rather than concentrating on one part, such as medical electronics.

To meet this need, the Biological Engineering Society was established two years ago. Its purpose is to bring together biologists, medical scientists, engineers and physicists on an equal footing, for the advancement of their subject and for the discussion of problems of mutual interest. The requirement for Full Membership is a qualification in physics, engineering, medicine or biology, and at least three years experience in biological engineering. The requirements for Associate Membership are less strict, and there are also Corresponding and Affiliate Membership. The latter is for commercial organizations, and permits them to send any two suitable representatives to meetings of the Society.

Scientific meetings are held four times a year, at hospitals and medical schools, research institutes, biological departments of universities, regional physics departments and commercial laboratories. Physicists and engineers, biologists and medical scientists are all represented among the membership and there are ample opportunities for free interchange of opinion. The Society does not have its own journal, because a new periodical *Biological Engineering and Medical Electronics* is about to appear. This is the organ of the International Federation for Medical Electronics, and it is edited by a member of the Council of the Biological Engineering Society.

Education

The formation of this Society does provide a much needed link between biologists and engineers, and that is what it was created to do. It could be argued, however, that more is required.

The professional status and the training of biological engineers is nobody's business, and at present it is woefully haphazard. There is no course available at any of our universities or technical colleges leading to a degree or diploma in Biological Engineering.

A partial exception to this statement is a course starting this year at Imperial College. It is organized by the Department of Electrical Engineering; lasts for six or for ten months (according to choice) and will be available to a dozen or so participants who already hold qualifications in engineering, medicine or biology. It is designed to cover the applications of electrical engineering to medicine and biology. This venture is greatly to be welcomed, and those who have taken it will probably have no difficulty in finding posts where their skill and knowledge can be put to use.

But Biological Engineering is a rapidly advancing and expanding field, and its development ought to be guided by a proper academic and professional body. What is needed is an Institute of Biological Engineering, linked with a University, employing a well qualified academic staff and adequate classroom, laboratory and workshop facilities. It would have to enjoy the confidence of the medical and the engineering professions and have the approval of the Ministry of Health. It might have as its Director a man qualified in both disciplines or, failing that, a physicist of sufficient eminence and experience. Its staff should be drawn from medical scientists, engineers and biologists of sufficient calibre. One difficulty might be the assurance of sufficient properly remunerated openings in Britain for its graduates (though opportunities are increasing), but there would be any number of openings overseas; and indeed many of its graduates might come from the Commonwealth countries. Apart from its training programme, it would engage in research and development on projects of its own, and on those referred to it by the Ministry of Health, by other medical and biological organizations, and possibly by industry.

A comparable Institute already exists in Moscow; departments of biomedical engineering are being set up in Universities in North America and a research institute has recently been set up in Paris by the International Federation for Medical Electronics. It is to be hoped that we will be able to grasp the opportunity in Britain.

Acknowledgements

The author wishes to acknowledge the help of Dr Alfred Nightingale of the Physics Department, St Thomas's Hospital, London, and Mr W. J. Perkins of the National Institute for Medical Research, London, who kindly read this paper and made many helpful suggestions.

The idea of applying the principles of mechanics to the living body dates back at least to Aristotle. In the 17th century, under the name of iatromechanics, it became discredited for a period owing to the all-embracing character of the crude mechanistic approach of its protagonists. Today, both engineering and medicine have reached a sophistication which not only facilitates but necessitates combined operations.

Biomechanics in the Modern World

by R. M. Kenedi, BSc, PhD, FRSE, MIMechE

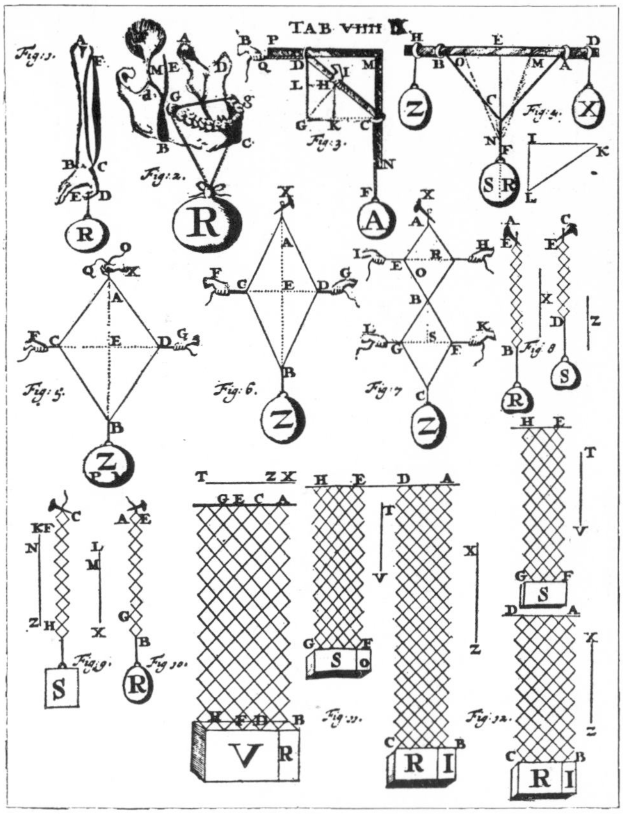

Fig. 1. The first to associate movement with anatomical leverages, the 17th century mechanist Borelli did much to clarify the true functions of various organs, such as the stomach and the heart. This is a page from his book 'De motu animalium'

Biomechanics, the 'study of organic bodies and their components on a macroscopic level by the application of engineering principles', is an inter- or (more correctly) multi-disciplinary activity. Many excellent surveys of it[1],[2] have appeared. It is traditional in this field to begin by referring to Leonardo da Vinci's *Notes on the Human Body*. It tends to be forgotten (and perhaps it is better so) that Aristotle (384–327 BC) wrote about the movements of animals and tried to apply geometric analysis to their actions. In my view these studies were motivated purely by a commendable curiosity and were not intended to be applied. It was left to the iatromathematicians or iatromechanicians of the 17th century to introduce a more practical aim and to show just how completely abortive a purely mechanistic approach could become when applied to living entities.

The concept of man as a machine arose to some extent as a reaction to the system of medicine practised by the iatrochemists which turned man into a retort. Chief among them had been Johann Baptista van Helmont (1577–1644) and Franciscus de la Boe, or Sylvius (1614–1672). They assumed that the human organism was governed exclusively by chemical processes, therapy being simply the addition or removal of appropriate chemicals.

Crude mechanicists

The level of all-embracing arrogance reached by the iatrochemists produced an equally exaggerated reaction. This manifested itself in the creation of an all-embracing mechanical system in which the physiological happenings in the human body were treated as rigid consequences of the laws of physics.

This approach in its mathematical form had been prepared by René Descartes (1596–1650) whose *Traité de l'homme* described man as a self-contained machine. The real founder of this school of thought was Giovanni Alfonso Borelli (1608–1679) of Naples, a pupil of Galileo. Apparently of a quarrelsome disposition, he taught mathematics in Messina, Pisa, Florence and eventually in Rome. There he was for a time supported by Queen Christina of Sweden for whom he wrote *De motu animalium*, a page from which is reproduced in Fig. 1. Within the concepts available in his time, Borelli's work did much to clarify the functions of the muscles, the lungs and the stomach from a mechanical point of view. He was the first to associate movements with muscle-powered leverages.

Borelli had many followers in many countries. Some pursued his ideas to absurdity, for example, Giorgi Baglivi (1668–1706) in Rome, who pushed the mechanical analogy to the point of regarding the human machine as consisting of innumerable smaller machines such as scissors (teeth), bellows (chest), etc.

As would be expected, such a medical 'theory' had to be utterly divorced from medical practice. Baglivi was, in fact, a most successful physician, which he only

This is a condensed version of his nominated lecture to the Institution of Mechanical Engineers in 1966

70

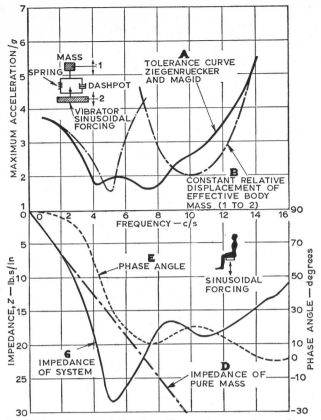

Fig. 2. Typical results of vibrating table experiments to discover human tolerance for accelerations at various frequencies: the suggested equivalent mechanical system is shown top left and the lower curves indicate the effect of phase angle

achieved by dropping, on entering the sickroom, all the theories he taught his students. His writings convey a shrewd and wordly mind and he is the author of a number of maxims. Two, in particular, are most pertinent even today: "the mania for forming new words checks the beginner in his successful advances". And again:

> To frequent academies, to visit libraries, to own valuable but unread books, or shine in all the journals, all these do not in the least contribute to the comfort of the sick.

The Reverend Samuel Haughton of Dublin must be mentioned in respect of his *Principles of Animal Mechanics* (1873). He is best remembered for the discovery that an egg emerges from the hen blunt end first and for his formula for the length of slack to be left in the hangman's rope! Contemporary with him was Muybridge who made the first photographic studies of human and animal gait, using a series of sequentially triggered cameras—before the days of cinematography.

The reason for the more recent interest in biomechanics is on the one hand the increasing pace of engineering theory and practice. The materials of engineering are becoming more and more sophisticated and its theoretical outlook is becoming better equipped to deal with unconventional problems. New fields have been opened also by today's sophisticated instrumentation.

On the other hand, medicine is placing more and more emphasis on rehabilitation and reconstruction. Surgery is now emerging from its traditional primary phase of almost entirely destructive removal of diseased and damaged tissue. Excisional surgery demands little knowledge of the structural characteristics of the tissue involved; the biological property of healing is usually sufficient for a successful result.

Reconstruction and repair have advanced a long way by trial and error alone but have reached the stage where much more exact knowledge of many factors must be gained if precision is to replace empiricism and further major advances are to be made. Much more knowledge is required of the structural and mechanical properties of normal and diseased human tissue, the problems associated with the use of prosthetic replacements, etc.

Tolerance of acceleration

Interest in body response to environmental influences, particularly of a vibratory and impact kind, has been triggered off by the man/machine interaction problem. Modern technology has produced a variety of new transport and stationary machinery. Vibration is apt to be an unwanted side-effect. It not only shortens the life of the machine but may also affect the user. Both from the point of view of the operator and the customer, a knowledge of the human body's response to such an environment is required.

The body's initial response to mechanical forces is passive, manifested by tissue deformation. This initiates, as a secondary effect, either a physiological response (increased blood flow, etc) or tissue damage.

The eventual aim of research is a unified theory of bio-dynamic response, covering tissue deformation, physiological response, injury mechanism and behaviour responses. This, however, is very far off. Meanwhile three kinds of alternating forces on the body are receiving attention. Steady-state vibration and buffeting, impact, and air-transmitted noise.

At present, experiments on the whole body are largely directed at elucidating the effects of steady-state vibration in the biomechanically significant frequency range of 1–20 c/s. Below 1 c/s physiological, rather than mechanical, effects predominate, while above 20 c/s protection by mechanical damping is relatively easy.

Typical results of vibrating table experiments for human subjects in the erect sitting position are shown in Fig. 2. Curve A gives experimentally derived values[3] of the tolerable vertical accelerations. The subjects, seated on a platform, underwent forced vibration for periods ranging from 206 s at low acceleration to about 18 s at the highest values. The individual was able to switch off the equipment when the acceleration imposed became unbearably painful. At 1 c/s the acceleration tolerated was around 3·75 g.

Curve G represents the variation of the mechanical impedance with frequency. This convenient dynamic para-meter is defined as the ratio of the instantaneous value of transmitted force to the corresponding instantaneous velocity of the point where the force is applied. Curve E gives the corresponding variation of the phase angle between force and velocity. Curves G and E are based on Coermann's[4] results.

It is seen that G has two significant resonance peaks and that there is deviation from the pure mass impedance (line D). Coermann has shown that the appropriate mechano-dynamic model of the body in this case could be effectively approxi-mated to by two simple mass-spring systems, of the kind shown in the upper part of Fig. 2. The characteristics of these systems are computed to correspond to the respective resonance peaks and phase angles of the impedance-frequency variation.

Once these simple mass-spring systems have been deduced from the experimentally obtained impedance and phase angle curves, it is possible to construct corresponding acceleration tolerance curves. These are obtained by assuming that the limit of tolerance is achieved when a particular tissue stretch is reached, irrespective of frequency.

This, in terms of the equivalent mass-spring systems, means plotting forcing acceleration against frequency for a constant relative displacement between the mass and the point of application of the sinusoidal forcing action. These are the curves labelled B in the upper part of Fig. 2. Vertical 'matching' with curve A is obtained by taking the tolerable acceleration at 1 c/s, the same for B as it is for A.

Significance of findings

The double peaks of the impedance curve G have been tentatively identified with those of the two resonances of the pelvis and with that of the spine which has a resonance at around 5 c/s. The experimentally determined acceleration tolerances of curve A are much lower (around 5–9 c/s) than those of the comparable curve B, deduced from the experimentally determined impedance variation. The subjects indicated that, in this frequency range, their acceleration tolerance was limited by severe pains in the chest which became maximal around 7 c/s.[4] It has been suggested that some of the organs in the chest, probably the heart, have a resonance of around 7 c/s and, by literally bouncing up and down, stretched certain tissues to produce intolerable pain.

The mass of these organs in comparison with that of the whole body is so small that their impedance will have no measurable effect on the total body impedance and would, therefore, not show up on curve G nor on its derivative curve B.

Structurally, the body consists of a hard bony skeleton, held together by tough fibrous ligaments embedded in a highly organised mass of connective tissue and muscle. The soft organs are contained within the rib cage and the abdominal cavity. Thus all body tissues are essentially complex —yet in current investigations for dynamic responses the body is considered to be a linear passive mechanical system containing elastic and viscous resistances and inertias. This assumption appears to be adequate for small deformations.

The most generalised mechano-dynamic model for use in describing whole body behaviour when subjected to vibration, impact, and external pressure load due to acoustic effects, blast and decompression, is shown in Fig. 3, devised by von Gierke.[2] The quantitative assessments of acceleration tolerance limits in practical terms are closely related to the ultimate strength of body tissue. A great deal of work has been done on this, for example, by Nickerson and Drazic,[5] Yamada,[6] and Stech and Payne.[7]

Two somewhat thought-provoking features emerge: the mechanical strength of tissue tends to decrease with increasing age beyond the age of around 40; and the extension of the human life span tends to uncover unexpected mechanical weaknesses of body parts, the effective life of the whole, so to speak, outlasting that of the component. Thus, in the age range of 65–75, fracture of the hip has become a major affliction; extrapolation of existing data[7] also shows that, if the average human life span were extended to 119 years, the ultimate strength of vertebrae would drop so low that a 50 per cent probability of spinal fracture would obtain under the normal load of gravity.

One point worth touching on in passing is the effectiveness of the anthropometric dummy. Elaborate attempts have been made, particularly in connection with impact experiments, to produce a dummy whose mass effects are the same as that of the human body. This has been reasonably successful[8] but, as a comparison of curves G and D in Fig. 2 shows, body response to dynamic action is by no means mass response only. In consequence there are very severe limitations on the validity of these dummies and experimental results obtained by their use should be interpreted with extreme care.

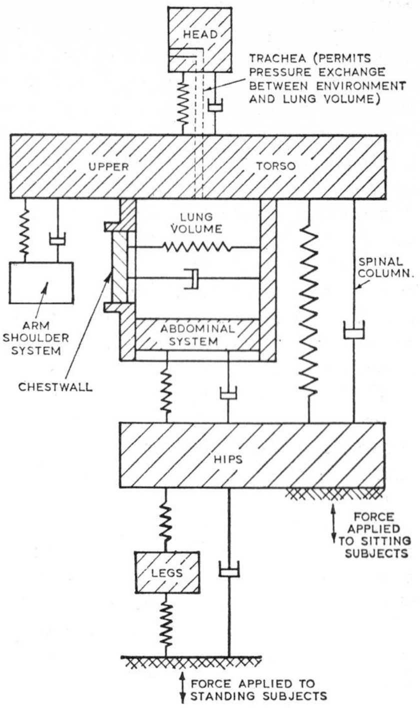

Local application

The same biodynamic techniques have also been applied[9] to the study of localised areas of tissue in the body in the frequency range of 0–20 kc/s. Vibration was applied through a piston, the circular base of which was attached to the body surface;[10] or an air-filled tube was pressed to the tissue surface and the air column inside it was excited at one end by a membrane, driven electro-mechanically. In all cases the vibration was applied at right angles to the tissue surface, on the thigh or on the upper arm.

The impedance Z, defined as before, was now considered in terms of its resistance R and reactive Q components, so that $Z = R + iQ$. To analyse the results von Gierke and his collaborators employed Oestreicher's[11] vibration propagation theories in a semi-empirical fashion. These apply to a homogeneous, isotropic, elastic, viscous and compressible medium when a rigid, fully submerged sphere is vibrated within it.

Von Gierke et al showed experimentally that a vibrating disc, applied to the surface of such a medium, is equivalent in its effects to half that produced by the submerged theory of Oestreicher.

The resistance R up to about 10^4 c/s is primarily viscous (ie shear wave transmission) while beyond this, compressibility effects are likely to predominate.

This type of biodynamic tissue excitation applied at right angles to the surface involves a response not only from the superficial tissues, such as skin, but also from the supporting tissues underneath. The results obtained, therefore, are integrated effects due to a complex system. Biodynamic assessment of individual tissues requires a modified technique, such as is currently under development at the University of Strathclyde.

Fig. 3. A generalised mechanical model of whole-body behaviour capable of simulating reaction to vibration, impact, acoustic effects, blast or decompression

Self-energised actions

The obvious example is locomotion. Effective studies of its kinetics began only at the turn of the century.[12] In recent times they have been principally concerned with evaluating the force transmission characteristics of joints, particularly those of the hip joint.[10-18] This is virtually a ball-and-socket joint, controlled by 22 muscles, the rounded hemispherical head of the hip-bone (femur) forming the ball and the corresponding spherical bearing surface in the pelvis, the acetabulum, forming the socket.

The reasons for the particular interest of the hip joint are that normally it carries the greatest body load, fluctuating from zero to maximum; and its mechanical failure and that of the neighbouring bony structure is one of the major problems of orthopaedic surgery.

The Bioengineering Unit at the University of Strathclyde is one of the centres where locomotion kinetics is being actively studied. Various aspects of the relevant instrumentation and techniques have been described in detail.[15, 18, 19]

Fig. 4 shows a test in progress. The subject, with joint centres marked, equipped with electrodes over selected muscle sites and accelerometers on the lower leg, walks along a path which incorporates a platform, supported on four strain-gauged pillars, by means of which all the components of the interaction between the foot and the ground are measured. This type of platform design was originated by Cunningham and Brown.[20]

The electrodes pick up the variation of electrical potential of the selected muscle during activity. The record of these variations (the electromyograph), while not capable of direct quantitative interpretation as muscle tension, does permit the phasing of muscle actions to be correlated in time with the kinetic characteristics of the walk cycle. During the test walk, the subject is also filmed in, and perpendicular to, the plane of progression at 50 frames/s. Films, force plate, electromyographic and accelerometer output are all synchronised. In addition, the anthropometric characteristics are determined.

In the analysis of the forces transmitted by the joint, the highly complex system of 22 muscles had to be simplified to make the problem tractable, by neglecting certain muscles and grouping others. Such grouping, shown to be permissible,[18, 21] resulted in a reduction of the system to six significant muscles or groups. All the resulting data are handled by a computer programme.

Such studies have now been carried out for a variety of normal, arthritic and amputee subjects, walking on the level, up and down ramps and stairs. Typical results of joint force for a normal subject are shown in Fig. 5, curve A. The two maxima are both essentially in the stance phase, the first following heel strike for the leg considered, the other in the region of heel strike for the other leg.

The absolute maximum occurs in the second region and is surprisingly high, 4·1 times body weight. Paul[16] records that in 12 normal subjects of ages 18–36 years the peak joint force to body weight ratio varied from 2·3 (female subject walking very slowly with a short stride) to 5·8 (heavy male subject walking very energetically) with an average for all subjects of around 5·0. This high value is due primarily to the joint stabilising action of the hip muscles, not unconnected with the adaptation of an essentially four-legged creature to the inherently unstable two-legged posture!

Curve B presents results[22] obtained from a 51-year old patient who, as a result of an accident, had to have a complete replacement of the femoral neck and head implanted. The prosthesis (of stainless steel) was strain gauged and six months after the operation the forces on the prosthetic head were directly measured during walking. The pattern is broadly the same as that for the normal subject, indicative of relatively normal muscle actions, the two regions of maxima being again in evidence.

One of the applications of the Strathclyde studies has been the design of a new type of surgical 'nail' for hip fracture fixation.[23] This has been undergoing clinical trial since December 1964 and more than 40 of these nails have been implanted to date.

Another important application of the walkpatch technique is the study of amputee gait. Hughes[19] has shown in a preliminary study that the probable joint forces in the hips on the natural and artificial leg sides were significantly dissimilar in an above-knee amputee.

This lack of kinetic symmetry is not surprising when one considers that present day artificial legs reproduce the articulated layout of the natural limb but without its musculature.

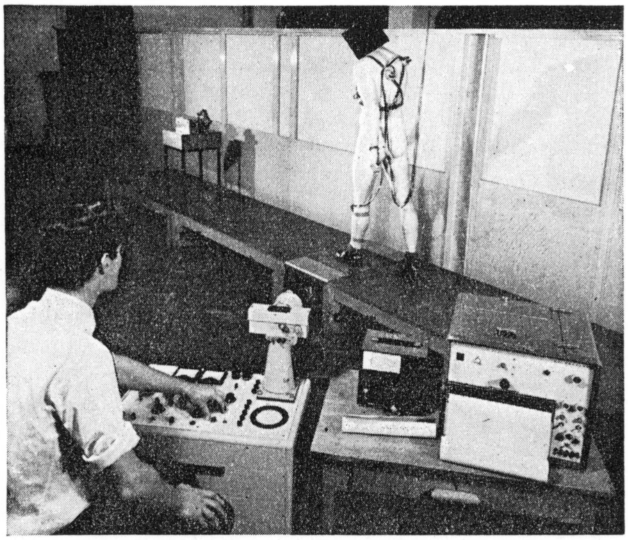

Fig. 4. In the 'walkpath' test the subject festooned with electromyographic electrodes, is filmed from front and side, and walks on a strain-gauged platform. All the resultant data are correlated by computer to evaluate the forces acting on the human joints under various conditions

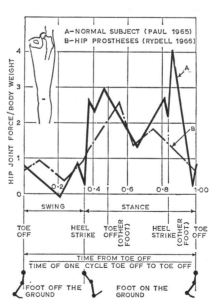

Fig. 5. The forces acting on a normal hip joint during walking, as evaluated by the technique illustrated in Fig. 4: the maximum can be several times the static body weight

Since in the natural limb it is only the muscles that compensate for the gross inefficiency of the articulated skeletal layout,[24] the choice of this very layout for lower limb prostheses seems somewhat less than inspired from an engineering point of view. Furthermore, the mass of the artificial limb is usually different from that of the natural limb so that an amputee trained to produce kinematic symmetry of gait (a natural walk) is bound to create unsymmetrical forces in his two hips and to alter significantly the original distribution and magnitudes of the forces in the remaining natural joints.

This appears to be an undesirable feature, particularly when considered in conjunction with certain clinical evidence[19] of significantly higher incidence of joint arthrosis in amputees than in normal subjects. It seems that a basic reappraisal of artificial limb design is necessary.

I believe that a prosthesis designed to produce the same forces as the limb it replaced is practicable and would find ready acceptance by the amputee, provided that functionally it was at least as good as the limb it replaced.

Body tissues

Modern surgeons, to put it crudely, 'construct' in human tissue, therefore the engineering characteristics of these tissues are acquiring increasing importance. For instance, the webbed skin scar formed as a result of a burn may be subjected to considerable tension produced by the local contraction of the damaged tissue. Such a tension field scar on the frontal part of the neck is shown in side view in Fig. 6. The corrective surgical procedure is shown in the centre picture where the same web of skin is shown in a frontal view. The arrows indicate the self-locked tension field with the webbed scar *BD* marked out together with the outlines of two triangular skin flaps *ADB* and *CBD*. First the scar tissue along the line *BD* is excised. Then the two triangular flaps, of flap-angle θ, are loosened from the underlying tissue, remaining attached only along their bases *AB* and *CD*.

Due to the release of tension consequent on incision, particularly along the lines *AD* and *CB*, the two flaps will tend to become transposed, rotating clockwise. If the original outline is well designed, the flaps will transpose naturally and, when stitched in place, appear as shown in the right hand picture, with an effective increase in the distance between points *B* and *D*.

The geometry of this so-called Z-plasty has been worked out [25, 26] assuming the skin to be inextensible. In the practical range of flap angles, extensions of the order of 25–125 per cent are predicted. In a more complex form of flap design for a similar purpose, the split Z-plasty, four triangular flaps are interdigitated on transposition. The percentage extensions and contractions were predicted by Devlin,[27] again on the inextensible basis.

The above concepts assume the plastys to be planar; Furnas[28] produced similar computations for three-dimensional flap forms. All of these predictions, however, tend significantly to underestimate the results actually obtained by disregarding skin extensibility. In clinical terms this effect manifests itself by an increase in the flap angle on raising the flaps, due to the release of the locked-in tensile system.

The actual extension for example can be 30 % greater than that predicted when assuming inextensible behaviour, for a flap angle of 60°. This emphasises the clinical need for knowledge of the stress/strain characteristics of tissue; in this instance skin. Many other clinical examples could be cited.

Experimental investigations of tissue require specialised and sophisticated techniques, particularly in view of the time-dependence, temperature and humidity-sensitivity of the mechanical characteristics.[29-33] Biological tissue in general is a multi-component material. Microscale studies of skin, for example, have shown it to consist of apparently randomly oriented networks of collagen and of elastic fibres, permeated and enclosed in a gel-like matrix, the 'ground substance' (Fig. 7). As a tensile load is applied, an increasing number of fibres are oriented in bundles parallel to the load, as shown in Fig. 8, culminating in a fully oriented and closely packed structure with the elastic fibres lying sandwiched between these collagen fibre bundles.

Fig. 9 shows typical results of a tensile test on skin at controlled extension rate on the macroscale. The righthand portion of the figure gives the load/direct strain curve. As has been pointed out,[34] the macro- and microscale behaviour correspond and when the terminal straight line is reached, the collagen fibre bundles themselves are being strained.

The left-hand side of the figure shows the variation of lateral strain with load. In magnitude these are comparable to the direct strains and, as can be predicted from the microscale test, the specimen width tends to a limit defined by the geometry of the fully compacted fibre structure.

Skin exhibits viscoelastic features such as hysteresis, creep and stress relaxation. Stabilisation of the load/extension curve appears to obtain after a few cycles at low loads. This implies some form of stabilisation of the initially random arrangement of fibre bundles. Daly,[33] who has shown this experimentally, also discovered that significant stress relaxation is exhibited when the collagen is substantially oriented and is undergoing direct stressing; therefore it seems reasonable to suggest that the significant viscous mechanism is associated with the collagen. The fibres consist of finer fibrils and it is conceivable that, when the collagen is stressed, these begin to slide, giving rise to a viscous effect.

This has suggested an elementary mathematical planar network model incorporating viscoelastic features, which gives, for unidirectional tension, calculated load/extension and lateral contraction relationships confirmed by experiment.

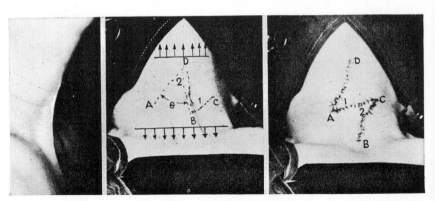

Fig. 6. The scar tissue shown in the profile on the left, contains locked-up tension. To relieve this, triangular flaps are cut as shown in the centre and transposed as on the right. Design of the flaps is complicated by lack of skin extensibility data

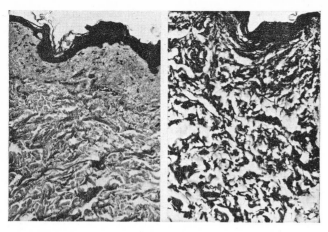

Fig. 7. (left) Collagen and (right) elastin networks in the unstressed human skin (× 65)

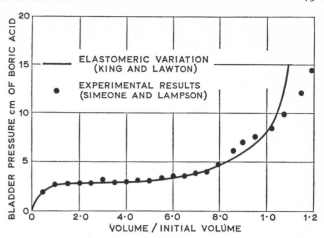

Fig. 10. The bladder behaves as an elastomer over a considerable range showing practically no hysteresis

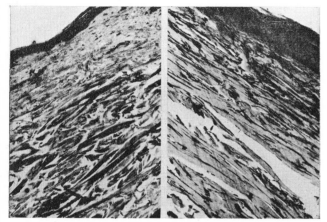

Fig. 8. As soon as the skin is partly stretched (left) fibres begin to align themselves with the force: wholly stretched skin on the right (× 65)

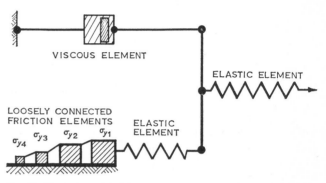

Fig. 11. Rheologic model of bone to represent uni-axial stress behaviour

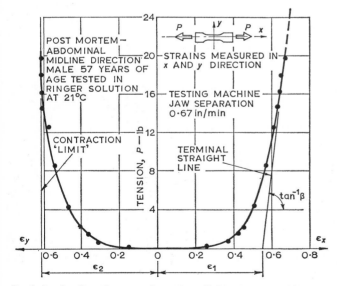

Fig. 9. Results of tensile test on skin at controlled strain rate: load/direct strain on the right, lateral strain on the left

All tissues of the body exhibit in a similar manner the features corresponding to their particular multi-component structure. Thus, the *ligatentum flavum* (a ligament in the spine) which has a virtually fully oriented structure, predominantly of elastic fibres, shows practically no hysteresis or stress relaxation and behaves as an ideal elastomer.[35] So also does the bladder,[36] the quasi-static pressure/volume relation for which is shown in Fig. 10.

Bone also contains the various components mentioned but possibly in a better-defined physical structure. It is possible,[37] qualitatively, to describe its deformation response to load by the rheologic model shown in Fig. 11 which contains elastic, viscous and plastic elements.

In the cartilaginous connections of the ribs to the breast-bone, collagenous and elastic components are present but it is the gel-like ground substance which predominates. Abrahams[38] has shown that such cartilage is linearly visco-elastic at small deformations. Because of the predominance of the ground substance, the stress/strain relationship, corresponds more to that of a homogeneous material and exhibits linearity in the fully stress-relaxed state.

Plastic surgery

Such cartilage is used extensivley in reconstructive plastic surgery to repair a collapsed nose bridge and generally restore facial contours. In these cases the cartilage taken from the rib endings is carved to shape, prior to transplantation. Gillies[39] first noted that distortions could result from carving.

Others[40, 41] have shown that these are due to the fact that such cartilage is prestressed in its natural state so that it may act as a torsion spring and return the ribcage to its neutral, resting state after exhalation. The stress system in these cartilages has been evaluated[30, 38] by adaptation of established engineering techniques and a carving procedure for producing distortion-free grafts has been developed.

Such adaptation for function is very common throughout the body. Skin, acting as a container of the other tissues, is subjected to tensions which vary in magnitude and direction from point to point. Cartilages in the nose, the ears, etc, which preserve their shape are all prestressed corresponding to function. One most interesting example is that of the inter-vertebral discs of the spine. These essentially consist of a thick tough annular sheath, the centre of which is filled by a pulpous substance contained under pressure.

In engineering terms such prestressing can only result from differential deformation, *ie* differential growth.

Another associated tissue feature brought to light as the result of biomechanical engineering has been the histo-chemical effect of stress on collagen. Craik and McNeil[42, 43] have shown that the collagen fibres in skin change their staining reaction under stress. It is now known that similar effects obtain in other tissues (bone, cartilage, etc.) which contain collagen. This discovery is of some importance since, in addition to providing an easy technique of identifying stressed collagen, it is also the first recorded example of a mechanical trigger producing a biochemical effect.

Education and training

The Biological Engineering Society is fostering the development of bioengineering training and education but otherwise there has been relatively little effort to examine the problems in a systematic manner. The time seems particularly opportune now when Britain is likely to experience an explosive expansion of bioengineering.

In the USA there were only three centres with academic programmes in the subject in 1961 but there were 61 in 1965.[44] In Britain, at present, I know of three centres with postgraduate courses, *viz* Imperial College, the Universities of Strathclyde and Surrey. The course at Imperial College, which was the first, is basically medical electronics, mainly for students qualified in the life sciences; at Strathclyde the course is biomechanical, mostly aimed at students qualified in the physical sciences, while in Surrey the course appears to encompass a catchment area covering both physical and life science graduates.

Arthur[45] gives a definition of bioengineering as consisting of four major categories: Bionics—the application of the mechanics of biological systems in engineering to create hardware; Applied Biology—the use of the biological systems themselves on an industrial scale to create new products; Biomedical Engineering, which covers the application of engineering to provide replacements for damaged structures, diagnostic and therapeutic instrumentation; and finally Environmental Health Engineering—the use of engineering to create and optimise environments for life.

It will be seen that the bioengineering field is enormous and touches upon all existing and contemplated fields of engineering. Even Biomedical Engineering alone (which is usually meant by bioengineering in this country) covers a quite intimidating range.[46, 47]

It is therefore essential that, prior to designing a course, a reasonably clear idea should exist of its aim. Biomedical engineering is a multi-disciplinary activity, conceived as a

Prof. R. M. Kenedi *was born in Budapest and educated and trained as a civil engineer in Scotland, graduating BSc (Glasgow) in 1941. He joined the staff of the Department of Mechanical Engineering of the then Royal Technical College as a temporary Assistant Lecturer in 1941 and has been there ever since. His main research activity, prior to becoming interested in bioengineering, was in the field of thin-walled structures in which he obtained his PhD in 1949. Since 1960 he has been active in bioengineering and was appointed Research Professor in this subject in 1963. He now heads the postgraduate teaching and Bioengineering Research Unit at the University of Strathclyde. He has spent periods of secondment in South America and the United States; is the author of some 40 papers and editor of two symposia.*

team effort. Education should produce a team member (rather than an all-knowing hybrid), highly competent as an engineer and equipped with the appropriate multi-disciplinary background. This should assist the bioengineer not only to appreciate the medical problems that arise and to communicate with the other members of the multi-disciplinary team but to meet the ethical and social problems associated with medicine today, some of which arise due to the very applications of the technology with which he is concerned.

An intake of first degree students in the physical sciences must be given (i) an appreciation of, and the ability to communicate in the life sciences, (ii) their engineering know-how must be increased by the relevant up to date techniques of analysis, design and manufacture; and (iii) the methodology and practice aimed at the solution of practical problems in biomedical engineering must be taught. These concepts are developed at Strathclyde in the context of a multi-disciplinary team which includes the patient as an essential member.

In a full-time, one calendar-year course, around three-quarters of the time would be roughly equally divided between these three aspects. The remaining quarter is devoted to a project arising from a practical clinical problem. If the course is to be effective, the importance of such clinical association in the form of direct contact with doctors, patients and the general hospital atmosphere cannot be overemphasised.

Philosophical problems

Human problems arise from a number of sources. For example, some of today's new medical techniques are so expensive that at present society is not ready to accept their general application. As replacements of vital organs, in particular, are likely to be in short supply, how is the individual beneficiary to be selected?

Again, longevity, apparently highly desirable, is in danger of extending only an increasingly dependent part of life. With the increase of life span, the human body appears to exhaust its resources for independent existence. Tissue appears to deteriorate in a mechanical sense from the age of 40 onwards. Thus as medicine pushes ahead it creates an ethical problem: should human life be extended if all this does is to expose further weaknesses, conferring the questionable gift of longevity wholly dependent on others?

There is also the macabre, yet very relevant, question as

to when an individual receiving artificial replacements and/or natural transplants ceases to be the same individual.

To these questions and problems there are no easy answers today. Ultimately they may resolve themselves by a blurring of the distinction between life and death and by a definition of life beyond our present comprehension and social acceptance. But all those entering the field should be made aware of these considerations and of their own responsibilities.

Apart from the unique regenerating and self-repairing ability of tissue, body characteristics are, on the whole, inefficient from an engineering point of view. Not every component has a function; and many are freely adapted to serve functions for which they may or may not be basically suited

The present mode of life, directed to serve the convenience of the body as it is, tends on the whole to suppress, rather than encourage its development. Thus as bioengineering activities expand and intensify they may influence greatly in a manner as yet uncomprehended the essential qualities of human life itself.

REFERENCES

1. CONTINI, R., GAGE, H. and DRILLIS, R. 'Human gait characteristics', *Biomechanics and related bioengineering topics*. 1965. Ed. R. M. Kenedi. Pergamon Press, Oxford.
2. VON GIERKE, H. 'Biodynamic response of the human body' 1964, *Appl. Mech. Review*, vol. 17, no. 12.
3. ZEIGENRUECKER, G. H. and MAGID, E. B. 'Short time human tolerance to sinusoidal vibrations', *W.A.D.C. Technical Report* 59–391, Wright Air Development Centre, Wright-Patterson A.F.B., Ohio.
4. COERMANN, R. R. 'The mechanical impedence of the human body in sitting and standing position at low frequencies'. 1963, *Vibration research*: Ed. S. Lippert. Pergamon Press, Oxford.
5. NICKERSON, J. C. and DRAZIC, M. 'Young's modulus and breaking strength of body tissues'. 1964, AMRL-TDR-64-23. Wright Patterson A.F.B., Ohio.
6. YAMADA, H. 'Human biomechanics', Personal communication from Department of Anatomy, Kyoto. University of Medicine, Kyoto, Japan, covering the period 1950 to date. Now being prepared for publication in the U.S.A. (1963).
7. STECH, E. L. and PAYNE, P. R. 'Dynamic models of the human body. 1964, AMRL-TDR. Wright-Patterson A.F.B., Ohio.
8. DEMPSTER, W. 'Space requirements of the seated operator', *WADC Tech. Rep.* 1955, 55.
9. VON GIERKE, H. E. OESTREICHER, H. L., FRANKE, E. K., PARRACK, H. O. and VON WITTERN, W. W. 'Physics of vibration in living tissues', *J. Appl. Physiol.* 1952 vol. 4, 886.
10. FRANKEL, V. *The femoral neck* 1960. Alnquist and Wiksells, Uppsala, Sweden.
11. OESTREICHER, H. L. 'On the theory of propagation of mechanical vibrations in human and animal tissue', *U.S.A.F. Tech. Rep. No.* 6244, 1950.
12. FISCHER, O. 'Der gang des menschen'. 1895, *Abk. K. Sachs. Ges Weis. Math-Phys.* vol. 21, 153; 1899 vol. 25, 1; 1900 vol. 26, 87; 1902 vol. 25, 471; 1904 vol. 28, 321; 1904 vol. 28, 533.
13. HIRSCH, C. and FRANKEL, V. 'Analysis of forces producing fracture of the proximal end of the femur'. 1960, *J. Bone Joint Surg.* vol. 4213, no. 3, 633.
14. PAUWELS, F. 'The importance of biomechanics in orthopaedic surgery'. 1963, *9th Congress of the International Society of Orthopaedic Surgery and Traumatology* (Vienna).
15. PAUL, J. P. 'Bioengineering studies of the forces transmitted by joints, Part 2'. 1965, *Biomechanics and related bioengineering topics*. Pergamon Press Ltd, Oxford).
16. PAUL, J. P. 'The biomechanics of the hip joint and its clinical relevance'', *Proc. Roy. Soc. Med.* (In press) (1966).
17. MERCHANT, N. 'Hip abductor muscle forces'. 1965, *J. Bone Joint Surg.*, vol. 47A, no. 3, 462.
18. SORBIE, C. and ZALTER, R. 'Bioengineering studies of the forces transmitted by joints—Part 1'. 1965, *Biomechanics and related bioengineering topics*, Ed. R. M. Kenedi. Pergamon Press Ltd.
19. HUGHES, J. 'Dynamic gait studies applied to the design of lower limb prostheses'. 1966, *Proc. Int. Symp. external control of human extremities*, September. Dubrovnik, Yugoslavia (In press).
20. CUNNINGHAM, D. M. and BROWN, G. W. 'Two devices for measuring the forces acting on the human body during walking'. 1952, *Proc. Soc. Exp. Stress Analysis*, vol. 9, no. 2, 75.
21. MARKS, M. and HIRSCHBERG, G. G. 'Analysis of the hemi-plagic gait'. 1958, *Ann. N.Y. Acad. Sci.*, vol. 74, 59.
22. HIRSCH, C. and RYDELL, N. 'Forces in the hip joint—Parts I and II'. 1965, *Biomechanics and related bioengineering topics*, Ed. R. M. Kenedi. Pergamon Press Ltd, Oxford.
23. ROSS, D. S. 'A new tubular form of hip nail'. 1965, *Biol. Eng. and Med. Electronics*, vol. 3, 301.
24. KENEDI, R. M. 'Bioengineering—concepts, trend and potential'. 1964, *Nature*, vol. 202, no. 4930, 334.
25. LIMBERG, A. A. *Skin plasty with shifting triangular flaps* 1929. Leningrad Tech. Inst.
26. MACGREGOR, I. A. 'The theoretical basis of the Z plasty'. 1957, *Brit. J. Plast. Surg.*, vol. 9, 256.
27. DEVLIN, P. 'Biomechanical aspects of flap design in plastic surgery'. 1965, B.Sc. Thesis. University of Strathclyde, Glasgow.
28. FURNAS, D. W. 'The tetrahedral Z-plasty'. 1965, *Plast. Reconstr. Surg.*, vol. 35, 291.
29. RIDGE, M. D. 'The rheology of skin'. 1964, Ph.D. Thesis. Department of Medicine, University of Leeds.
30. ABRAHAMS, M. and DUGGAN, T. C. 'The mechanical characteristics of coastal cartilage'. 1965, *Biomechanics and related bioengineering topics*, Ed. R. M. Kenedi. Pergamon Press Ltd, Oxford.
31. KENEDI, R. M. GIBSON, T., DALY, C. H. and ABRAHAMS, M. 'Biomechanical characteristics of skin and costal cartilage'. 1966, *Federation Proceedings*, vol. 25, no. 3, May, June, Pt I) American Societies for Experimental Biology.
32. EVANS, J. H. 'The mechanical characteristics of human skin'. 1965, M.Sc. Thesis. University of Strathclyde.
33. DALY, C. H. 'The biomechanical characteristics of human skin'. 1966, Ph.D. Thesis, Bioengineering Unit. University of Strathclyde, Glasgow.
34. KENEDI, R. M., GIBSON, T. and DALY, C. H. 'The determination, significance and application of the biomechanical characteristics of human skin'. 1965, *Digest of the 6th International Conference on Med. Elec. and Biological Eng.* Tokyo.
35. NACHEMSON, A. and EVANS, J. H. Verbal communication on 'Work in progress on the biomechanical characteristics of the ligamentum flavum'. 1966 Bioengineering Unit. University of Strathclyde, Glasgow.
36. KING, A. L. and LAWTON, R. N. 'Elasticity of body tissues', *Med. Phys.* Ed. O. Glaser vol. 2, 303. Year Book Publishers, Chicago.
37. SEDLIN, E. D. 'A rheologic model for cortical bone (a study of the physical properties of human femoral samples)'. 1965, *Acta Orthopaedica Scand.* vol. 36, suppl. 83.
38. ABRAHAMS, M. 'The mechanical properties of costal cartilage'. 1965, Ph.D. Thesis, Bioengineering Unit. University of Strathclyde, Glasgow.
39. GILLIES, H. D. *Plastic surgery of the face* 1920. Oxford University Press, London.
40. GIBSON, T. and DAVIS, W. B. 'The distortion of autogenous cartilage grafts; its cause and prevention', *Brit. J. Plast. Surg.* vol. 10, 257.
41. KENEDI, R. M., GIBSON, T. and ABRAHAMS, M. 'Mechanical characteristics of skin and cartilage'. Oct, 1963, *Human Factors J.*
42. CRAIK, J. E. and MCNEIL, I. R. R. 'Histological studies of stressed skin'. 1965, *Biomechanics and related bioengineering topics*, Ed. R. M. Kenedi. Pergamon Press Ltd, Oxford.
43. CRAIK, J. E. and MCNEIL, I. R. R. 'Microarchitecture of skin and its behaviour under stress'. 1966, *Nature*, vol. 209, no. 5026, 931.
44. TRUXALL, J. C. Quotation in editorial comment, 'Bioengineering boom forecast'. January, 1966, *Biomedical Engng*.
45. ARTHUR, R. M. 'Biology at Rose. April, 1964, *Rose Technic*.
46. ALT, F., HARLOW, W. G. and METZ, H. D. 'Biomechanical instrumentation engineering'. 1965, *Biomechanics and related bioengineering topics*, Ed. R. M. Kenedi. Pergamon Press Ltd. Oxford).
47. DAVIES, F. E. 'Medical instrumentation'. September, 1964, *Inter. Sci. Tech.*

Economics as an Aid to Policy-making

by Austen Albu, MP, BSc(Eng), MIMechE

Applied economics is an art rather than a science; but it may influence important decisions by Government and industry. Engineers and economists should know more about each other's work and the important differences in outlook conditioned by it.

Very few engineers will have met an economist in the flesh and not many could say how they are employed. Equally, although economists can be presumed to have at least a theoretical knowledge of industrial procedures, few of them will know anything of the daily work of professional engineers. This is not surprising, because most engineers are employed in industry but, until the last few years in Britain, very few economists were to be found there. Now the situation is changing; several university-trained economists are at the head of large industrial organizations: for instance, Mr Paul Chambers of I.C.I., Professor R. S. Edwards of the Electricity Council, Lord Piercy of the Industrial and Commercial Finance Corporation.

Perhaps more significant is the current growth in the employment of economists by companies and nationalized industries in forecasting, market survey and planning departments. The sudden renewal of interest of both Government and industry in economic planning and the setting up of the National Economic Development Council have led to a demand for economists which the universities may well have difficulty in filling during the next few years. If the N.E.D.C. follows the French pattern and sets up a number of committees representing the main branches of industry to act as their channel of communication with the individual firms in industry, these will themselves need to employ economists to process the information they receive from the firms and to work out the implications of any national plan for their own industries. Already many of the larger firms have realized the need to employ economists.

Another new motive for the employment of economists has been the development of the European Common Market and the probability of Britain's entry into it. Firms who have not felt the need for economic information in the past, relying, as too many have done, on traditional markets, now want to find out what the effects on their business at home and their chances of exports to Europe will be if Britain joins.

Decision-making

Economists, therefore, are likely to play an increasing part in the decision-making processes of business management over the next few years; and engineers had better learn something of their methods. Equally important is it that economists, who in the past have tended to prefer ivory towers or government departments, should become fully acquainted with the technological aspects of the phenomena with which they deal. To be fair to the economists, they have not, so far, been exactly welcomed into industry. And the Civil Service, including the Treasury, with its traditions of amateur administration, has always relegated the economist,

as it has most other professionals, to a backroom advisory position. The N.E.D.C., whose present staff is headed by, and largely consists of, economists, is independent of the Treasury and has direct access to the Chancellor of the Exchequer. Professional economic advice is, therefore, likely to play a more direct part in Government policy-making·in the future than it has in the past.

Applied economics, which can lead to the taking of decisions by government or business, is an art; just as engineering design is an art. But, whereas engineering is an art based on the physical sciences, economics is an art based on the far less precise social sciences. In most firms the conduct of business, that is to say the taking of those decisions which form the basis of economic activity, is determined by hunch and experience. It is still in the pre-scientific stage which engineering was in at the time of the founding of this Institution.

The majority of businessmen learn their trade entirely by practice; generally without even the formal apprenticeship of the early engineers. Salesmen and accountants in Britain rarely have any economic training. This situation is now changing as the advantages of economic forecasting and planning become more widely recognized and as the scope of assistance which a professional economist can provide becomes better understood.

Differences of approach

There is, however, a substantial difference between the scientific basis which an engineer uses to arrive at decisions on

— they have not, so far, been exactly welcomed into industry . . .

78

Mr A. H. Albu, *M.P. was educated at Tonbridge School and the Imperial College of Science and Technology where he obtained his BSc in engineering. He was for many years a Works Manager at Aladdin Industries, Greenford, and later served in a senior position in the Control Commission for Germany. In 1948 he was appointed Deputy Director to the British Institute of Management and in the same year he was elected Member of Parliament for Edmonton. From 1965 to 1967 he was Minister of State at the Department of Economic Affairs. Among the Parliamentary offices he has held are Chairman of the Parliamentary & Scientific Committee, Member of the Select Committee on Nationalised Industries, and Member of the Select Committee on Procedure. Mr Albu was a Member of the Board of Governors of the Imperial College of which he is now a Fellow. He is also a Fellow of the City and Guilds of London Institute and a Doctor of the University of Surrey. He has written, lectured and broadcast extensively, chiefly on industrial and economic matters, and he is the author of a book entitled 'The Young Man's Guide to Mechanical Engineering'.*

which to construct a design and that which an economist can employ when drawing up a plan for economic or commercial action.

Both professions are concerned with predictions; the engineer predicts the performance of a mechanism or structure, and the economist of the performance of an individual organization, of a nation, or of a group of nations. The engineer's task is always clearly defined and he can base his prediction on well-established scientific laws and empirically verified generalizations. He is continuously improving the efficiency of his products as scientific research enlarges his knowledge of the behaviour of materials under different conditions; but he usually safeguards himself against unforeseen phenomena by the use of safety factors in the design of components; and even then his predictions are sometimes wrong; as when he under-estimates the rate of increase of the efficiency of thermal power stations when they grow in size; or when the design for a new diesel locomotive fails on its first trial.

The task of the economist is rarely so clearly defined and he is expected to determine objectives without the kind of hard information available to the engineer. The social sciences on which the economist must rely are in a much less advanced stage than the physical sciences and, because of the nature of the phenomena with which they deal, it is probable that they will ever remain so. The infinite varieties of human behaviour, whether in individuals or in groups, the impossibility of more than the most limited forms of experimental research and the continuously changing context of behaviour must limit the degree of precision that the social scientist can hope to achieve.

In the past the great economists, such as David Ricardo, (an ancestor of that great mechanical engineer, Sir Harry,) while observing, and even participating in, business activity, based their theories on deduction from assumed hypotheses which they had no means of testing. Among such hypotheses were the laws of supply and demand and of human motivation by self-interest which formed the staple of the ideas of the Manchester School in the last century.

Although many of these hypotheses represented obvious rough approximations to the facts of economic life, there are few economists today who do not realize that they are inadequate, by themselves, as guides to economic behaviour in an increasingly complex world. It is not only that the scale of industrial activity has become so much greater, that a few vast concerns can often dominate a market, and that the cost of research and development has brought governments into economic activity in a way that would have horrified the early economists; but political ideas, based on human experience in the industrial countries, have led to an increasing degree of government participation in economic activity. The free play of competition among a horde of equally effective economic units and the free choice of consumers, exercised on the basis of individual taste and experience, are obviously no longer—if they ever were—unimpeded factors determining economic activity.

There are still economists who believe that the object of government economic policy should be to withdraw from all intervention in the economic process and leave it to the operation of natural laws; but these will hardly be found, except perhaps in times of war, assisting in the purposeful activities of government or business planning departments.

Another development which has led to greater sophistication in economic thinking is that of psychology; particularly of social psychology. At its lowest level this has enabled businesses to operate more rationally by basing their sales and production policies on market research conducted by social survey methods. But it has also made economists realize that human beings do not always act in what appears to be their own immediate self-interest and that, under certain economic conditions, for instance that of mass unemployment over a prolonged period, economic insecurity reinforces inherent psychological insecurity and can lead to outbursts of mass irrationality with dangerous political results.

Models of the economy

Much economic research and writing deals with global abstractions: national product or income, balances of trade or payments, investment, consumption and government spending rates of interest or wages. Economic thinking is largely concerned with trying to establish relationships between these global factors and, more recently, with trying to determine parameters by means of which mathematical models of the working of the economy can be built. Such models can then be used in an electronic computer (or even

— *the free choice of consumers ... is no longer an unimpeded factor ...*

in an hydraulic model) so that the effect of changes in some factors on others can be examined. Professor Richard Stone of Cambridge has been working on such a model. These methods, with their appearance of mechanical precision, will obviously appear attractive to the engineer and there is no doubt that they have a great potential as aids to national economic planning; just as the rather similar methods of mathematical operational research have in industrial planning.

It must always be borne in mind, however, that the parameters fed into the computer are not natural laws, verified in the laboratory, but relationships discovered by the examination and comparison of past statistics. They are, therefore, no more accurate than the original information on which they are based and, in a dynamic economy, may well be out of date by the time they are used. It is easier for economists to predict the direction of movements in the factors with which they are concerned, for instance prices, production and demand, than it is for them to put a figure on the extent of such movements.

Economists are well aware of these limitations of their methods and are constantly trying to refine them. Along with other social scientists they have been developing new research techniques. In spite of the great diversity which is characteristic of human behaviour, research is disclosing statistical regularities which assist in its prediction. In the field of economics statistical surveys, sometimes of future intentions, make possible the estimation of trends in, for instance, personal consumption or investment in new buildings and machinery.

In industry

Whereas, in the past, economists were mostly concerned with the overall performance of the economy, nowadays more and more of them are directing their attention to the behaviour of the individual firm. This has led them to try to find out the basis on which investment decisions are taken, the conditions which encourage or discourage technological innovation and the kind of management which leads to business success.

Often these studies are carried out in conjunction with other social scientists, generally sociologists or social psychologists.

Examples of work of this kind are the studies of science and industry, made by Professors Carter and Williams of Manchester University and on innovation by Burns and Stalker of Edinburgh.

It cannot be claimed that these studies have led to the discovery of any clearly defined scientific laws governing business conduct, but they are of great assistance to economists in helping them to adjust the parameters in their economic models to take account of more intangible factors. These include the rate of technological advance in particular industries; the effects of better educated managers or more qualified technologists; and the handicaps due to traditional attitudes in workers or managers.

Today the more sophisticated economic model builders try to feed into their computers factors which take account of likely alterations in the rate of change of the various economic indices, overall or in particular industries, based on research of this type and on careful enquiry in industry. In this way they are increasing the possibility of being able to produce more accurate forecasts.

Apart from the differences in method of working between the two professions, the nature of the work they perform and their education for it engender a difference in attitudes

— appearance . of mechanical precision will obviously appear attractive to the engineer . . .

of mind between the economist and the engineer. Engineers are primarily concerned with the performance of individual mechanisms or structures, or of single undertakings. Many engineers, because of their primary interest, put technical excellence first in their assessment of a design or construction; although no engineer in a senior position can neglect economic considerations. Generally, such considerations are concerned with the profitability, under given commercial and financial conditions, of an isolated project. Knowing certain prices of raw materials, rates of wages and rates of interest, the engineer can calculate the cost of construction and of operating a machine or complex of machines. To do so he should make estimates, based on his experience, of the life of the project and of the cost of its maintenance over that life.

In arriving at the best compromise between technical perfection and economic performance the design engineer must plan for an optimal, not a maximum, life for the project and for reasonable maintenance costs. To seek perfection is always too expensive and it is by achieving this compromise in his design that the engineer demonstrates his art. In the economist's language, the engineer is seeking marginal utility, when he endeavours, often unconsciously, to find the point in the development of his design where either a little more, or a little less, would reduce the return on the investment.

In making his calculations, however, the engineer has to assume definite figures for a number of economic factors about the future, of which the economist is probably a better judge than himself.

Although the economist is constantly improving his information about the way the national economy or individual businesses work, economics still remains more a way of looking at life and of posing the questions on which administrative decisions have to be taken, than a science in the sense that the engineer understands that term. As Keynes wrote many years ago:

'The Theory of Economics does not furnish a body of settled conclusions immediately applicable to policy. It is a method rather than a doctrine, an apparatus of mind, a technique of thinking, which helps its possessor to draw correct conclusions'.

Best use of resources

The foundation of this way of looking at things is the economist's belief that all resources are scarce and that policy is concerned with making choices between their use. The problem of choice remains even when, as in the United

States today, resources no longer appear scarce; for scarcity is relative to needs or desires. There remain, even for Americans, such choices as putting more men more quickly on the moon; or increasing leisure by reducing the hours of work; or re-building the slums of Harlem. Obviously there is a limit beyond which all these things cannot be achieved at the same time and it is with the implications of choosing one or other of them that the economist is concerned.

It is not for the economist to say that a nation should aim at the production of maximum wealth; but if that is the national choice, however expressed, he is especially trained to make clear the implications of such a choice for the way in which resources are used, whether they be of labour, material, land or finance. The same applies to the problems of the individual business, whose management is continually faced with the problem of the best use of the resources available, normally in order to maximize profit. The economist is concerned with the relative costs and returns of alternative policies or of alternative methods of carrying a policy out.

Wealth and well-being are not identical, but economists endeavour to deal with both as quantifiable factors by using the price (or the tax) which people are prepared to pay for them. The fact that many people are not prepared to pay anything for amenities they protest they value a great deal often leads to the sort of difficulty in which the Central Electricity Generating Board finds itself, when it wants to run a grid line over Snowdonia; or the Manchester Corporation when it wished to use Ullswater as a reservoir.

But the economist is not generally concerned with individual projects, unless they are on such a scale that they affect a whole area of country or a whole industry. For this reason, what may appear a reasonable project to an engineer may not appear worthwhile to the economist. On the national level, the economist is concerned with advising Government or criticising it on overall economic policy. It can be assumed that one of the objects of all modern Governments is to increase wealth, but on the means by which this is to be done there is no universal agreement even among economists; and even if there were, governments would not necessarily use them, because they have to take account of political considerations.

Nevertheless all good economists will look at the national picture first, before they pay attention to the problems of particular regions or industries; and their views will be based not only on the immediate situation, but on such factors as rates of investment and production, and on the changing pattern of world trade by which, over a period of years, all national economies are affected. It is for this reason that the advice they give may often seem hostile to the interests of engineers in a particular industry or a particular firm.

Assessing an engineering project

An example of the sort of question which is approached differently by the engineer and by the economist, is whether or not to build a hydro-electric power station on some remote river where enormous quantities of power are flowing to waste. The engineer sees the possibility of harnessing this power and can make estimates of the costs of its generation which make his mouth water. He will no doubt also estimate the costs of distribution.

What he is less likely to do is to make a study of the possible economic development of the country concerned, of the number of likely consumers of electricity and of the amount of power that could be sold. To do so he would have to know a great deal about the population of the country and its

distribution; about the national income per head; about the rate of industrial investment and of exports and imports. He would then have to estimate the effect on all these things, and therefore on the country's production, of the building of the proposed power station and he might come to the conclusion that smaller local thermal stations were more economical; or even that bullock power would be cheaper! The economist would not only consider these matters; but he would also try to find out whether a similar capital investment would not show a greater economic return elsewhere.

This is not to say that the decision to build the station might not be taken in the light of political or social considerations, or of governmental plans which would themselves have the effect of changing the relevant factors in the situation. The decision would be more likely to be the correct one, however, if the economist's considerations were taken into account when it was made.

Governments, today, have to make many decisions on highly technical matters; such as whether or not to build more nuclear power stations, whether to build a supersonic air-liner, whether to build more new towns or more motorways. In all these questions the economist will emphasize the elementary facts involved in making a choice: that resources used for one project cannot be used for another, that plans in excess of productive capacity will force up prices and wages, that resources used in the home market cannot be used for export. These facts are often overlooked by administrators and by engineers, but the economist's emphasis on them can help to ensure that they are given full weight when decisions affecting large-scale expenditure are made.

Statistics

Applied economists are generally skilled in the use of statistics; a subject which increasingly links them to engineers. Their education makes them expert, both in interpreting published statistical material and in the use of statistical mathematics for purposes of analysis and programming. Reliance on published statistics can be a dangerous matter for one not versed in the methods by which they are arrived at and the conventions underlying their use.

Statistical methods are becoming of increasing importance in operational research, where they are employed to translate the economists' concept of choice in the use of scarce resources in order to maximize output or profit, into practical industrial programmes. It is obvious that in this activity the closest contact is necessary between the specialist economic statistician and the engineer or business man. As Professor Amey of Bristol has pointed out, with the growing size of the modern business and of the scale of individual capital projects, the cost of mistakes, both to the firm and to the country, becomes very much greater. Managements today find themselves planning for longer periods ahead and are anxious to obtain greater control over their destinies by enlarging the area in which they have to make conscious decisions.

The statistical economist is the one professional man trained for this purpose. Even if he cannot help the businessman to make definite decisions now about a future which can only be dimly foreseen he can advise on the sort of action that should be taken to cope with each future contingency as soon as more information becomes available.

Economists in advisory or decision-making positions are concerned with the future. Engineers may be concerned to calculate the stresses and strains in a proposed mechanism;

economists with the stresses and strains involved in a proposed rate of economic expansion.

The factors which the engineer uses are unlikely to change substantially before his design is completed; although new discoveries and inventions may render it obsolete before the end of its planned life.

The economist is often trying to forecast over longer periods and to advise on courses of action which take many years to come to fruition. Examples are rates of expansion (or contraction) of railways and roads; the rate of investment in particular industries; the number of places required in universities and colleges of technology for different subjects; the trend in the need for skilled labour; the future demand for different fuels; the effect of changes in the size and composition of the population and of rising (or falling) national incomes.

—— *engineering students . . . take some lectures in economics . . .*

Clearly, many of these things are very much affected by innovations due to scientific discovery and technological invention. It has been said that the extraordinary phenomenon of modern economic growth in industrialized countries has primarily been due to technological innovation based on the scientific revolution of the seventeenth century. Technology is, in this sense, at the very heart of economics and today the pace of technological change is very fast indeed. It is, therefore, of the utmost importance that economists concerned with forward planning should work closely with scientists and engineers so that they can be kept aware of the technological changes that are in the making and, perhaps even more important, of those that could profitably be made, if sufficient resources were now devoted to their development.

Educational links

There remains the question whether there should not be some more formal educational link between the two professions. As professional engineering education finally leaves behind its original basis in craft apprenticeship and is built more on the teaching of the engineering sciences, mathe-

matics and statistics play increasing parts in the curriculum. This gives the engineer the possibility of assimilating without difficulty the methods of the statistical economist. In view of the fact that a high proportion of professional engineers find themselves, sooner or later, in managerial positions or as consultants on large-scale projects, there is no doubt that an understanding of the economist's approach would be of great assistance to them. Already many engineering students in Britain, and more in other countries, take some lectures in economics during their degree courses. This tendency should be encouraged.

At the Imperial College a postgraduate course in technology is run for economics students from the L.S.E. It would be a good thing if every economics graduate should be required to have knowledge of at least one of the physical sciences up to at least G.C.E. 'A' level, and take a course in his degree syllabus aimed at giving him an understanding of current technological developments, and of the methods of applied science and engineering.

At the Massachusetts Institute of Technology there is an undergraduate course which combines engineering and economics for those going into engineering business management.

This is a field in which one of the Colleges of Advanced Technology might well experiment. In a sandwich course the student might spend his first two industrial periods in the engineering departments and his last two in sales, market forecasting, operational research and production planning.

No doubt recognition of a course of this type would run up against the ingrained conservatism of the institutions which guard the purity of our professions; but industry and government might find that a man or woman trained in the two methods of weighing evidence and of arriving at decisions, able to understand both the possibilities of technological achievement and its economic limitations, had an important role to play in a rapidly changing and competitive world.

The magnetic tape filing system of a modern computer—one of the regions where engineers and accountants have a joint interest

[Courtesy of I.B.M. Ltd]

The Accountant's Function in Industry

by J. A. Goldsmith, IMA, FCA, ACWA

During the past few years the importance of the accountant in industry has grown and at the same time the character of his job has completely changed.

The old Dickensian concept of him seated at a high desk in a small dark office and writing up a large leather-bound ledger has all but vanished although such offices could still be found up to a few years ago. The routine work they did in keeping records ensured that the firm's money was properly received and paid; while this work is still an essential part of the accountant's duty, drudgery is now largely removed by mechanization.

The accountant is thus responsible for wages calculation, invoice preparation, maintaining records of debtors and creditors, and summarizing all receipts and payments to ensure that the accounts can be presented each year to the shareholders with a certificate by the auditors that proper books of account have been kept. It is a vital service which, however, runs as a routine, only calling on other departments for specified information at specified times.

Control by budget

It was only about 25 years ago that the costing side of the accountant's job began to develop, with the wide title of 'Management Accounting'. Up to that time nearly all costing had been confined to the analysis of expenses, first to departments and then to the products made during the period the expenses were actually incurred. The result of this was a comparison of the actual cost of each product with the selling price which purported to show management which products had made a profit. But in the end there was nothing

that could be done with all this information for so much analysis had been done that nobody could tell why a product cost had gone up, and anyway the method of analysis was decided solely by the accountant. Furthermore, the analyses usually appeared long after the event, and so were only fit for immediate filing.

The new concept of management accounting recognizes a need to plan sales, production and expenditure beforehand in detail so that a coherent blueprint can be created for the future operations of the firm. To enable this to be done means that:

1. responsibilities of every person must be defined so that the part he plays in earning or spending the firm's money is clearly laid down;

2. budgets are set in accordance with those responsibilities and agreed by each person concerned;

3. as frequently as may be necessary for each item the actual production and expenditure is summarized, compared with the budgets and the report passed to the person responsible.

If this is done fairly and methodically there is no doubt that control through budgets in this way will, with good management, weld the individual members and managers into a team, giving each a real knowledge of the part that he can control.

To set the budgets in sufficient detail may involve considerable effort for the first time, but this alone invariably points to potential improvements and savings in operation. By applying a financial control of this type to every aspect of a firm's operations the accountant has been able to provide a real service to management, and is qualifying as one of the executives to be consulted before any decision is made.

Control of other activities

As an accountant, one frequently wonders why the principles behind management accounting couldn't be applied to every facet of a company's activity. The principles are very simple; they involve first, a detailed plan, then a series of measurements and finally a feedback which highlights deviations from the plan.

This comment particularly applies to the work of the design engineer. For example the total cost of a design project is normally estimated from quite broad assumptions. The project is then broken up into parts for detailed design, leading on to detailed manufacturing plans. If the cost of each part of the project could be budgeted before detailed work started and checked again after design and before manufacture, there would be opportunity for over-elaborate designs to be pruned before it is too late to make any changes and before a loss is inevitable. However, this sort of approach means a tedious procedure, the estimates are difficult to make, extra staff are needed and there is often the argument that it inhibits good design. So it is all too seldom done.

However, as an example of what can be done, one firm manufacturing heavy steel works plant introduced a new department into their design office for this additional work, and surprised even themselves by saving £100,000 on one contract alone.

Similar considerations apply to the planning of new capital projects. If the preliminary work is only done in outline, with the idea that plans will be detailed as erection proceeds, it is probably inevitable that the ultimate cost will be higher than expected and the completion date put back. As the accountant, one all too often finds it difficult to persuade the designers to spend sufficient time and effort before the first brick is laid in making a detailed estimate of costs. Experience shows that, from this point on, one is faced with the choice of meeting any unexpected charges, or—almost unthinkable—abandoning a half-finished project.

These considerations do not apply only to costs. For example, it often seems that preliminary work on new capital projects concentrates far more on getting the job started than on trying to foresee all the snags that will arise. Usually a chart is prepared which shows the dates on which each section must be started and completed, but the planning sometimes goes little further than this. The contract then often proceeds by way of a series of crises with the completion date being slowly put back. At each crisis one wonders why it couldn't have been foreseen—it is invariably due to some late delivery which could have been known about days or weeks earlier, had a detailed plan been made.

The answer again is to spend time and money on initial planning, which by itself is likely to show where the crises are likely to arise; and then, by continual checking, to trace any deviations from the plan and inform all those concerned.

One successful solution, where a company is controlling its own capital development, is to create a formal liaison between the engineers and other members of management through a steering committee. In this way the designer concentrates on the design—the job he does best. The accountant, as another member of the committee, can make the best use of his skills in analysing, planning and reporting. This principle has been most successfully applied by my own firm where we have recently completed the first part of an £11 million project to install a new electric steel melting shop. A series of steering committees have met regularly to control different aspects of the project with the result that it has been completed six weeks ahead of schedule, and at a cost almost exactly equal to the estimate.

Use of computers

Any attempt to use computers is certain to entail closer co-operation between engineers and accountants, if the new possibilities are to be exploited. Instead of having each department taking responsibility for its own clerical staff, often repeating work done by other departments, the central computer will be able to integrate the processing and provide what each manager needs. The control of plant maintenance expenditure is a typical case where information may be needed by the wages department, cost department and maintenance planning department, each analysing and summarizing the data in different ways to enable them to carry out their work. If integration is to be successful it calls for respresentation by all departments, including engineers and accountants, on the team that is to design the new system; coupled with the determination and ability to set down and agree what really should be provided.

A major advantage of a computer is that it can check completely through a large file of data such as a list of stores accounts, apply a set of rules or budgets, and extract and print out automatically only those items that do not conform to the rules and therefore call for action. This opens up enormous possibilities of developing the techniques of budgeting and planning discussed earlier and using the computer to delve through the masses of resulting data.

A typical development in the engineering field is Critical Path Analysis which has been used under the names of P.E.R.T. and P.E.P. in the United States to control contracts like the *Polaris* missile. Very detailed study of a development contract is carried out before work is started, to determine the time that every step should take, and to consider what operations must be done in parallel so that two parts (for example the fuselage and wings of a prototype aircraft) will be ready simultaneously. With a big contract this becomes extremely complex; it has invariably been found that some small item such as a special bolt was likely to be critical in delaying the contract and thus would need the most detailed progressing. Following this planning phase, the computer is used to check actual progress day by day, update the plan and report on any new critical paths needing attention.

Computers must in future lead to a far greater insistance on planning by every executive and the man who is able to do this will be preferred to the managers one so often sees today, handling one crisis after another with the utmost energy. Put another way, many of the decisions now made by middle management will disappear, and those executives must face the choice of drifting down to less responsible jobs or increasing their sphere of influence and working with a far broader outlook.

Inevitably the accountant will play a major part in this change. In many companies he will lead the team, and be responsible for the computer department itself. As a major processor of data he is at least certain to be deeply involved. Engineers and accountants developing these techniques may become a formidable team.

Training

If it is accepted that planning should be a part of the everyday routine, it means that training is essential, both to inculcate the necessary outlook, and to enable the techniques to be learned. Up to the present the initial training of engineers seems to have been limited to mathematical or technical facts and figures rather than management problems. Any university or technical school course is already too full with technical subjects to allow other problems to be considered in any detail. My personal experience at university included three lectures on industrial problems, which were generally considered of little importance. A subsequent apprenticeship taught no more, though perhaps this was due to my own lack of interest at that time.

It is true that today, management problems are adequately covered in courses designed to fit a man to become a manager, but one still feels that the engineer who continues to ply his art in industry perhaps gets too little opportunity or incentive to broaden his knowledge. This seems strange when it is nearly thirty years since Mr T. G. Rose, himself an engineer, first put forward his ideas on management reporting. He wrote his first book *Higher Control in Management*[1] in 1934 with a sixth edition in 1957 and the ideas and techniques set out still hold good today. Two passages can be quoted:

> Higher control can be defined as the general management of a business on a planned basis, the adherence to the plan being watched by a monthly survey, made from the business, technical, trading, and financial viewpoint.
> My first claim for higher control is that it goes far to making any business manageable. My last is that it makes the responsibilities of a director a burden that any man of intelligence and some practical experience can safely undertake... Higher control is an assurance that the proofs of mismanagement are so unmistakably and swiftly disclosed that errors can be retrieved before their consequences become 'irreparable'.

In contrast to the engineer, the accountant is able to gain a broad experience of management problems right from the start of his training, which can follow one of two avenues. The first of these is through serving articles in a professional office to become a chartered accountant. The articled clerk starts with routine checking and adding of figures but later has opportunities of visiting a variety of clients and taking part in the preparation of their annual accounts, and possibly joining discussions of the accounts with the client. With luck, therefore, these early years provide a wide background of experience in different firms before taking up an industrial appointment.

The second method of training is to join an industrial firm from the start as a junior clerk, and reach the senior posts through promotions while obtaining an accounting qualification such as membership of the Institute of Cost and Works Accountants through evening studies. This method involves joining in the day-to-day work of the company from the start, thus providing experience of a different kind

from the more superficial contact of the articled clerk. At the same time the examinations, particularly those of the Cost and Works Accountants, demand knowledge outside accountancy, including management organization and statistics.

Then, following training, the accountant in his new capacity as provider of management information is likely to take part in every major decision and thus have an opportunity to gain a wide experience of both management and technical problems. Probably he has far more opportunity of broadening this experience beyond his accounting skills, than a purely technical engineer has of transcending his own sphere. Moreover, because every company has quite similar accounting problems, the accountant can readily increase his experience still further by changing employment over widely differing industries during his early years; this is not nearly so easy for an engineer.

The accountant as manager

A great deal is written about the difficulties of engineers becoming managers but, even with his close contact with management, the odds against the accountant making a good manager are probably greater. His training is one of caution and attention to detail, his duty is to point out the risky policies and acclaim the financially sound ones, and not to do this would be to fail in his job. A good manager needs, in addition, a flair for picking the right policy on the slimmest evidence, and a belief in his own judgement.

In fact, despite publicity accorded to a number of accountant-controlled companies in the past few years, fewer accountants than engineers have reached the ranks of top management. Mr Bosworth Monck published an analysis of the position in 1952[2] which showed that in 725 companies selected from a wide variety of industries, there were 4148 directors of whom 670 were chartered engineers, 390 were accountants, 215 had some other technical qualifications and the remainder had no specified qualification.

An analysis of the list of member of the Institute of Cost and Works Accountants for 1959 showed that only 266 out of 5189 members in Britain held directorships.

So, like the engineer, most of us will expect to follow careers as experts in our own field, each serving management according to his skills. But surely there must be scope for some integration between the activities of engineer and accountant. In this new field of management accounting, the accountant is frequently delving into technical problems, and it may be that an equal or even more effective service to management could be given by a trained engineer with appreciation of accounting skills. One can quote from American experience where many companies have a 'Treasurer' to deal with the routine accounting work and the management accounting service is provided by the 'Financial Controller' who may have trained as an accountant or an engineer or may be a business school graduate who has reached that position through the budget office which agrees the standards, and prepares management reports.

It is certain that there is a need in British industry for engineers who are interested in joining accountants in management accounting and this could form the basis for a new branch of qualified engineers, perhaps with its own post-graduate qualification—'Engineer Accountant'.

Mr J. A. Goldsmith *graduated in Mechanical Sciences at Cambridge, and subsequently spent several years working as an engineer, first in the Royal Navy and later in the aircraft industry. He then served articles and qualified as a chartered accountant. From 1952 to 1961 he was a consultant with Robson, Morrow and Co. He was a member of the O.E.E.C. team which visited the United States in 1960 to investigate developments in the use of computers for data processing. In January 1962 he joined the United Steel Companies to become Secretary of Steel, Peach and Tozer.*

REFERENCES

1. Rose, T. G., *Higher Control in Management*. 1957 6th ed. Sir Isaac Pitman and Sons Ltd, London.
2. Monck, Bosworth, 'The Eclipse of the Engineer in Management.' *Engineering*. September 10th 1954, Vol. 178, '2', p. 329.

Government, Parliament and the Law

by Sir Lionel Heald, PC, QC, MP*

Engineers have long been associated with lawyers in many fields. Today there is a great need for more scientists and engineers to take an active part in the job of governing the country at all levels of administration. This may require changes in the structure and procedures of Parliament. The weight of the scientific and engineering institutions in Britain should be placed squarely behind any measures designed to secure closer co-operation. Since this was written in 1962 the Select Committee on Science and Technology has been set up but it has had little impact on the points made here.

Some difficulty of understanding one another is always likely to arise when practical men first come into contact with lawyers and politicians, who work with words rather than things. Thus, when George Stephenson was invited to represent South Shields in Parliament, he declined, saying 'Politics are all matters of theory; there is no stability about them, and I should feel quite out of my element'. James Locke, Stephenson's pupil, did eventually become the member for Whitby, but he seldom spoke in the House, and when he died in 1860 his obituary notice remarked, 'The House of Commons is a place above all ill-suited for a thoroughly practical man'.

The average engineer might not necessarily accept the title of this article as justifying any serious consideration. 'What have we in common', he might well ask, 'with those who are engaged in the making or administration of the law?' Yet I believe he would agree that, unless Society can harmonize its way of life with technical progress, civilization itself can hardly be expected to survive. And how can we hope to make the necessary adjustments for this purpose, unless there is real understanding and co-operation between scientists and those engaged in government? The pure scientist and the layman are drifting further apart every day, and direct contact between them is becoming almost impossible. Somehow a bridge must be built across the gulf, and the man who can do this most appropriately is the engineer.

In this context, of course, I am using the term 'engineer' as including all practical scientists, though many may not strictly be entitled to it. The engineer is, if I may be allowed to borrow the words of H.R.H. The Duke of Edinburgh, 'the means by which the people are able to enjoy the fruits of science'.

Science and the legislators

Others, far better qualified than I, have discussed and deplored the great gulf between scientists and non-scientists in its particular relation to higher education. The academic scientist usually takes very little part in public affairs; he is not in touch with politicians or lawyers, and they, in turn, are apt to regard him as a remote being with whom they could hardly hope even to carry on a brief conversation. It is thus only too easy for the layman to form an entirely wrong impression of the scientist and his work; apart from retarding progress, this could involve a very real danger to society.

Recent scientific achievements have conferred immense potential benefits on mankind, but unfortunately what has most impressed many people is the perfection of means whereby the human race can destroy itself. It is, therefore, by no means inconceivable that a demand might arise, if not for the actual suppression, at least for some drastic limitation of scientific activities. No sensible person would wish to see this, but it is no use pretending that public uneasiness does not exist.

When we think about the vast power which scientists and engineers have acquired over matter and the serious implications all this may have for the rest of us, the possibility of a 'revolt against the experts' ceases to be a mere Wellsian fantasy.

It might even provide a cure for electoral apathy: in Samuel Butler's *Erewhon* the Professor of Hypothetics proved that 'machines were ultimately designed to supplant the race of man'. It was, therefore, decided to make a clean sweep of all those that had not been in use for more than

---

*This is a condensed version of the author's Graham Clark lecture: *The Engineer and the Law.*

271 years 'and to prohibit any further inventions under pain of the worst penalties of the law'. Truth is, in this case, not far behind fiction, for we find Professor K. Mannheim suggesting in his book *Man and Society* that inventors should be compelled by law to place their models on show before a panel of experts, with power to decide, in combination with a scientifically conducted public ballot, which of them should be permitted to be mass-produced!

In all seriousness, is it not time to make up our minds about our attitude towards scientific developments? It may be later than we think. For no government, of whatever party, will grasp a nettle of this kind until compelled to do so, and today the House of Commons simply does not possess the necessary technical equipment to enable it to exert the required pressure.

Sir John Russell has described this difficulty as follows:

'Of all mental adjustments those in political thought are the slowest, because the pace is set by those of lowest mental calibre. At the base of much of our present-day political thought, lie the ideas of the nineteenth century, and these unfortunately are inadequate to carry revolutionary science'.

The man in the street sees no evidence today of any policy designed to avoid the dangers of science and this is beginning to disturb him. For example, the noise made by jet aircraft is becoming a very live issue. Thousands of people in the Greater London area complain that they are woken up and shaken in their beds, because 100 passengers in one jet must be able to fly to Timbuctoo in five hours. We have been warned authoritatively by Lord Brabazon that the supersonic plane will be even more noisy. How much interference with life and comfort must be tolerated by the ordinary individual? What are the priorities? Sooner or later we must begin to consider how they are to be enforced.

For this purpose the House of Commons will have to become better informed on technical matters. How is this to be achieved? Can we hope for more Members with technical qualifications? At present the proportion who have any engineering background, even in the widest sense, is minute. One Cabinet Minister is a member of the Institutions of Mechanical and of Electrical Engineers, and another member of the Government is a practical scientist and a member of the Institution of Electrical Engineers. I am sure the presence of such men in the Government must be a satisfaction to engineers, but they are rare birds in the Westminster aviary. We must get more of them to help in the business of government, by actual membership of the House; by their employment in the Civil Service, not only as technicians but also in important administrative posts; by the constitution of unofficial working groups to which engineers of all kinds can be co-opted from outside; and by any other means we can devise, even if this involves introducing radical changes in our traditional Parliamentary machinery.

First encounters with the law

In considering the engineer's present relations with Parliament, it is helpful to trace them back in history. Until the Industrial Revolution got into its stride, the engineer seldom had any special contact with the law. Statutes concerned directly with his activities were comparatively rare, but it may be of interest to mention one or two examples. The earliest I know of, dated 1541, was called *An Act Concerning The Conduits At Gloucester* and its language provides a model of clarity and brevity.

'The Mayor of the City of Gloucester, and the Dean of the Cathedral there, may carry water in pipes of lead, gutters and trenches from Marstone Hill, and from time to time repair them, satisfying the owners of the ground for the digging thereof'.

London had a similar Act in 1543. Two important 17th century Acts were those for the drainage of the fens and the rebuilding of London after the Great Fire, the latter being notable in providing for the recovery of 'betterment' from all who benefited from the improvements. The most interesting Statute, however, for present purposes, was the River Thames Act of 1714, providing for the stoppage of a mile-wide gap in the bank at Dagenham, which threatened to be 'utterly disastrous', and also for a toll of 3*d.* per ton on every ship entering the Port of London, to pay for the works; for this resulted in the first recorded appearance of an engineer before Parliament. He was a remarkable character—John Perry, a former naval captain who, after an unfortunate encounter with some French privateers off Cape Clear, had been fined £1,000 and sent to the Marshalsea prison for ten years. After four years he was released and, contriving to meet Peter the Great during the latter's sojourn in London, Perry spent 14 years in Russia, completing a number of canals and other projects.

The Thames Committee, to which Perry explained the technical details before being entrusted with the execution of the works, were greatly impressed by his evidence, the Chairman saying to him at its conclusion 'You have answered us like an artist and a workman; it is not only the scheme but the man that we recommend'.

The Common Law had no specific concern with technical matters as such, but the traditional rights of action of trespass and nuisance were always available to the opponents of any novel project which involved interference with private rights of property.

Such rights could only be overridden by a Private Act of Parliament, giving statutory powers to the so-called promoters and even as late as the 1830's this could still be a 'desperate enterprise', as George Stephenson's friend and employer, Edward Pease, found in the Parliamentary battle for the Stockton and Darlington Railway Bill.

Canals and railways led the way to Westminster, due to the insistent demands of industry for the supply of raw materials and the distribution of manufactured goods. But soon many other so-called public utilities followed, such as those supplying water, gas and electricity, and later telegraph and telephone services. Public authorities also entered the field with applications for powers in all the ramifications of public health and local government. Here, as with the railways, there was at first strong opposition to such novel ideas as sewage disposal, when many M.P.'s seem to have rated cleanliness very far below godliness. But the objections were gradually overcome, and particular reference should be made in this connection to a great civil engineer, Sir Robert Rawlinson, originally employed by Robert Stephenson. In 1848 the General Board of Health was established, with Rawlinson as its Chief Engineer: this was not a Government department, and after some years of successful operation Whitehall caught up with, and duly abolished, it. Re-created in 1872, it became the Local Government Board, again with the indefatigable Rawlinson as Chief Engineer.

As applications for the grant of statutory powers became widespread, Parliament's approach to them underwent a marked change. Starting in most cases with an inborn objection to technical progress, the legislators gradually abandoned their hostility and began to adopt a new and positive attitude, aimed at making new inventions and techniques available to the public, subject to proper safeguards and compensation for all interests directly affected. It needs no emphasis that this change could never have come about but for the determination of the early pioneers and their patient exposition of the technical considerations.

What may not be so obvious is that it was also essential for the technical men promoting such enterprises to secure the co-operation of able lawyers, in order to produce the legal framework without which Parliament would never have granted such powers. Thus our two professions were drawn closely together at Westminster and came to realize that, far from engineers and lawyers having nothing in common, they could in fact work together happily as highly efficient teams. Some even combined the two professions.

John Smeaton, founder of the Society of Civil Engineers, was destined for succession to his father's extensive practice as a solicitor in Leeds, but he was much more interested in science than law. Even as a boy he showed 'marked mechanical ability', especially when he designed a special pump and emptied a valuable fishpond in the garden! Smeaton senior fortunately took this very well, and eventually agreed to young John abandoning the law and becoming apprenticed to a London instrument maker. Before completing his time, he was already writing papers for the Royal Society and soon struck out on his own as an engineer.

It was the initiative of Lord Lindsay in assembling men of science from all over Europe to confer at his laboratory in London 'in an informal and informative fashion' that led to the foundation of the Society of Telegraph Engineers in 1871. A dining club was also instituted, known first as the 'Arcangels' and later as the 'Dynamicables'. It brought together parliamentarians, lawyers and engineers in friendly conviviality, and one of its practical achievements was a report on the Electric Lighting Act of 1882. This was largely the work of Lord Bury, a tireless worker for closer co-operation, and John Fletcher Moulton, F.R.S., who was called to the Bar in 1874, elected an associate member of the Society in 1877, Member of Parliament, a Lord Justice and Lord of Appeal and Director General of Explosives during the 1914–18 war.

Lord Moulton was the most famous engineer-lawyer. A famous lawyer-engineer was Lord Armstrong, F.R.S., who was originally a practising solicitor; he was President of the Institution of Mechanical Engineers in 1861, 1862 and 1869, and President of the Institution of Civil Engineers in 1881.

Parliamentary committees

The engineer's need for the assistance of the lawyer became pressing as soon as the former's activities began to threaten the rights of land-owners; this frequently brought the two professions together before Private Bill Committees in Parliament. The story of their collaboration would fill many volumes, but I should not be doing justice to the pioneers if I failed to give some account of their courage and determination in the face of highly organized and quite implacable opposition. One notorious blimp is reported to have said that he would rather have a highwayman or a burglar on his premises than an engineer; and Samuel Smiles' life of George Stephenson contains a vivid description of how the first appearance

of a mysterious-looking instrument, the theodolite, excited such alarm and fury among the natives of Lancashire that, although carried by a noted bruiser, it was destroyed in a pitched battle.

Nor was it only the mere interference with rights that had to be justified. There was also the fear of actual damage to property; and the Common Law treated such things as artificial flooding on the same legal basis as the escape of a wild animal, where the owner had an absolute liability for all the consequences. It must, therefore, have been a remarkable scene when, in 1762, James Brindley, a practical millwright and canal builder, appeared before a committee at Westminster to support the Duke of Bridgewater's revolutionary proposal for the Barton Aqueduct scheme, to carry his canal over the River Mersey. It is recorded that Brindley was hardly able to write, quite unable to spell and made no calculations; but that he had nevertheless 'a sort of intuitive perception which enabled him to assess the project before a survey was made and an estimate prepared'. Brindley himself said that, when he had a puzzling bit of work to do, he would go to bed and think it out.

He duly appeared before the committee and explained his proposed methods for constructing the canal. The opponents made great play with the danger of escaping water but, surprising to relate, Brindley obtained permission for a demonstration. They may not have appreciated what they were inviting, for Brindley proceeded to construct, before their very eyes, an ordinary earth embankment which he filled with water, whereupon the bank burst and the Committee Room floor was flooded. Brindley then repeated the experiment, but used puddled clay to produce an impermeable wall, and this contained the water.

The Duke's lawyers may well have been apprehensive about the Committee's reaction to this unconventional behaviour, but the Bill was approved on the very next day.

George Stephenson was less fortunate, perhaps because by 1825, when the Manchester and Liverpool Railway Bill was introduced, the obstructionists had really gone into training. The Bill went into Committee on 21st March 1825 and was discussed almost continuously until 30th May, when

the main clause was defeated by 19 votes to 13. The promoters returned to the charge in the next Session and succeeded in getting the Bill through, this time after only ten days' hearing.

During the first hearing, Stephenson spent three whole days under cross-examination, particularly on his plan for 'floating' the permanent way over the Chat Moss bog, and was unmercifully harried and abused, both by the opposing Counsel and by the members themselves. He complained strongly of their 'quirks and quiddities'; some, he said, had even suggested he was a foreigner, presumably because of his north-country accent; and he had been variously described as an ignoramus, a fool and a maniac. But it is pleasant to be able to add that he bore no malice and kept his sense of humour.

On the second occasion a desperate attempt was made to wreck the Bill at the last minute by a member having the appropriate name of Sir Isaac Coffin. He said, among other things, that he 'would not consent to see widows' premises invaded', and asserted as a fact that any pheasant flying over the locomotive would instantly drop dead from poisoning by the fumes.

I cannot, unfortunately, do justice to the many eminent men who have sat in the witness chair at Westminster but, by way of contrast with the 18th and early 19th century figures, I would like to refer to one leading and greatly respected technical witness of modern times—Mr Harold Gourley.

As technical adviser to the Office of Works and Under Secretary of State, Mr Gourley appeared on many occasions before Parliamentary Committees and subsequently provided an outstanding example of the engineer-politician, who has been all too rare in the past.

Engineers and the courts

Professional men have frequently played a most important part in the actual administration of justice as assessors, referees or independent experts. They are always greatly in demand as arbitrators or umpires in technical disputes, where the parties are happy to leave the actual decision in expert hands, subject to an appeal to the court if any important question of law arises. Engineers are also appointed to make inquiries and reports for Ministers on a variety of subjects.

But the capacity in which they come into the closest touch with lawyers in the courts is that of the expert witness, above all in actions involving patents. The first requirement in a difficult technical case is that Counsel should work in the closest and most sympathetic association with his expert advisers and that he should be very humble in his approach to technical matters. Thirty years ago it was by no means uncommon for eminent lawyers to attempt to lay down the lines on which expert witnesses should approach their task, sometimes with quite disastrous results. Today it is generally recognized that the essential thing is team-work, the expert supplying the technical foundation and the lawyer adopting this as the basis of his preparation.

The two men who will always stand out in my memory in this field are the late Sir James Swinburne, F.R.S., and Sir Harry Ricardo, F.R.S. Swinburne, who lived to be over 100, was very well known and I will not attempt to describe him or his remarkable career. I will 'stick to a few simple facts' as he advised me to do, the first time I met him. An eminent judge once said of him that he 'gave his evidence with that complete disregard of the consequences which makes him the ideal expert witness—at least from the point of view of the court!

Sir Harry Ricardo has never, I am sure, been forgotten by anyone who heard his evidence in a very important case about the cylinder head patent which bore his name. In those days it was regarded as hazardous for any patentee to give evidence, but in his case it was a striking success. He even prompted counsel cross-examining him by saying at one critical moment 'I think what you really ought to ask me is so-and-so', thereby making the question much more difficult for himself. At the conclusion of his evidence the cross-examiner paid a remarkable tribute to his 'eminent fairness' although speaking in his own cause and the judge warmly endorsed this.

There were, of course, 'giants' in the old days, notably Sir Frederick Bramwell, F.R.S., President of the Institution of Mechanical Engineers, 1874–5. He must have been a very formidable figure in the witness box, where he was once described as 'the most powerful lay advocate of his time'.

His family provided a notable example of contact between engineers and lawyers, since his brother George became a Lord of Appeal, and a story, probably apocryphal, has it that Lord Bramwell once said, 'There are three kinds of expert witness—liars, damned liars and my brother Fred!'

Strong and sometimes heated encounters are inevitable between counsel and opposing experts. But it is almost invariably taken in good part and many friendships have grown up between those who started by being antagonists in some great legal battle. The engineer plays a vital part in litigation today, with the much greater complication of technical cases, and the courts must rely to an ever increasing extent on his skill and integrity.

Government policy

On many occasions engineers have been members of Royal Commissions or departmental committees and have advised Ministers personally. The institutions are also, of course, represented on advisory boards and committees of all kinds, notably those concerned with technical education.

But consultation has been mainly on an *ad hoc* basis, and on many important subjects there is no guarantee that it will ever take place. This has often resulted in things being done, or left undone for many years, which are quite unacceptable to informed technical opinion. No doubt the situation is much better than 100 years ago, when the institutions had to work for more than 30 years to induce Parliament to provide an efficient Patent Office. But we must not be too censorious of our predecessors. Taking into account the developments of science since those days, can we claim that Government or Parliament are, relatively, any better equipped today to deal with technical matters?

While the chief responsibility for answering this question must rest on the Government, the engineering institutions and other learned bodies have a most important part to play. It should be much more widely appreciated that technologists are prepared to accept the responsibility for a real share in the conduct of the country's affairs.

They are, in fact, already taking part in a most important organization at Westminster—the Parliamentary and Scientific Committee. This is an unofficial and non-political body composed of some 200 Peers and M.P.s of all parties, as well as a small number of very distinguished honorary members, and nominated representatives of over 100 scientific and technical organizations. It brings together that very combination of informed experience which is so desirable for the purpose I have indicated.

The future

If we are to meet the challenge of the coming century, our machinery of government must be much better equipped to deal with technical subjects, and the practical scientist has an indispensable contribution to make in this field. The problem remains—how is he to be encouraged and enabled to play his part? I do not see how it can be solved without considering the whole future of Parliament.

It is obvious that a modern House of Commons ought to contain a really significant proportion of members who have both technical qualifications and current experience of industrial and scientific development. But, to quote a most informative article, published in *The Observer* by Mr T. F. Peart, M.P., a leading Opposition spokesman in this field,

There are no practising scientists in the Commons. Those who have qualifications are unable to follow their scientific pursuits or to combine Commons activities with other work.

Nor do I believe it will ever be possible to persuade such men to enter Parliament in sufficient numbers unless very substantial changes are made at Westminster. For example, it would be out of the question for them to be present in the mornings, except in some emergency, and regular afternoon and evening meetings of committees would have to replace some of the sittings of the whole House. Business-like arrangements for voting and pairing, officially approved and open to all members within reason, would also be essential. Busy professional men simply cannot be expected to wait

about until a late hour, night after night and in far from comfortable surroundings, during debates in which they are neither directly interested nor able to take any useful part.

The formation of specialist committees is very desirable, and these should have power to examine orders and regulations as well as Bills. Under the present system, technical provisions go through, more often than not, without any effective scrutiny from private members. A mere speech from a departmental brief, rejecting a specific amendment, provides no real scope for informed criticism, and in the case of delegated legislation there is not even a committee stage.

Given the availabililty of sufficient qualified members, those responsible for new proposals could be required to appear and give evidence justifying them and explaining their effect. This 'inquisitorial' system is used in the Public Accounts and Estimates Committees, where members with financial experience are able to cross-examine the responsible officials. It also applies to the so-called Select Committees to which public Bills are occasionally referred, but this is not frequently done. In the Standing Committees to which most of them are sent, members are confined to speeches, and the technical man has no real opportunity of probing the details.

Another possibility is the formation of special groups or working parties to consider particular problems or subjects, with the assistance of suitably qualified officers from government departments and, where appropriate, expert advisers co-opted from outside.

Sir Charles Snow has advocated the employment of scientists at all levels of government, and in administrative as well as technical posts. This is clearly a desirable objective and such officers would be of great assistance to specialist committees as well as to Ministers, but they would have to be fully integrated in the existing system.

It is inevitable, of course, that any proposals which would have the effect of enabling private members to exercise more control over technical legislation would meet with determined opposition, quite irrespective of their merits, both from the Government Whips and from the Departments. Such obstruction must be firmly overcome if Parliament is to be able to function efficiently. If minor reforms, such as I have indicated, fail to produce the necessary increase in qualified members, resort might have to be had to much more revolutionary changes. Provision might be made for technical members to be elected on a special franchise, or even nominated. But these can only be regarded as long-range projects. We must try to get some results by immediate action. For this purpose the Parliamentary and Scientific Committee might be invited to study the whole problem of the handling of technical legislation, and to advise Parliament on it.

If the consideration of possible action were delegated to one or two small sub-committees, each representing the collective experience of the membership, their recommendations could hardly be ignored, even by the most reactionary opponents of progress.

A hundred years ago, Michael Faraday, one of the greatest of engineers, told the Royal Commission on Colleges and Schools: 'Science is knocking at the door; there is a prospect of it being opened, and it must be opened, or England will stand behind other nations'.

Since he spoke, there has certainly been progress in the field of education but, as regards Parliament itself, the door is still no more than ajar, and bears heavily on its hinges. A real effort is needed to move it and I venture to appeal to the engineers of today to place the weight of their institutions squarely behind any measures directed to that end.

Why We Should Enter Parliament

by E. R. Lubbock, BA, MP, MIMechE

In dealing with those aspects of Parliamentary government which are of particular concern to engineers, the author calls for an enlarged Parliament containing more engineers and scientists. Since few of today's politicians have enjoyed a scientific education, they may well be biased against selecting and promoting technical men who enter politics. But, in view of the technical nature of many of today's vital decisions, this makes it all the more necessary that engineers should put themselves forward as candidates. Besides their technical knowledge, they could bring their trained minds to the solution of problems too often dealt with in an atmosphere of emotions and politically conditioned reflexes. This article was written in 1963.

'If, as we well may, we regard politics as the struggle for power, engineering is not a political activity. It aims at something quite different—the control and exploitation of natural resources to give a greater degree of human welfare.' This view was expressed by Sir Maurice Bowra in his Graham Clark lecture, and it is with some diffidence that I have to begin by crossing swords with such a distinguished authority. It is perfectly true, of course, that engineering is not primarily a political activity, but on the other hand the politicians are not seeking power for its own sake, but for the opportunity which it gives them to mould society according to a particular set of ideas. To a very large and increasing extent, these ideas are concerned with human welfare and are dependent on the work of engineers for their fulfilment.

There are certain fundamental political propositions which are almost universally accepted today. We all agree that continued technological advance is desirable and necessary, that the State has a duty to help the less fortunate members of the community, and that Britain is uniquely dependent on her competitive position in world markets for her survival as a major economic power. The domestic policies which the Parties follow all have to be contained within this framework of axioms, and the differences between them are largely a matter of the allocation of resources. As these resources are created by engineers in the first place, their disposition must be of more than superficial interest to the whole profession.

Engineers in Parliament

My first conclusion, then, is that engineering and politics cannot be separated into watertight compartments, but are inter-related in ways which will repay detailed study. Surprisingly little has been said or written on the connection between the two, however, perhaps for the simple reason that there are so few people who are involved in both professions. There are today (1963) six Members of Parliament who are members of the professional institutions, and if we look back over previous Parliaments we find that the number is lower now than before, and just after, the 1939–45 war.

This may appear surprising when one reflects that over the period covered by these figures, engineering has been of steadily increasing importance to the national economy. The switch from traditional industries into those with an engineering base has been accelerating, and the last twenty years have seen the appearance of totally new industries, of which atomic energy, jet engines, computers and television are examples. One might have imagined that these changes would have been mirrored by an increase in the number of engineering M.P.s rather than the reverse, and it is interesting to speculate why this has not happened.

On taking a closer look, we find that certain professions and occupations provide the vast majority of candidates and hence of M.P.s. Out of 1462 candidates at the General Election of 1959, 256 were lawyers, 171 were teachers, 103 were publicists or journalists, and only 22 were engineers. There seem to be two fairly obvious reasons for this situation, but I offer these as opinions rather than as *ex-cathedra* pronouncements. The professions which are most heavily represented are those which are most easily combined with the work of an M.P. and they are also those which confer on their members a thorough training in the art of communication. In contrast, an engineer's job is not very easily combined with other activities, nor does it give much practice in writing or speaking.

Most engineers are salaried employees, and it is asking rather a lot of any employer that he should allow the time off which a candidate must have if he is to fight a successful election. If an engineer can persuade his employer to do this much for him, he still has to face the prospect of losing his job if he is elected, with a consequent drop in income. The expenses which a member of Parliament has to bear reduce his gross income of £1,750 to below the £1,000 mark, and the vast majority of members, therefore, supplement their income from some other source—no easy matter for an engineer.

As for the problem of communication, that is a simpler one to solve; the presentation of data in verbal and written form ought to be a compulsory subject in every degree course. This I do not advocate solely for the help it would give to the would-be politician, but because the lack of experience in communication is often a grave handicap to the engineer in industry.

Table 1—Professional engineers in the Commons

Inter-war average	1945	1950	1951	1962
13	11	10	9	6

Bias in selection?

There remains yet another barrier to the engineer who wishes to stand for election to Parliament: in the 1959 election, more than a third of the candidates were educated at public schools and no less than 106 had been at Eton, a school which cannot claim to have produced many engineers, whatever its other virtues may be. Similarly, Dr Ross has shown that, in the 1951 House of Commons, 15 per cent of the members had aristocratic connections, and this again is not likely to coincide with an engineering background.

We do not know the extent to which these patterns are imposed by constituency selection committees. It would be interesting to have an analysis of the occupations of rejected would-be candidates, to show whether or not some bias exists. As selection is generally based, in part, on a speech made by each short-listed candidate, it is possible that fluency of expression may outweigh other valuable qualities, and that polished manners may count for more than real merit. On the other hand, the explanation may be that engineers have not hitherto offered themselves for selection in any numbers.

It may be noted here that the constituency selection committees play a part in the electoral process, the importance of which it is almost impossible to exaggerate. In these days of psephologists and opinion polls, the result of an election is known with reasonable certainty in advance. A very small number of people—in relation to the electorate as a whole, that is—are in fact responsible for picking the Member. There is scope for a fascinating analysis of the way in which the decision is reached and the difference of approach between the three parties, but the investigation would be a difficult one by reason of the secrecy surrounding the vetting of candidates. The factors influencing the choice, if they could be determined, might show whether or not the qualities required differ in some fundamental way from those which an engineer is likely to possess. However, it may be that, even if one had all the facts, no common pattern for all constituencies would emerge. The characteristics which would suit Stepney are obviously not those which would appeal in North Cornwall, for instance.

In any case, it must be self-evident that, the number of engineer candidates is ludicrously small compared with the importance of the profession to Britain's future position in the league table of industrial nations. This by itself might be a good enough reason for thinking that it ought to be increased if possible. But to reinforce the argument, we might consider how they could play a valuable role in the House in complementing the Arts-trained men and women who now form the overwhelming majority. Engineers would certainly agree that any political questions which are amenable to a logical and scientific examination should be so treated. One might think this so obvious that it could be said of non-engineers as well. That this is far from true has been shown by Professor Susan Stebbing, who quotes many leading politicians to show that the English are not, in general, a logical nation. One typical instance is taken from a speech made by Baldwin in 1937:

'One of the reasons why our people are alive and flourishing and have avoided many of the troubles that have fallen to less happy nations, is because we have never been guided by logic in anything we have done.'

Lest it should be thought that such opinions are no longer held, consider Lord Salisbury's remark about the Parliamentary and Scientific Committee at the Annual Luncheon of that body: '. . . it is illogical, and it is successful. In fact, it is a typically British institution.'

Lord Hailsham, on the other hand, speaking on the same occasion, said that he was profoundly disturbed by what he called the flight from reason in the modern world. He was talking about the world in general and not specifically about Britain, but we are by no means immune from the disease of unreason.

Words and phrases like 'nationalization', 'landlord', 'sovereignty', 'unofficial strike', seem to have the power of arousing predictable reactions—one might almost say conditioned reflexes—in men who are otherwise quite reasonable. It would be easy enough to give plenty of illustrations taken straight from the pages of Hansard, if it were not for the necessity to be impartial. Most of the obvious examples one can think of demonstrate a split of opinion on party lines, and it is no part of my case that the fault lies all on one side. The point is that an engineering training would help the politicians to avoid becoming mere party automatons.

The engineer approaches any problem with the question 'What are the facts?'—not 'What are the politics of this question?'—and it is certain that many of our actions would be the better for this objective approach, which is by no means limited to engineering or scientific problems. In addition, there are quite a large number of subjects on which engineers could make valuable contributions as a result of direct experience, and in the limited space of this article I must concentrate on these.

The supply of engineers

The future of Britain as a major industrial power is dependent on the continuous application of new technological advances, to keep ourselves at least abreast of our major competitors in world markets. This, in turn, depends on the level of efforts to increase the supply of trained manpower. Sir Winston Churchill, in 1955, drew attention to the inadequacy of our efforts in this connection, compared with that of the U.S.S.R. and the warning has been repeated on many occasions since.

This is by far the most important subject on which engineers would have something to say in Parliament. Nobody can be satisfied with what has been done over the last few years. It is obvious that if the number of science and technology students is to form a larger proportion of the total, expenditure will have to rise much faster than *pro rata*. If, however, one has been brought up on Latin and Greek one might well pay lip-service to these needs, while secretly viewing them as unnecessarily extravagant.

The financial experts at the Treasury seem to have overlooked the fact that technological education is necessarily far more expensive per student than Arts courses. The importance of a low ratio of students to staff would hardly need stressing to anyone more familiar with engineering. According to Sir Keith Murray, the Chairman of the University Grants Committee, this ratio has deteriorated since before the war from 7·6 to 10·7.

In the debate on Scientific Manpower of 21st December 1961, the Parliamentary Secretary to the Minister for Science, Mr Denzil Freeth, admitted that for scientists and engineers in purely vocational posts this was not the whole story: one of the chief reasons for the backwardness of large sections of British industry is the presence of amateurs in the boardrooms and in senior management positions, and no assessment of future requirements which ignores this factor can be valid.

Even in the field of research and development, the outlook is far from satisfactory. According to a report of the Federation of British Industries there is an average vacancy rate of about 13 per cent overall in industrial research and development departments. The Government can exert considerable influence on the pattern of research since it spends two-thirds of the total R and D expenditure of nearly £500 million itself, either through Government-financed institutions or through contracts with private firms.

Important decisions

Unfortunately, when these subjects are debated in the House of Commons the attendances are often poor—deplorably so, considering the role science and engineering play in the modern economy. Perhaps the laymen are frightened of science and engineering, and feel that their forensic abilities are no asset in these debates, although one might have imagined that they would show greater willingness at least to listen.

To give just one example, there was the question of the Government's decision to go ahead, in co-operation with the French, with the development of a supersonic airliner. The amount of money involved was at least £75 m for the British contribution alone, although, if past experience in these matters is any guide, it is more likely to be £1000 million before we have finished. Yet the responsibility for sanctioning the expenditure rested on a collection of people, 90 per cent of whom couldn't tell you the difference between a turbine disc and a flame tube.

The Minister of Aviation, in announcing the decision, gave no information on which the House could decide whether the project was likely to be a profitable one, nor any assurances that the technical·problems could be satisfactorily solved. I suppose that, if he had done so, there would have been very few Members who could have understood.

This was a particularly interesting question to me, because I think it is one on which we should have examined the technical pros and cons, before thinking of the political aspects. In the absence of technical knowledge, we could only score debating points about whether or not this is the best use of available resources.

At any rate, I hope I have said enough to illustrate the contention that there are many issues on which the voice of engineering should be louder in the House. The interconnected subjects of higher education and the supply of scientific manpower are the most important of these, I think, but the list would be a long one if it were to include every subject of direct interest to engineers.

I speculated at the beginning about the reasons why so few engineers are to be found in national politics. The same question arises in relation to local government, and at least one of the answers is similar. There is in local government a bias in favour of certain groups of people, and this is particularly marked in authorities which meet in the afternoon. This practically rules out people who have to work regular hours for a living, and favours the self-employed, retired persons, married women whose children are grown up and those who have 'independent means'. Even where the Council meets in the evening, many people would have difficulty in attending up to a dozen committees a month at 6 or 6.30 p.m.

The result is that it is becoming harder to find potential candidates for Council vacancies, and the number of uncontested elections shows a continual increase in all types of authority. Yet local authorities are responsible for spending millions of pounds of the ratepayers' money every year, much of it on engineering works.

If this trend continues there might well be greater centralization of some powers which are at present delegated, and a transfer of other powers from directly elected bodies to *ad hoc* committees composed largely of professionals.

The parties

Not everybody can take a direct part in the processes of government, either local or national—but all are responsible, in theory at least. This means that the very minimum political activity ought to be the making of an informed choice at election time; but if one believes that the collective result of those decisions is important, something more is necessary. Politics today are entirely party-based, and the policies of each party are discreet pictures of the type of society at which its adherents are aiming. No individual can hope to influence policy by acting in isolation, but only through his membership of a political party. The choice of party is a subjective one which is in no way based on scientific principles and thus we find engineers in all major parties.

But, whatever their views may be, engineers would probably agree that their own influence on twentieth century civilization has been at least as important as that of the politicians. Our history books tell us more of Disraeli than they do of Brunel, yet there is no doubt that the evolution of today's Britain owes as much to the one as to the other. The state is based on the technological revolution, and Governments have merely handed out the prizes which have been provided by the engineers. We have also perfected means whereby not only the state but the whole of civilization might be destroyed. Politics has become altogether too dangerous for us to leave it to the professional politicians.

Sir Lionel Heald, Q.C., has already called on the engineering profession to take a more active part in the processes of legislation. He has said that 'under the present system, technical provisions go through, more often than not, without any effective scrutiny from private members.'

New procedures needed

When one considers that, in the last session, no less than five important Bills were subjected to the guillotine procedure, one can appreciate that it is not only the technical provisions which are being treated in this way. Parliament is having to cope with ever-increasing masses of legislation, and when we go into the Common Market in the future there will be more still. Yet the number of M.P.s has been more or less constant for the whole of this century. Obviously, the House of Commons has got to be enlarged if it is to do its job.

Secondly, the Standing Committees which deal with legislation ought to be given the power to call for advice from technical experts. This would enable engineers to play a more direct part in the processes of legislation, and although it would probably lengthen the proceedings, it would ensure that technical considerations had been properly taken into account before legislation finally reached the Statute Book.

Sir Lionel Heald also called for the extension of the Select Committee procedure, under which M.P.s already are allowed to cross-examine the responsible officials. If the House of Commons were enlarged, as I propose, there would be an opportunity of creating Select Committees to examine many technical and scientific matters which at present get pushed through almost 'on the nod'. I have already mentioned the supersonic *Concorde*; other examples of equally important matters which require direct contact between engineers and politicians are atomic power and the transport system.

It is true of course, that engineers have taken part in the work of Royal Commissions; but this does not really achieve

Mr E. R. Lubbock, M.P., was educated at Upper Canada College, Harrow and Balliol College, Oxford. While at Oxford he won a boxing blue and obtained a degree in engineering. In 1949 he joined the Welsh Guards and after two years' service he began work in the Aero Engine Division of Rolls-Royce Ltd, Derby, in the Foundries and Sales Department. In 1955 he became Management Consultant for Production Engineering Ltd, and in 1960 he joined the Charterhouse Industrial Development Company Ltd. From 1962 to 1970 Mr Lubbock was the Member of Parliament for Orpington, Kent.

the object of bringing together the legislators and the technologists so that each understands the other. We are failing to educate our masters in a sense which the originator of that phrase could never have foreseen.

The Parliamentary and Scientific Committee does its best to rectify this situation, but the trouble is that it has no official status, nor do its activities attract much attention among people who are not immediately concerned with politics. If it could be given definite responsibilities instead of acting as a mere forum, there might be great opportunities for engineers to influence scientific legislation.

I am not pretending to say just how this should be done—that might be rather presumptious after less than a year as a Member of Parliament—but we have got to begin thinking about how the machinery of government can be adapted to deal with technical subjects in one way or another, and it seems to me that the Parliamentary and Scientific Committee at least provides us with a starting point.

More than technical issues

But while engineering and science are most inadequately dealt with under the present system, I would certainly not accept the idea that an engineer's interest in politics should be limited to technical matters with which he is directly familiar through his work.

We no longer refuse to concern ourselves with the important problems of our time; many of these such as the cold war or the Common Market, are not technical problems in the narrow sense. Nevertheless, they would not have arisen in their present form had it not been for the technological progress which has been brought about by engineers. The cold war would not have been nearly so dangerous if nuclear weapons and ballistic missiles had never been developed; and the Common Market might not have come into existence at all but for the advent of mass-production techniques which demand a large home market. Thus on both these issues the engineer cannot absolve himself from the responsibility of deciding how his work should be used.

I realize that in this article I have probably raised more questions than I have answered. It is one thing to deplore the lack of engineer M.P.s or to wish that engineers would play a more active role in the parties, but quite another to suggest the remedies. It seems to me that the first requirement is for engineers themselves to begin thinking about how their talents could be used in politics and to decide if that is what they really want, or if they would prefer the fruits of their work to be regulated, as hitherto, almost entirely by members of other professions and callings.

Sir Joseph Rawlinson obtained the degree of M.Eng. at Liverpool University. He spent nearly six years with civil engineering firms and contractors. Late in 1924 he joined the Liverpool Corporation and was appointed Assistant City Engineer in 1932. In 1936 he was appointed City Engineer of Westminster and in 1947 he became Chief Engineer and Surveyor to the London County Council, a position which he held until his retirement in 1962. Sir Joseph was awarded the C.B.E. in 1953 and created a Knight Bachelor in 1962. He is also an Honorary Citizen of Winnipeg, Manitoba.

What are the engineering functions of local government and how does the work of its engineers differ from similar activities in industry? The author, a former chief engineer to the L.C.C., discusses these points, the problems of liaison between public authorities and their contractors, as well as other points about local government which are of interest to the engineer. Here, as in Parliamentary politics, there are many engineers among the employees but very few sit on the Councils that employ them.

Local Government and Engineers

by Sir Joseph Rawlinson, CBE, MEng, FIMechE

The Right Honorable J. S. Maclay, M.P., Secretary of State for Scotland, in a preface to the *Municipal Journal* for 1960, said: "Local Authorities today are deeply involved in the social developments by which the pattern of life is being determined in our time". The social developments for which local authorities are responsible cover an enormous field and a full list would be a long one.

The Local Government Acts of 1933 and 1939 describe the constitution of several classes of public authorities concerned with local government. The country is divided into areas known as counties, county boroughs, boroughs, urban and rural district councils.

All these local councils have two features in common; they are elected periodically by the people and they all depend to a large extent on local taxes (rates) levied on occupiers of property in the area.

The sizes of the councils vary considerably, some of the smaller ones having as few as twenty members while some of the largest cities have over a hundred.

Local councils deal with many of their affairs through committees which form a special feature of English local government practice. It is impossible to describe here in detail the constitution of these committees with a list of their responsibilities, but two committees should be of particular interest to engineers: those dealing with town planning and with public works.

Today the Town Planning Committee has wide powers and exceptional responsibilities. The pattern of our future cities, and therefore the welfare of the people, is largely in the hands of its members and yet it is surprising how little attention it attracts outside the precincts of the local town hall.

The Works Committee is generally responsible for the construction of roads, bridges, pumping stations, sewage disposal, refuse collection depots and markets.

There are, of course, other committees whose work should appeal to engineers. Probably the most important of these deals with housing.

The housing problems facing the country today are immense and, as far as one can judge, are not likely to decrease in the foreseeable future.

Essential services

History clearly shows that the need for certain services brought into being those public bodies which have evolved and developed over the centuries into the present form of local government. The early stages of this development took place when the principal industry of the country was agriculture, in the form of smallholdings, which needed little more in the way of machinery or tools than the plough. From these beginnings many services vital to the continuance of society, as we now understand it, have developed into highly technical organizations, requiring the best brains which local government and industry can attract. The generation of electricity, for instance, was initiated in many towns by local authorities, who were also responsible for building many

The London County Council's development project at Roehampton which has its own comprehensive heating scheme

Table 1—Income and Expenditure for 1959–60

	Rate Fund Services (£'000)	Trading Services (£'000)	Special Funds (£'000)	Total (£'000)
Revenue:				
Income from rates, grants, etc.	1,334,834	186,591	34,374	1,555,799
Expenditure	1,313,301	184,592		1,497,893
Capital Account:				
Receipts from loans, grants, etc.	499,032	42,614		541,646
Expenditure	510,091	44,950		555,041

Table 2—Engineering expenditure

Purpose	Annual expenditure (£'000) approx.
Sewerage and disposal	36,710
House and trade refuse	36,855
Baths and washhouses	9,080
Housing—part—	262,748
Town planning etc.	11,095
Land drainage	11,392
Highways and bridges	105,679
Public lighting	16,842
Water supply—part—	60,444
Passenger transport	75,033

docks and harbours. The list could be extended almost indefinitely. It is, however, worth referring specifically and in some detail to one of the earliest of the public services, which helped in the middle of the last century to transform London, where outbreaks of the plague were all too common, into one of the healthiest cities in Europe. Typical of the perhaps prosaic but vitally important services to the community is the very old one of main drainage, which is also of considerable interest to many engineers.

The drainage system of London has evolved over the centuries. Its earliest function was to carry surface water from the land and streets to the rivers. So long as these streams contained only clear water, no trouble arose; but it was a very different matter when they became polluted with sewage. In 1388, in the time of Richard II, the first Act of Parliament which dealt with the purity of the Thames and other rivers was put on the Statute Book. In 1531, in the reign of Henry VIII, the Bill of Sewers was passed which remained in force for nearly 400 years. Commissioners of sewers were appointed who had power to levy rates and who carried out a number of useful works, e.g., improving the rivers and constructing sewers.

This is an early example of the work carried out by publicly appointed bodies which has now been taken over by local authorities. Such services have grown in importance and have been extended from time to time to keep abreast of the development of the country, particularly during the last hundred years.

The Ministry of Housing and Local Government* is responsible in a general way for a wide range of work done by local authorities, including housing, water supply, town planning, sewage disposal, smoke abatement, and many others. The Ministry of Transport* more particularly since the expansion of the road programme in the middle fifties, has placed considerable responsibilities on local authorities for developing the highways of the country and this has brought the engineers of various industries into much closer touch with those of local authorities.

The total annual expenditure of local authorities is shown in Table 1, and indicates the measure of their responsibilities. These funds come both from the local rates and from Government grants. The approximate annual expenditure for which engineers are principally responsible, including administrative and running costs, is shown in Table 2. In addition there is, of course, an expenditure against Capital Account.

There are 201 000 miles of roads in Great Britain which are maintained by some 1288 Highway Authorities. Public passenger vehicles operated by local authorities carried no less than 4800m passengers during 1960, which represents about 33 per cent of all passenger journeys made on public service vehicles—including railways—for the whole country.

The above figures illustrate the very . great part local authorities are playing in the social development of Britain.

Relations with industry

Orders placed by local authorities with private engineering firms during the last hundred years have contributed in no small measure to this development. In fact, there are large and thriving engineering firms which depend to a considerable extent upon the work they carry out for local authorities to keep their establishments open. These are the firms specializing, among other things, in construction of roads and bridges, water supply, sewage disposal, street lighting, etc. The list could be extended, but sufficient has been said to show that the activities of local authorities have a very considerable impact on British engineering industries.

The power house, opened in 1960. of the GLC's Northern Outfall Works' one of the most modern installations in the country

* Now part of the Dept of the Environment

local authorities on getting goods of a quality for which they are paying is absolutely essential in a democratic country such as ours. When differences of opinion do arise, it is important that some method should be readily at hand which can be used to obtain a quick and fair solution. To resort to Arbitration or to the Law is, in some cases, the only way to settle serious differences of opinion, but both methods can be costly and generally take a long time to set in motion.

The number of cases which are allowed to find their way into the courts are now, fortunately, relatively small, which is a sign that the differences that do crop up are being settled during the performance of the contract.

Notwithstanding these changes, much can be done to improve relations between local authorities and industry. For instance, when it is necessary for a local authority to enter into separate contracts—e.g., civil and mechanical, for the building of a power station, particularly when there are many sub-contractors and employees on both sides, it is my experience that, in some cases, improvements are required in site organization and supervision.

The mechanical and electrical plant is manufactured, examined and tested at modern works and almost without exception the results are excellent. This expensive plant is then packed and dispatched to site; in some cases, all the care and attention which has been bestowed on it ends once it has left the works. When it arrives on the site, the planning and organization so necessary for efficient erection and testing is too often left in the hands of a single individual, sent down from the works with little authority and few resources at his disposal. It is his job to find local labour, judge the condition of roads leading to the site, to find any other equipment needed there during erection, and so on.

It is understood of course, that local labour must be employed but this side of the work should be organized beforehand and not left to the representative sent from the works.

It is interesting how firms differ in their methods of supervising the work on site. I know some who insist on a director paying a weekly visit to each outside job with the result that many local problems are dealt with quickly and efficiently; while other firms rely on the 'site representative' who has much less authority than a director and, in consequence, spends a great deal of time asking for instructions or travelling to and from head office.

The erection and testing of plant is a very expensive business and, on the larger contracts where there are many sub-contractors, a first-class organization is necessary to ensure that each sub-contractor is doing the right job at the right time, thus causing as few delays as possible to other trades. The accumulation of these troubles often results in claims which lead to long arguments and delays in settling accounts.

This often brings engineers in public service into contact with their colleagues in private industry and the relationship between them is of great importance. During my own long experience of working for local authorities I have enjoyed the happiest relations with engineers from a great many industrial firms. When differences do arise they are generally the result of financial troubles and not because of disagreement on engineering gounds. This is not to suggest that finance is only of secondary consideration: indeed, £ *s. d.* is the ultimate yardstick of all engineering efforts, though not necessarily their purpose.

However, many of these differences would be avoided if each side understood better the many problems and difficulties facing the other.

Contracts

One of these difficulties arises in the drawing and interpretation of contracts. The regulations governing the award of public contracts are rigid: it is customary to advertise all invitations to tender in the press and, on the whole, manufacturers and consultants understand the procedures involved. But there is, perhaps not unnaturally, a tendency on the part of local authority engineers to endeavour to cover themselves and their authorities in all contract documents against every possible contingency. When this happens, engineering firms complain—not without reason— that if they in their turn covered themselves against all contingencies and conformed strictly to all the conditions of the contract, it might be difficult for them to finish the job.

All engineers should appreciate that the insistence of

Conditions of service

A great number of engineering appointments by local authorities are—and have been for nearly a century—permanent and pensionable positions, but it should be remembered that the employee pays something like 5 per cent of his salary towards his pension, which is calculated on the number of years of pensionable service. Fifty years ago it was unusual, but today it is common practice, for privately employed engineers to receive pensions after long service with their firms. In fact, many large concerns pay at least as good pensions as do local authorities and provide additional fringe benefits. Any gap, therefore, which might have existed in this respect in the past between public and privately employed engineers has now disappeared.

It would be fair to say that industry and private employment offers greater scope for engineers in the more specialized branches of our profession. The opportunities to keep abreast of the changes which are continually taking place in the science of engineering are more readily available to engineers employed in industry than to those in local government service. But this does not mean that the latter can rely on outside sources to keep them informed as to what is happening in the engineering world. On the contrary, they are expected to keep themselves informed on those matters which might affect their departments, and be in a position to advise their authority accordingly.

The Greater London Council's engineering service, for example, covers a very wide variety of work, ranging from radio and television for schoolchildren to the design and maintenance of large storm-water pumping stations and supervision, control and maintenance of a small fleet of ships. The GLC is also responsible for ten Thames bridges and is the road improvement and main drainage authority for the area of Greater London. Similar responsibilities rest on the shoulders of other County Councils and the authorities of large cities and towns.

Keeping in touch with specialists

Do these divers responsibilities call for a class of engineer different from that finding its way into industry? By training, I would say, no: by experience, yes!

It is now a common feature of advertisements for staff to call for degrees or similar qualifications so that, up to the stage when young engineers obtain their qualifications, their education and training is generally on a fairly broad basis in civil, mechanical and electrical engineering. After the initial training period of five to seven years they branch off into more specialized forms of engineering.

The impact of modern technology on the administration and structure of local government is, I believe, fairly well understood by local authorities in Britain. But during the last two or three decades the engineering sciences have developed so quickly that local government engineers now have to depend to a greater extent on the specialist in industry or outside consultants for the latest information on certain engineering matters. It is true that in some branches of engineering the local government engineer is his own specialist; but when complicated plant is involved, the advice of an outside expert is often called in.

It is probably also true that manufacturers know more about the technical work of local authorities than vice versa. I think, however, that more could be done to keep the local government engineer better informed about progress in those branches of engineering in which he is particularly interested. The reply may well be that industry is spending substantial sums each year in advertising its products and it would help manufacturers if the local government engineer would indicate a little more clearly and precisely where his particular interests lie. There are obvious difficulties on both sides but some organization might be set up to keep local authorities informed of the latest engineering practices in industry.

There is a feeling among many engineers employed in industry that the local government engineer is inhibited and restricted in his activities because of his official position, a position which is largely controlled and directed by politicians. He does, of course, hold a public appointment and consequently his activities, both during and after office hours, are perhaps more restricted than those of his counterpart in industry and elsewhere. For example, he is not permitted to take on any other remunerative employment of any kind without the permission of his council. Now, this rule is a strict one and dispensation from it is seldom sought and very rarely granted.

Few engineer councillors

There are, I understand (although I have no personal knowledge of them), a few cases of engineers employed by an authority being allowed to sit as councillors for other authorities. But perhaps it is not surprising that few local government engineers find an outlet for their political activities in this way. What is surprising, however, is that so very few engineers employed in industry find their way into local councils. Why should that be?

I have heard it said that the committee system of administering and controlling the local government machine is cumbersome, outmoded and a waste of time. The actual time taken to get through the procedure of a committee, is to a very large extent, in the hands of the members themselves. But the general system of local government is laid down by Acts of Parliament and local authorities have no choice in this matter.

Some professions are more widely represented on the councils of local authorities than others. Many lawyers give freely of their experience and devote a great deal of time to this work, but the two largest categories of members are housewives and business men. I am not at all sure that I can explain why this should be, but a possible explanation is that business men show a greater willingness to take part in outside activities and enjoy the rough and tumble of local political life. It is also perhaps true to say that the average business man takes a greater interest in local affairs than his professional counterpart.

Whatever the reason, the number of professional engineers on local authorities is very small indeed: far too small, in fact.

It is a sad comment on the interest the public take in local affairs, that less than 50 per cent, on the average, vote at local elections. For example in 1962 only 23 per cent of the eligible voters took the trouble to record their votes in an important by-election in London. This dreadful apathy is interpreted in different ways at different times by different parties. It has been suggested that a small poll shows that the inhabitants are satisfied with what they receive from the local council. Whatever the reason for this indifference, the work of local government must go on, and the successful candidates—however small the vote—are for the time being the makers of that policy which can affect all or part of those services for which the local council is responsible.

Town planning—to take only one example—is a permanent responsibility of many local authorities, including a review every five years of the development plan. This is only one sphere in which the engineer councillor could be of great value to his council, in offering advice on the location of roads, housing and industry.

The actual work of the local authority is of course done by

permanent officers but they must carry out the policies and instructions of the elected councils they serve. Each council can—apart from certain statutory requirements—accept or reject proposals or schemes submitted to it. It is at this vital stage that the experience of the engineer councillor would be invaluable in shaping policy.

Future trends

The responsibilities placed on local councils by the Government tend to increase year by year and, as most services involve engineering of one kind or another, the responsibilities of the engineer have assumed formidable proportions.

It is impossible to forecast political trends over the next decade or two, but it is safe to assume that a general improvement in the standard of living will continue with the possibility of a big increase in all the public services. There will be a corresponding increase in the movement of traffic and this change will place still greater responsibilities on local authority engineers.

Will the very rapid strides in the industrial progress of foreign countries, particularly on the Continent, in America, Russia and Japan, have repercussions on local government here in Great Britain? Personally, I believe the effect of this great surge forward will be felt in every town hall in this country. The present keen competition from abroad—with no holds barred and with short shrift for the loser—can only be met by a great effort on the part of all of us, including those in local government.

Industry, like other sections of society, has to pay local rates and the inhabitants of industrial areas are sensitive to the services which are offered by local authorities: e.g., education, transport, public health, water, drainage and so on. In this way, local government does and must play an important part in helping industry to compete successfully in world markets.

There are unmistakable signs of a serious shortage of highly trained engineers in Britain, with little hope of improvement for some time to come. With the trade barometers set at 'Fair', industry can offer better salaries to engineers than local authorities; nevertheless it is vital to the interests of both industry and the country at large that the services of the local authorities are not impaired in any way. A serious setback here would be unthinkable.

Engineers outside local government must take a greater interest in local and national politics. In this scientific age the country has great need of all the engineering brains it can muster and both national and local politics are greatly in need of their services. It is to be hoped that, in spite of the demands made on their time, they will manage to shape the policies of local government which, in importance, are second only to the work of Her Majesty's Government.

Acknowledgement

The author wishes to thank the Clerk of the Greater London Council and the Chief Engineer for permission to publish the illustrations to this article.

In helping the engineer to assess the difficult problem of how to weigh possible loss and danger to life and limb against the benefits of new designs and advanced techniques, the insurance underwriter, in effect, represents society as a whole and helps to spread the risk of innovation as widely as possible. Good faith on the part of all concerned is essential if the system is to work for the responsibility of the inspector is a heavy one.

Fig. 1. Collapse of the main jib of a dockside crane due to an overload when operating at maximum radius: accidents such as this are covered by an engineering crane policy

Insurance and Inspection

by J. Eyers, BSc, MIMechE

A common characteristic of the great members of our profession in the past was a passionate interest in problems ranging far beyond those associated with their everyday activities. But of late we engineers in general have become ever more absorbed in the fundamental aspects of our work, to the exclusion of other factors, even when these may have a direct effect on our function which is, or should be, the production of some amenity at the most economical price associated with the greatest benefit to our fellow citizens. During the past decade in particular there has been a tendency to narrow specialisation, encroaching even on the technical sphere so that, for example, the young modern mechanical engineer may have only a cursory knowledge of structural design outside his own circumscribed field which, if not corrected, may be greatly to his disadvantage at some crucial point in his career.

Fortunately the dangers of a too restrictive training are being realised, as is evident from the repeated reference to this in the technical press and from the increasing attention given to the broadening of basic education by the professional institutions and scholastic authorities. Certainly a determined effort must be made to overcome the almost pathological distaste of many of our young colleagues for such subjects as administration, finance and insurance. It needs only a short visit abroad for the great advances being made in industry by our foreign friends to be appreciated. If we, as a nation, are to retain a leading position in production, the industrial manager, accountant, engineer, insurance expert, salesman and others must learn to act as a team, appreciative of each others problems; and not, as is so often the case, to work in enclosed compartments.

The broader outlook can best be inculcated during college instruction and it is interesting to observe that many foreign universities devote the final year of their engineering degree course to subjects connected with management, a practice which might be followed with advantage in this country.

Of all the business functions affecting the activities of the engineer, be he consultant, designer, works manager, or field operator, that of insurance is normally the most dis-

regarded, even though the costs of this are directly reflected in the selling price of the ultimate product; and in spite of the fact that a wrong decision concerning what cover to effect might bring financial ruin to his concern. The usual attitude of one of our profession to insurance contracts is that these are the sole responsibility of the company's insurance manager or secretary and if, apart from his personal car or household policies, he has any knowledge of the subject at all, it is usually relative to the cover and services given to his plant by an engineering policy; and this only because such a contract provides him with an independent inspection service by specialist surveyors who can speak his own language. Nevertheless, from the standpoint of an evaluation of industrial hazards, it is the engineer who is, or who should be, in the position best to appreciate these, and he should be prepared to take his share of responsibility for the decisions made about insurance. By so doing, however, he is automatically involved not only in his professional capacity but also in his relations to society, for in addition to protecting the interests of his employer he is ethically obliged to consider the welfare of the general public, his staff and colleagues and, indeed, his underwriters.

Nearly 2000 years ago Seneca remarked: "Fortune plays with her gifts; she takes away that she has once given and gives back that she has taken away; and it is never safe to put her to the proof when she does not give rise to misfortune". The truth of these words needs no emphasis when relevant to our own interests, but should be even more important when the fortunes of others are involved.

Public relations

Perhaps it has to be admitted that the engineer's distaste relating to insurance matters has in the past been largely the fault of the insurance industry itself in that it has not taken steps to educate the public on its activities and, indeed, until lately such measures would have been regarded as bad form. Yet for hundreds of years insurance has played a vital role in the industrial economy and without its protection the present stage of business development could not have been achieved, a fact recognised even by the totalitarian states.

One of the major complaints voiced against insurance practices is that concerning the exclusions and conditions detailed in the so-called 'small-print' on the policy. Since, however, this document details the text of a business contract, the scope and limitations of the latter must be clearly defined and should be studied and understood by the insured. In the past many policies were worded in archaic legal form which could not be readily interpreted by the layman but of late these have been condensed and framed in everyday language so that censure in this respect is no longer valid. In any case there is no excuse for the insured not seeking clarification from his broker or insurer if he is in doubt on any matter.

Further criticism made against insurance companies concerns their prosperity, as evidenced in their business investments and property holdings which, it is implied, must result from an excessive levy on society. The answer to this is that the insurance industry, like any other business, has the right to make a reasonable business return to its owners and that profits are maintained at an equitable level by the natural effects of free competition at home and abroad. A large premium income automatically implies a correspondingly large potential liability and wise investment in business and property is essential in the building up of reserves to meet the possibility of catastrophic losses (which may have to be made good at any time at short notice). Furthermore, large reserves are the basis for the eventual lowering of premium rating.

Legal aspects

The provision of adequate insurance cover is intimately connected with the liability an executive or employer may have to sustain following legal action and all professional engineers should therefore have at least an elementary knowledge of statutory legislation, and of the laws of tort and contract. However, jurisprudence develops with changes in society and it would be foolish not to make insurance provision for some matter on which liability might be incurred, simply on the basis of the findings of some previous complicated legal case history. Furthermore, legal rulings vary in different countries and in some areas decisions may be biased against the foreigner. The comments in this article apply to the situation in the British Isles and do not necessarily hold good abroad.

The majority of engineers are concerned with the provisions of the Factories Act and kindred legislation which embody definite regulations, absolute in their context. Even if an accident is not suffered, failure to comply with such provisions may lay the offender open to possible legal penalities which, of course, cannot be covered by insurance. If, by infringement, an accident does occur, any injured party may take action under Common Law against the person held legally responsible. Except in the case of culpable action of the latter however, any damages recoverable, together with costs incurred, may be reimbursed under the terms of an insurance policy.

There are two main aspects of the law of tort which affect the engineer in his relation with his fellow men; those of nuisance and of negligence. The interpretation of 'nuisance' frequently involves the unlawful interference with a person's use or enjoyment of land or his rights thereon and the Courts may be expected to interpret any claim for damage arising from this in its absolute sense. If, for example, a manufacturer uses on his land a pressure vessel equipped with a bursting disc which operates in an emergency and thereby prevents explosion, but in so doing discharges noxious products about the neighbourhood, then, although the owner may have taken all reasonable precautions regarding the safety of the plant, he may still be liable for losses suffered due to the consequences of the event.

Apart from absolute liability arising from the use of land, claims for indemnity are most often based on negligence and the award of the court will be assessed on the extent to which this is considered as attributable to the defendant, relative to possible contributory negligence on the part of the plaintiff. It is recognised that society must progress and that, consequently, use must be made of objects which involve latent dangers; it would therefore be illogical to make the manufacturer, retailer or owner of such objects liable for every accident which might arise from them. Any automobile constitutes a danger in itself because of the possibility of accident consequent on failure, for instance, of the steering gear. If it can be proved that the owner has maintained the car in adequate running order and that the failure occurred due to hidden defects which could not be reasonably detected; and if he were driving carefully at the time, he would rarely be deemed liable for the results of the accident. On the other hand, if it were proved that he knew that a defect existed in the steering gear prior to the accident, he would almost certainly be held responsible. Except where the absolute effects of statutory legislation apply, the same broad principle of responsibility for negligence applies for all losses sustained by third parties.

In this connection the engineer should remember that a minor omission on his part may involve major liability. The case may be cited in which a ship was destroyed by fire due to a spark originating from the fall of part of a hatch cover. It was argued that the disaster was due to accidental fire but ultimately, the stevedores were held to blame because of their negligence in not ensuring that the hatch cover was properly secured.

Again, third-party reliability may arise to the ultimate user of products which could not reasonably be inspected by the original purchaser.

Employer's liability

The engineer is, by reason of his position, particularly concerned with the matter of employer's liability which involves the legal relationship between master and servant. Basically the employer owes the same duty to avoid injury to his servant as he does to any other citizen. But his responsibilities are more extensive and may be summarised as:

1. the engagement of a competent staff;
2. the employment of safe apparatus and material;
3. the provision of a safe system of work, together with adequate supervision.

Furthermore, an employer is responsible for the acts of his servants arising out of their duties between themselves and as affecting other members of the public. He is subject to the absolute provisions of Statute Law (ie; laws laid down by Parliament) to avoid accidents taking place; and to Common Law (the traditional laws as upheld by the courts), which ensures that a servant may obtain adequate compensation if an accident does occur, subject to the court's ruling the employer to be at fault.

The terms of Statute Law are absolute. For instance, under the provisions of the Factories Act, plant used must be of sound construction and if, for instance, a crane fails (Fig. 1) and causes an accident, employer's liability cannot

be avoided by putting forward a defence that it had been
subjected to every available test to ensure safety. Such
absolute duty may also extend to the system of working.
For example, under the Coal Mines Act it is laid down that
mine owners, unless they possess the necessary technical
knowledge, must not assume the duties of technical management; a competent mine manager must be appointed. In
one incident resulting in legal action this had been done but,
due to a fault in administration on the part of the manager,
an accident occurred. In spite of the fact that the owners
had fulfilled their obligations under the Coal Mines Act
by appointing a qualified manager, they were still held
liable for the accident on the grounds that their duty to
provide a safe system of work for their servants was absolute
and could not be entrusted to another person, no matter
how competent.

The engineer is particularly affected by the law of contract
in the matter of warranties, conditions of sale, specification of
performance and dates of contract delivery. It is hardly
necessary to remark that the utmost care should be taken in
the wording of the relevant documents (on which legal
advice is advisable) for, if any action arises from them, the
courts may be expected to base their judgment on a strict
construction, irrespective of what the contracting parties
actually intended to imply.

A further matter in which the engineer may have a vital
interest is that of his relations to a client in a professional
capacity. Claims for compensation for losses suffered by
professional negligence have of late become increasingly
common. Such claims must be sustained under the law of
contract in contrast to losses from accidents which are
subject to Statute Law or to the law of tort. The law's
attitude to professional negligence is flexible and it does not
maintain that the judgments of a professional man must be
infallible. It is summed up in a ruling given many years ago:

> Every person who enters into a learned profession undertakes to
> bring to the exercise of it a reasonable degree of care and skill. There
> may be persons who have higher education and greater advantages than
> he has but he undertakes to bring a fair, reasonable and competent
> degree of skill.

Until a recent ruling a professional man was only liable
to his client but in future third parties may be able to claim
against him for negligence.

The function of insurance

Nowadays insurance in some form or other impinges on
the life of every citizen but it must be acknowledged that
even its fundamentals are little understood. An old and
frequently quoted definition is 'a means of ensuring an
equitable distribution of losses so that the loss falleth lightly
upon the many rather than heavily upon the few'. This is
perfectly true but its ramifications go much beyond this.

One of the most important functions of the insurance
industry is to permit wealth to be used for the development
of society rather than being kept uselessly in reserve to meet
possible losses. If every owner or person built up an individual
fund for such purposes the resultant locked up capital
would be enormous and industrial growth would be greatly
retarded.

If technical development could suddenly be halted, it
would be possible to arrive at an exact analysis of the daily
losses suffered by the community and to establish the correct
insurance premium accordingly. Such normal losses do in
fact constitute the basic layer of any insurance premium but
superimposed on this will be a second layer relating to the
unknown risks associated with developments. Since insur-

Fig. 2. The disruption of this 30 MW turbo-alternator resulted in damage
amounting to approximately £250,000. It was covered by an engineering
machinery breakdown policy

ance expenses form part of a product selling price, this results
in the whole community sharing in the costs of progress.

A third function of insurance is the reduction of losses by
the compilation of case histories and the employment of
experts in specialised fields who are able to give advice and
by so doing prevent the repetition of past errors. An example
is a generator explosion shown in Fig. 2. The investigation
revealed that the disaster was due to structural miscalcula-
tions and to new concepts which had never previously been
applied. The knowledge gained from this investigation will
certainly prevent similar disasters occurring in the future.
The price of such services forms the third stratum in the
average insurance premium paid.

The cost of insurance for any industrial risk is built up
from equitable premiums for a series of conventional
hazards such as fire, employer's liability, products liability
and other classes of risk. On each of these a team of experts
have spent a lifetime of specialisation. Individual policies
can be framed to cover one or more risks but the principle of
the summation of equitable premiums for each class remains
the same.

Table 1 lists the types of policy of interest to engineers.

Obligations

The community has to pay a daily toll for accidents and
losses in connection with the plant on which its progress
depends and which more often than not originate in faulty
planning and construction decisions. After a disaster,
the conscientious engineer will be tormented by such
questions as whether he had been justified in developing
a new design which, in the event, proved faulty; whether he
had been morally right in economising on features which
would have rendered the plant uneconomical but would
have improved its safety.

The matter as a whole involves the integrity of the engineer
as a citizen. If he is pressed to design a plant which, in his
expert judgment, incorporates features infringing fundamental safety principles, it is his clear duty to refuse to do
this, whatever the personal consequences may be. But every

Table 1—Principal Types of Policy for Engineers

1. The fire policy.
2. Engineering policies covering accidents to boilers and vessels and machinery breakdown (may include damage to owners of other property, third party liability etc.), and which are unique in providing an inspection service by engineering experts. Fig. 3 illustrates a typical accident for which compensation is provided by such a policy.
3. Consequential loss policies covering loss of profits due to 1 and 2 above. One glance at Fig. 4 will show that effects of an accident is not necessarily confined to the plant to which it occurs.
4. Erection or contractors all risks policies. (Accidents due to the erection and dismantling and as in 1 and 2).
5. Policies covering the risks of transport.
6. Policies providing indemnification for losses incurred under makers' guarantees.
7. Employers liability policies.
8. Public liability policies.
9. Products liability policies.
10. Policies covering special perils, including the so-called conventional risks of flood, tempest, earthquake etc.
11. Contractual liability policies (fines under penalty clauses etc.).
12. Policies covering accidents to employees whilst travelling in the course of their duties.
13. Policies covering liability for professional negligence.

plant involves some hazard and the engineer must decide the degree of risk he thinks his fellow citizens should be prepared to accept. In most cases he can take practical steps in estimating the possible effects of a catastrophe and endeavour to ensure that at least financial indemnification will be available to the victims through adequate insurance coverage.

The engineer is normally in a key position in that he knows if any hidden unconventional features are involved in the construction or use of the plant under consideration; but the final analysis of the risk should be made through group consultation with insurance underwriters and the broker.

The insurance underwriters have the duty not only of quoting an equitable premium but also that of giving advice on the individual hazards on which they have specialised. A broker's function is to ensure that all aspects of the risk are covered and that the various policies dovetail together and do not duplicate cover, resulting in the payment of unnecessary premiums.

One of the most important objects of this group consultation is the elementary duty to society to prevent accidents occurring, for it is the community as a whole which ultimately has to sustain the physical and financial burdens of these. Furthermore, no insurance can protect the industrialist against the indirect effects of accidents, such as loss of goodwill through delayed orders, bad publicity and staff relationships.

The correct apportionment of the amount of cover required for any industrial risk is not a matter which can be decided easily and expert judgment is necessary. For instance, consider the explosion of a small vessel situated in the centre of an important chemical works, which might cause serious damage to adjacent gas containers with consequent widespread devastation to the adjacent countryside (Fig. 5). It would be quite wrong to arrange insurance cover solely on the replacement cost of the vessel, disregarding the immensely more important secondary dangers, which might mean ruin for the engineer's employer and also for members of the general public in the event of no funds being available for compensation.

Similar considerations arise in relation to consequential loss and personal accident, both of which entail obligations to society. The emphasis in the former is usually on material loss, but an employer may be faced with grave difficulties if, owing to failure to insure for the payment of salaries, he

Fig. 3. Multiple damage occasioned by the explosion of two economisers at a power station. Such accidents are covered by a boiler policy

Fig. 4. Destruction at a hydroelectric station due to the overspeed and failure of a turbine. Insurance cover for the outage losses by a consequential loss policy of this kind

Mr J. Eyers *served his engineering apprenticeship at Scotts' Shipbuilding and Engineering Co. and, after graduating, was employed by the Canadian Pacific Steamship Co. as a Marine Engineer for 5 years. After obtaining a Ministry of Transport extra chief engineer's certificate, he was appointed to the engineering staff of the Vulcan Boiler and General Insurance Co. in 1933. During the 1939–45 war he was commissioned in the Royal Engineers and was responsible for the conversion of merchant vessels into port repair ships and later the mechanical rehabilitation of the Middle East Railways. Subsequently, he rejoined his Company and became its Chief Boiler Engineer in 1948. Later he was entrusted with the development of its overseas connections and the work associated with nuclear developments. Mr Eyers has collaborated in the work of several rule-forming committees.*

loses key staff; to whom, from the ethical standpoint, he may indeed be under an obligation for past faithful service. Similarly, for employees who have to travel extensively, it is surely incumbent upon the employer to arrange adequate insurance cover in the case of accident, a provision which will improve the relationship between master and servant.

Relations with the underwriter

In 1552 the eminent Portuguese jurist Santarem, in his treatise on insurance, formulated two principles which are fundamental to the insurance industry. These are that it shall not be the means for the insured to get rich but simply to avoid loss; and that it shall be a contract made in good faith.

From the first it follows that the insured's post-accident condition must be no better than it was before the event. Any property damaged (unless insured on a reinstatement 'as new' basis, for which an extra premium will be required) will be restored to its pre-accident condition or corresponding indemnity paid. The owner must not expect the costs for improvements or maintenance repairs to be included in the settlement. There is an obvious loophole here for inflated claims which, if successful, would eventually result in a general increase of insurance premiums. In other words, the originator of an augmented claim is endeavouring to rob not only his insurance company but also his fellow citizens. Fortunately, at any rate in Britain, which is perhaps the most insurance-conscious country in the world, fraudulent claims are comparatively rare.

It is, nevertheless, through the implications of Santarem's second principle that the engineer will find himself most closely involved with ethical aspects. Any insurance contract is based on the details contained in the proposal form. Good faith requires that full particulars of the risk covered should be given in this document and the proposer should detail therein all matters having a bearing on the risk, even though he may know that, as a result, the underwriter may refuse to insure. The underwriter accepts that errors in judgment can lead to accidents and subsequent claims will be paid without question or reproach. But his signature on the insurance contract implies his implicit trust in the integrity of the proposer as a citizen. This is indeed the foundation stone of the insurance industry.

The integrity required by the 'good faith' principle extends far beyond the insured and his underwriter, for the latter has a relationship to his own reinsurers and he is morally bound to charge the equitable premium which, in his expert judgment, is warranted by the hazards involved. It would be quite wrong for him to gamble with fortune and, for the sake of obtaining the business, to charge too low a premium.

It has been mentioned that one of the functions of insurance is to aid the progress of society by underwriting the unknown hazard, and indeed any legitimate risk must in theory be insurable, even though, in practice, the premium may be so high as to render some risks virtually uninsurable. In the engineering industry this is nowadays of great importance since many projects involving new concepts are of such magnitude that they are, in effect, prototypes which can only be tested when put into actual service; in contrast to the situation prevalent a few years ago when most plant could be tested in the shops, prior to delivery. Furthermore, the engineer has no justification for complaint if he is charged a higher premium than normal due to the fact that he has elected only to insure certain selected risks, a practice of which the ethics is open to question.

The most important aspect of 'good faith' concerns the possibility of loss of life or injury. How far is the engineer justified in taking unknown risks involving the safety of his staff and the public, when balanced against the general benefit of the community through reduced costs of his product or improved amenities? in other words, what is the price of human life? A partial answer to this complex problem is that certainly no engineer is right in dispensing with any measure involving the safety of his fellows simply by evaluating its cost against that of an increased insurance premium. The engineer is morally bound to make sure that any action of his is in the overall public interest and he must conscientiously present all the facts for consideration by his underwriter who is, in effect, the representative of society. In his turn, the underwriter must consider the implications involved and if he considers these to be against public interest he should, without question, refuse to be associated with the risk.

The inspection service

So far we have dealt with the relationship of the engineer to insurance in general. This article would, however, be incomplete without reference to some of the functions of the

Fig. 5. Damage amounting to over £100,000 was occasioned by the failure of a small sight glass on a chemical pressure vessel

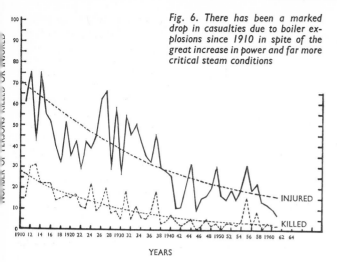

Fig. 6. There has been a marked drop in casualties due to boiler explosions since 1910 in spite of the great increase in power and far more critical steam conditions

British specialist engineering insurance companies which are sometimes not clearly appreciated.

Great Britain was the first country in the world to develop engineering inspection and insurance and to date still remains one of the few countries where this is administered by specialist companies, and not as a comparatively neglected branch of the general insurance concerns. The business was initiated in Manchester in the 1850's with the inspection and insurance of steam boilers and the principal object was the introduction of an inspection service to reduce the severe loss of life and damage to property occasioned by the explosion of steam generators which was then prevalent.

As an inducement to owners to have their plant inspected by independent experts, an insurance guarantee was given as an extension to the basic inspection contract. The business was a success from the start, in that the number of disasters diminished rapidly, a trend which has continued to the present day, as shown in Fig. 6. As time passed the scope of guarantees was extended to include all types of machinery and engineering contracts. Recognition was given to the engineering insurance inspection specialists as being competent persons for the duties of examining plant and for the compilation of the various official reports necessary to show that plant has been maintained in working order. These reports have to be retained for scrutiny by the Government inspectorate.

As a result, the British engineering insurance industry has had to engage a large team of specialists, trained in the art of inspection, backed by a body of underwriters who themselves have normally graduated from the practical inspection field. The latter must be qualified to appreciate all types of engineering problems and to discuss these on an equal footing with engineers in industry anywhere in the world.

In addition, this service has had to be provided with laboratories, staffed by metallurgists and chemists, for it is essential that the cause of any breakdown (however seemingly insignificant), be established in order to prevent a possible future catastrophe. All this has resulted in the specialist engineering insurance concerns having at their disposal a unique background of accident case histories and it is therefore not surprising that they play an important role on the various regulating committees.

The British industrialist has learned the value of independent inspection (whether required by legislation or not) as

a measure towards preventing unnecessary breakdowns. For if an accident does occur, no matter how well he may be insured, he will suffer indirect harmful effects. This fact is at length coming to be appreciated on the Continent where, until lately, independent inspection has been looked upon with distrust with the result that engineering breakdown claims and the costs are far higher than here.

The British system of engineering inspection and insurance imposes a heavy burden of responsibility on the certifying engineer for, if a disaster occurs, resulting in an official investigation, he must be in a position to justify the report bearing his signature. Any error or misjudgment on his part may be followed not only by financial penalties but by the ruin of his reputation. His fundamental duty is to ensure, as far as humanly possible, that plant will be safe.

Nevertheless, he must not restrict industry by unnecessary stipulations and, above all, the financial aspect must not be allowed to affect in the slightest degree the moral issues involved in certification.

Legislative requirements are absolute on such matters as periods of working between inspections, safety fittings and the like, but no fixed rules are laid down concerning design and construction of plant. This is left to the engineering judgment of the inspectorate, which results in flexibility. But in comparison with his Continental colleagues, the duties of the British certification engineer are far more onerous for the former may merely have to adhere to the code of design rules legally applicable in his country.

The charge is sometimes levied against British engineering insurers by the ill-formed that their methods are too restrictive and thereby impose a brake on the British manufacturer, particularly in relation to Continental competition. The truth is the reverse of this. The British system of plant certification permits the inspecting engineer to accept plant constructed in accordance with any recognised code of rules, which is certainly not the case in many other countries. An instance of this concerns the agreement by British engineering insurers to accept steam generators constructed in this country or abroad in accordance with the latest proposals of the ISO TC11 Code, an action followed so far only by France. Again, plant made abroad to an accepted code and examined during construction by the officially recognised inspecting body of the exporting country, is normally automatically accepted for insurance in this country; whereas certain Continental countries insist on plant exported from Great Britain being examined during construction by their own engineers.

The British engineering insurance organisations have traversed a long and difficult road but they now play a vital part in engineering, not only in this country but abroad. One of the most significant instances of this has been their role in the immense nuclear power projects in Britain and elsewhere. The severe losses which may have to be faced consequent on the breakdown of one of these plants or of ancillary installations, such as fuel fabrication plants, has necessitated the information of a British insurance pool in which the whole of the insurance industry participates and which effects reinsurance in connection with nuclear risks situated in many other countries. It is essential that each risk be technically analysed and reported upon by experts provided largely by the specialist engineering insurance companies.

It is difficult to envisage what new problems the British engineering insurance and inspection companies will have to face in the future; it is certain that there will be no respite and that their complexity will multiply as the years go by.

The creative qualities of engineering design have recently been the subject of critical enquiry both inside and outside the profession and the industry. It is pertinent, therefore, to examine the connections between engineering and the visual arts and to see how strong they have been, how strong they are today, and whether they amount to engineering being itself an art.

What the Artist Can Teach Us

by Fred Ashford, FSIA, Hon. Des. RCA

Fig. 1. 'Rockdrill' by Jacob Epstein: the body is seen as an assembly of engineered components

The question of whether engineering is an art or a science can be answered in two ways. By simple definition it is indisputably an art. The O.E.D. defines science as a knowledge of principles; art as skill in applying these principles in practice. The other approach to the question is: can we consider the visual manifestation of the engineer's work as art?

An essential component of art is style and in the appearance of the majority of engineering work there is often very little evidence of it. It may well be there, but concealed from view, like the gears in a gearbox. But in aesthetics we are concerned principally with what we perceive, so that if we use the second approach to the question the answer must be that what we see of engineering, in many cases, does not rise to the level of art. This is a pity because the skill, amounting to art, may be there but for practical reasons must be concealed. Only too often one comes across the most magnificent examples of mechanical engineering design but concealed behind covers of hideously insensitive design, often carrying controls and information which are unpleasingly and inconveniently arranged.

Here we have the heart of the matter. The satisfaction of mechanical function can often produce forms of great visual quality, for engineering is generally based on clean, geometric shapes, with necessarily high precision and surface quality. The engineer's responsibility for aesthetic quality extends, however, to the total presentation of his work and it is here, in areas usually less rigidly prescribed by mechanical function, that we can gauge the true measure of his artistry. A small proportion of engineering work achieves great beauty, but the mass of it does not: it is as though the art had gone out of it.

Beauty through function?

Visual coherence, by which is meant the relationship of the parts to the whole, is usually lacking, as are nice proportion and detailing. These criteria could be applied to a piece of sculpture; and what is a machine other than a piece of metal sculpture, in so far as aesthetics is concerned? It need not be capable of working at all, let alone efficiently, for us to evaluate it as an object. If good performance and appearance can be combined, so much the better; in the really satisfying examples of engineering design they obviously must be.

But confusion as to what are the aesthetic criteria applicable to engineering design exist in many people's minds, notably in those of the beauty-through-function protagonists.

The confusion was crystallized in a letter to *The Chartered Mechanical Engineer* claiming functional efficiency as the final criterion of any engineering product and that the product could be styled to suit current fashion. The implication being that the satisfaction of functional efficiency automatically guaranteed aesthetic quality which could be kept in line with current fashion by superficial changes. This view reflected the two fallacies of the Feilden Committee Report; that good engineering design in the mechanical sense would automatically provide good appearance and amenity; and that appearance and amenity can be divorced from function and regarded as a separate issue. These are wrong views and illogical ones.

Engineers generally appear to have lost the capacity to think liberally about aesthetics in connection with their work. They either favour the beauty-through-function theory or they concede that something superficial might be added to their work. The truth is that aesthetic quality and efficient function must be consciously considered together. It is necessary to think of what we are actually going to see, which will condition our aesthetic response, at the same time as we consider whether a machine will function efficiently.

Since we are concerned with objects of utility, some compromise may be necessary in order to achieve the optimum balance of qualities desired.

It is curious that this liberal and realistic view should find greater expression outside the engineering profession than within. This suggests that engineering thinking is becoming too specialized and inbred; that engineers have become indifferent to the visual arts and to the freer modes of thought associated with them. This has not always been the case.

Evidence from the past

It is fashionable to cite Leonardo da Vinci as sufficient proof of the unity of art and engineering. True, he gave graphic expression to many principles of mechanics and he invented many machines, although most of the latter seem to have remained on paper as unproved projects. The all-up weight of his flying machines would certainly have prevented any of them from leaving the ground, while his various engines of war would probably have inflicted heavier casualties upon their operators than upon their enemies.

While Leonardo was a great artist, his major contribution to engineering was possibly his uncompromising search for truth at a time when this was not popular, and the development of the technique of scientific enquiry. The unfortunate delay until the 19th century of the publication of his notebooks reduced the influence which he might have had upon the development of engineering and there is little evidence of any actual achievement by him of aesthetic quality in engineering.

We might look equally well to Cellini who was also a great artist and who certainly carried out many civil and military engineering projects. But we must look elsewhere than the Italian Renaissance for evidence of the true merging of art and engineering.

The 15th century provided examples of wind and water mills and firearms: later there were vehicles and some machines, but it is difficult to see in them any valid connection between art and engineering. Such examples were based upon craftwork, which might have been extended to heavy blacksmithing, but no more. It was not until the late 18th or possibly the 19th century that engineering emerged as something distinct from craftwork. Machine tools had been developed, black hand-wrought iron gave place to bright turned cast iron and brass, later steel. Wooden structural members gave place to cast iron but the ease of decorating the latter led to the incorporation of architectural and organic forms: still not any real evidence of art in engineering: today we should call it superficial styling.

It is not, in my view, until the late 19th century that we can see any influence of art in engineering, in that aesthetic quality became complementary to the satisfaction of engineering requirements. We can see it in much of the work of the great Victorian engineers, who were at their best when they were experimenting with new materials and methods of construction: at their worst when they were borrowing idioms from the repertoire of classical architecture, or introducing organic forms belonging to painting and sculpture of the time. Structures like the Crystal Palace, Paddington Station, the Saltash, Menai and Forth Bridges all exhibit aesthetic qualities arising directly from engineering requirements. They are not qualities borrowed from craftwork or classical architecture: we can see the same qualities in later locomotives, ships and some machines. We see in every case evidence of an overall control: we get the feeling that the designer was evolving his own art forms to suit the situation, not imposing visual characteristics arising from some other situation. We see aesthetic quality: the product of visual coherence, good proportion and sensitive detailing.

What is aesthetic quality?

It is not possible within the scope of this article fully to examine aesthetic quality in engineering: it is reasonable to accept that it exists. Aesthetic appreciation consists in reactions to certain stimuli: with engineering the stimuli are principally visual, although there may also be tactile and audio stimuli. In a machine, for example, not only its general form but the arrangement of every cover, nut and handwheel contribute to our reaction; these are the things we see and experience.

It is not unreasonable to suppose that, while leaving adequate room for the expression of personal taste, there exist some broad bases of agreement upon what are pleasing or unpleasing arrangements of the stimuli. This is the case with other visual arts, music or literature. For example, we may individually prefer Brahms, Mozart or Beethoven, but would agree that the works of all three are great music.

We may not always be able to lay down rigid rules or principles of aesthetic design but those who have talent and a knowledge of visual art tend to produce results which please the majority of mankind; and this, in the long term, is the criterion of all art. The diehard dialectician demanding measurement of aesthetic quality—so often heard in discussions —misses the point: if art were a science it would not be art.

The visual expression of many natural laws and mathematical formulae can be very pleasing, for instance the three-dimensional graphs in the Science Museum. But as yet, there is no formula for beauty.

On the other hand, many highly efficient expressions of engineering principles can be aesthetically displeasing. Few designs are in fact dictated completely by the function of the product: there is usually an element of choice between solutions of equal engineering merit and, therefore, the opportunity for some aesthetic decision by the designer. It is this which determines what we actually see.

Mutual influences

There is ample proof in the examples quoted, and others, that aesthetics and engineering are not irreconcilable, that there are factors in common and that they can and do influence each other. There is much evidence of engineering influencing the visual arts, beginning perhaps with the purely emotional effect evident in Turner's painting *Rain, Steam and Speed*, where a railway engine is part of the mood, rather than the subject, and continuing with a more profound influence upon modern art movements such as evidenced in Fig. 1.

The nostalgic romanticism of the Pre-Raphaelites provoked a revolt by such movements as Suprematism, De Stijl, Purism and Constructivism, which have their present day counterparts. They found in engineering a basis for expressing the fundamental principles of structure and movement. Not only does much of their painting and sculpture reflect general principles of statics and dynamics, but many works actually incorporate engineering components such as gear wheels, nuts and bolts and even mechanical movements. One look at Epstein's sculpture *Rockdrill* in Fig. 1 reveals the strength of this influence. Here is Man, seen entirely in terms of engineering components, as a completely mechanical concept.

An even clearer example is provided by the *Crankshaft* sculpture in Fig. 2, of which the very subject is an engineering component.

The influence of art upon engineering is equally strong. Fig. 3 shows that in many modern machines, computers and even domestic boilers we can see the indirect influence of the paintings and constructions of the artist Mondrian, who was preoccupied with geometrical abstraction involving formal structures of lines and logically arranged areas of primary colours.

Looking at many modern engineering products we can see the same geometric rather than sculptural forms; the same modular division of space; the same use of small areas of strong colour, often to serve some ergonomic as well as decorative purpose. Engineering designers can clearly learn a great deal from art.

It is significant however that conscious application to engineering design of the principles developed by artists like Mondrian and Brancusi has been principally due to industrial designers, rather than engineers. Not only is the work of today's engineers seldom directly influenced by that of painters and sculptors, but it seems rarely to achieve the full potential visual quality inherent in engineering function itself. At some period during the past 100 years a schism developed between art and engineering. Why should this be so?

The schism

It is sometimes suggested that if the engineer lacks interest in the visual arts it is due to the visually impoverished surroundings in which he works. This theory can have little foundation now that the 'dark satanic mills' are largely a thing of the past. A more fundamental reason for the schism between engineering and art can be traced back to the events following the Industrial Revolution, when scientific method and mechanization displaced intuitive judgement and craft skill.

The colleges of art set up after the 1851 exhibition to preserve aesthetic quality and guide industry towards better design and manufacture, soon relapsed into colleges of fine art, mainly for recreational and teacher training activities. One such establishment, the Royal College of Art, was not brought back to anything like its intended role until 1947. The colleges themselves were probably most to blame for this defection. It was more genteel and no doubt less arduous to provide instruction in figure drawing and studio pottery than to grapple with the more difficult and less understood problems of designing for commercial materials and manufacturing processes.

The technical colleges, on the other hand, had literally no time for art, and were increasingly concerned with scientific matters. Thus engineering and art drifted apart.

The intensification of the analytical and scientific aspects of engineering widened a gap which our system of general education has done nothing to close. The rigid division of young people in their formative years into artists and non-artists—the vast majority in the latter category—is still the general rule. The comprehensive school and recent changes in art education and recruitment have so far done little to

remove this artificial classification which in many people inhibits any expression of intuitive aesthetic feeling: it creates the idea that artists are a race apart.

The truth is that everyone has some capacity for aesthetic appreciation: indeed, working closely with engineers, one senses a greater understanding of form and aesthetic aims than one encounters in others whose background and training might be expected to produce greater rapport. It is not, however, an understanding which one frequently sees applied in their professional work.

The apparent gulf between engineers and artists is a mental, rather than a physical one. Many designers in industry work in close contact with production, and modern conditions often put engineers and creative artists into similar environments. It is true that many engineering works are situated in prosaic surroundings, and that the architecture and amenities of a number of older establishments, especially in the heavy industries, are unsatisfying. But many light engineering buildings provide an environment which, if not completely satisfying, is certainly no more sterile than that of other working, and even educational, premises. Fig. 4 shows part of a new steelworks where colour and form contribute to provide visual stimulation.

Engineering works, both old and new, frequently offer more visual stimulation than the average office or home: machine shops, for example, can offer a wealth of colour and texture in arrays of finished or part-finished components. Seen in some other environment, say a market-place or some natural setting, an equivalent display of colour and texture would stir the aesthetic consciousness of the average engineer. Why should it not stimulate him in the same way in his work?

Familiarity, the commonplace, an automatic association with work instead of pleasure; all these factors no doubt reduce the impact of the visual stimulus which is quite strong in most engineering environments. For example, areas where castings are weathered can exhibit the most fascinating displays of rust colours, ranging from bright orange to black. Foundries and forge-shops offer many visually exciting forms, some of which would stand exhibiting in their own right. They resemble works of modern art for the simple reason that the artist sees visual drama and interest in such

Fig. 2. 'Crankshaft' a sculpture by F. E. McWilliam. Like Fig. 1 it shows the strong influence of engineering upon modern art

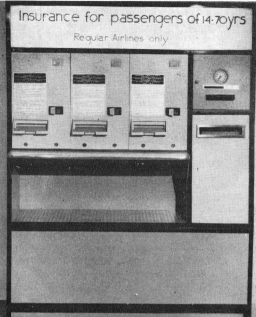

Fig. 3. On the left Mondrian's composition in red, yellow and blue. The indirect influence of his style can be traced in a vending machine, a boiler and many other engineering products of today

forms and consciously or sub-consciously seeks to reproduce them.

This hardly suggests a visually impoverished environment for engineers. Again, many products, such as electrical sub-assemblies, particularly those based on printed circuits, consist of the most exquisite formal arrangements of linear elements and colours: effects which many artists would be very satisfied to achieve. Steel works, rolling mills and construction shops all offer a great deal of visual interest and even drama.

Older works are not without visual interest. Part of an instrument factory has an area of wooden flooring with the remaining patches of a composition covering. Into these smooth, brownish ovoid areas are embedded, in delightful random arrangement, innumerable brass screws, gear wheels and other small parts. The whole floor being polished smooth by years of traffic, produces an enchanting effect which many artists strive consciously to create. Similar chance aesthetic effects can be found in many other old works, while newer ones, with planned colour and lighting, and equipment which is itself visually interesting, seem to dispose entirely of the aesthetically sterile environment theory.

A more valid explanation for the engineer's apparent indifference to aesthetic values is possibly to be found in his aesthetically impoverished training.

Aesthetics and education

What can be done about this dichotomy between engineers and artists, who clearly have a great deal in common? The possibilities of broadening the scope of engineering training and the difficulties attendant upon this have been well aired in recent years. The already crowded curricula and the increase in specialized technical subjects make the introduction of any worth while aesthetic studies difficult. A further difficulty lies in finding suitably qualified staff to supervise such studies. One of the many things revealed by the recent reorganization of industrial design training and reassessment of courses to meet the requirements of the new 'Diploma in Art and Design' qualification was a serious shortage of properly qualified teachers, particularly in the field of engineering design. The solution of the problem is therefore not simple, but more could be done.

Time is available for general studies but in very few colleges is any significant amount of this devoted to the appreciation of art and visual design. For example, in courses leading to the Dip. Tech., 20 per cent of the time should be devoted to non-technical studies. Given better co-operation between colleges of technology and art, or architecture, very useful courses of design appreciation could be jointly developed.

Their usefulness would be increased if they were the continuation of a wider aesthetic training beginning in secondary and even primary schools. At present there are no discernible links between the teaching of art and handicrafts in these schools, let alone between art and mathematics or scientific subjects. The graphic elements of many technical subjects could quite properly be used to show art and design as positive, creative elements of many activities. Such studies could lead to the development of a widely based visual awareness and aesthetic sensibility at a time when these capacities could contribute much to the pupil's understanding in other fields.

The few design appreciation courses at present being offered usually provide the first hint to engineers and design draftsmen in their thirties and forties of any connections between aesthetics and engineering. The earlier in life this

Fig. 4. Part of a new steelworks: colour and clean architectural design combine to produce a visually exciting environment

realization can come, the greater its value and its contribution to an improvement in the quality of engineering design.

It is difficult to suggest in detail how engineering training could be broadened but it is certain that it can be done, once the value of aesthetic awareness is appreciated. Some colleges have made a start. The Rugby College of Engineering Technology introduces third year Dip. Tech. students to an appreciation of the general principles of aesthetics and ergonomics, which they are able to apply to an engineering design in their fourth year. The colleges of art and technology at Leicester have arranged courses in aesthetics and engineering for 2nd year HNC students. At Salford the art school and the technical college have arranged courses forming an integral part of the engineering syllabus. Other institutions, for instance Imperial College, are including lectures on the various aspects of industrial design in their general studies course.

These are small beginnings but we can expect to see them grow fairly rapidly until an appreciation of the aesthetic aspects of his work is part of every engineer's training.

Whether the engineer will eventually become self-sufficient in aesthetic judgement is a matter for speculation. Technology tends to grow more complex and more technical specialization seems inevitable. The da Vinci type 'whole man' is unlikely to become more common, so that we cannot look to him for a general improvement of appearance, comfort and convenience in engineering products. There will certainly always be outstanding individuals with intuitive or acquired sensibility, capable of engineering work which could be considered as

art. But they are likely to remain a minority and organizational difficulties may prevent many of them from controlling the design in sufficient detail to ensure success.

Industrial designers fill the gap

At present the gap between art and engineering is being filled by industrial designers who may, or may not, eventually be rendered redundant. Only time can tell, but the mere existence of aesthetically conscious engineers would by no means solve the whole problem. Engineering design today is generally the work of groups, rather than individuals, and this group activity seems likely to extend. The contribution of individual members of the group can introduce visual problems which must be solved and their solutions coordinated to produce a unified result. It seems unlikely that many engineers would have the time to undertake the overall control of the visual and ergonomic aspects of the design, as well as to discharge their own specialized technical functions.

The very existence of industrial design specialists, working on a thoroughly practical and economic basis, seems to prove that the aesthetic and amenity aspects of design are a full-time job for somebody, whoever that might be. A properly trained specialist seems to be a reasonable answer to the problem and one which is gaining wider acceptance here and in many other countries. Whereas in 1960 there were fewer than ten companies making capital goods or equipment which employed industrial designers, the Council of Industrial Design estimates in 1965 that there were over seventy which used their services.

Many senior practitioners of industrial design in the engineering field are qualified by experience only, for the simple reason that no formal courses of training were available until 1939.

In 1965 there were four-year courses in industrial design (engineering) leading to the National Diploma in Art and Design, a first degree level qualification. The Royal College of Art offered a postgraduate course of three years and opportunity of further study in the form of fellowships, which were also available at Manchester.

These courses cover aesthetics, the appreciation of fine art and the history of art as a general background. Technical subjects include engineering drawing and workshop practice; engineering theory; the strength of materials; statics and dynamics; materials technology and manufacturing processes. Other studies cover design analysis; problem solving techniques; the analysis and presentation of information; ergonomics and the social aspects of design. As the course develops, proportionately more time is spent on design projects which can involve direct contacts with industry and which also often result in the production of working prototypes.

A wider knowledge of the content and the quality of industrial design courses would do much to dispel any suspicions which are due to their being held in colleges of art. Proper qualifications must of course depend on postgraduate study and practical experience, as in any other profession.

A qualified industrial designer would almost certainly be a member of the Society of Industrial Artists and Designers.

Some facilities for part-time training of design engineers, design draftsmen and others have already been indicated. In addition there are courses in design appreciation for executive engineers and engineering designers, held twice yearly by the Council of Industrial Design. Subjects include

Mr F. C. Ashford, *F.S.I.A., Hon. Des. R.C.A., began industrial design work in 1935. He joined the staff of Raymond Loewy in 1938 and at the outbreak of war he joined the R.N.V.R., serving in the Coastal Forces and later in H.M.S. 'Excellent' he worked on gunnery trials and development. After the war he went into private practice as Scott-Ashford Associates until 1955 when he became a Reader in Industrial Design (Engineering) at the Royal College of Art, establishing the department of that name. Since 1959 he has had his own practice, principally concerned with the design of capital goods. Mr Ashford is the author of the book 'Designing for Industry' and has contributed to 'Design' and other journals. He has recently lectured on industrial design and engineering for the Council of Industrial Design in Poland and Russia.*

ergonomics; aesthetic factors in engineering design; problems of form, colour and style. This kind of course could certainly be duplicated with advantage, possibly at evening classes, to make them available to larger numbers of students.

The formal part of the COID course covers five days and, brief as this is, there is abundant evidence that it is filling a real need. It has been the starting point in the improvement in the design organization of a number of principal engineering concerns and the establishment, for the first time, of recognizable design policies.

Intelligent co-operation

In the relationship of engineer and artist there is no partisan issue at stake. The perpetuation of differences based on bigotry or brashness, evident still at most conferences, or in forums where art and engineering are the subjects for discussion, is tedious and pointless. It is a fact that there are aspects of engineering demanding artistic ability; indeed,

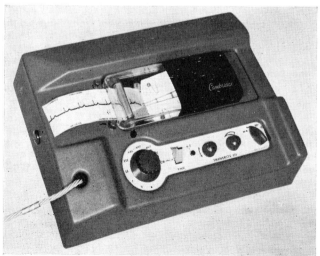

[Courtesy: Council of Industrial Design]

Fig. 5. In this electrocardiograph function, structure, and ergonomic requirements are co-ordinated and the result visually satisfying

without certain human qualities which art can give to it, engineering may be inadequate functionally, for even in this age of automation the ultimate user is likely to remain human.

The exercise of some art can thus be seen as essential to the proper discharge of the engineer's responsibilities. His task is to provide machines and equipment which are efficient, safe and economical. But implicit in this primary responsibility is a secondary one: to ensure that his products are in every way acceptable to those who have to purchase, operate, or simply live with them. This is a very considerable responsibility, bearing in mind the term of life of many engineering products.

That it can be achieved is witnessed by many well-defined products, such as the electrocardiograph shown in Fig. 5, which combine the structural, functional and ergonomic requirements with pleasing appearance. Where the engineer does well in these matters, he can be congratulated for having made good, either by intuition or acquired sensibility, what his training has failed to provide. Where he does badly, he demonstrates the need for a more liberal form of education and training for engineers or, alternatively, the case for recognizing the human aspects of engineering as a field for the specialist.

However the matter is finally resolved, it seems certain that engineers alone cannot at present design every product so as to withstand foreign competition and to retain our place in world markets.

Nor, indeed, can they alone ensure the necessary aesthetic quality in products for domestic consumption.

The Feilden Committee, in spite of its curious neglect of aesthetics and amenity, found engineering design to be lacking in many qualities. The Federation of British Industries is similarly concerned, but this body does not believe that a general improvement in mechanical design alone will put matters right. It supports the view that appearance and amenity are fields for specialized activity, but it is also in favour of extending, by every possible means, the interest of engineers and design draftsmen in the visual aspects of their work.

Committees and Reports are all very well but, by themselves, they do not result in action. What can we do now to improve matters? In industry itself there is an increasing awareness that design, in its fullest sense, is becoming more and more a factor in determining sales. Heavy engineering firms who have, for many years, enjoyed supremacy in their field, and a gratifying share of world markets, now find that they are losing ground to foreign competitors. They are also losing ground, in many cases, to local industries in countries which previously relied entirely upon imports.

All this is inevitable. What is equally inevitable is that we must add to the traditional virtues of British engineering products the qualities of better appearance and more convenient layout, which so often characterize the products of our competitors. We are only now beginning to do this seriously and, in recognizing past omissions and in taking the initiative in correcting the situation, the engineering industry is acknowledging the value of the contribution which industrial design can make.

Whether it comes from engineers or outside specialists, art is an essential factor in engineering. Given intelligent co-operation between engineers and industrial designers, and between those responsible for training both, we should, one way or the other, ensure that art and engineering, having come together, do not drift apart as they did once before.

How the Architect Sees Us

by W. A. Allen, BArch, FRIBA

If engineers had a definitive training in creative design, their collaboration with architects would be greatly facilitated and the result would be buildings which give more satisfaction to users and viewers alike.

Although I am an architect my work has always been involved intimately with scientists and engineers and I have spent a good deal of time thinking about the differences that develop among us, how we might diminish them, and how important for the national good it is to do so.

As a measure of the importance of this, let me first recall that the building professions and industry are between them responsible for half or more of the country's gross fixed capital formation, and that new building amounts to some £1600–£1800 m. annually, of which the engineering content is probably upwards of £500 m. If we do not work together smoothly, the scope for waste is formidable. It is troublesome enough at the personal level. Between us we are the country's main spending professions, and the building industry is the country's largest.

You will notice my reference to 'the building professions'. I am not sure to what extent the engineers who deal with building problems see themselves as a building profession but this is the view architects take of those with whom they collaborate in design. The differences between these two building professions are not inherent but are man-made and we can therefore bring them to an end if we wish; but because they are formulated in educational establishments where inertias are large, they will take some time to eradicate if and when it is decided to attempt this.

Some resolution of the differences is in fact taking place but among practitioners rather than in education, and no doubt because the realities of practice drive the process forward. The changes have been especially evident among structural engineers, partly because of the vigorous selection which architects have exercised in seeking out those who could participate creatively in design; and partly because, among structural engineers, there were some who had the vision to see this potentially greater fulfilment.

I think the search for creative collaboration with equipment engineers has been equally vigorous but has had less perceptive response. Perhaps it is harder for mechanical engineers to see a creative role in building design; but it is certainly there, waiting to be developed.

The nature of the differences

The most fundamental differences between architects and engineers are in values and outlooks, and perhaps motivations.

Let me assume that a projected building involves a fair amount of engineering work, let us say, a music college, which will need a multitude of insulated rehearsal rooms, a theatre and a concert hall to which the public must have access, dining facilities, and administration quarters. The site is noisy and tight, and the budget is strictly limited.

Mr William Allen *was educated in architecture in Canada. A member of a scientific family, he came to this country in 1936 in search of postgraduate opportunities for acoustical study. After a period of normal practice, he found what he was seeking at the Building Research Station, though he was also concerned with factory design. After this his interests widened and he became first Deputy Head of Physics, and then Chief Architect to, the Station. In 1961 he became Principal of the Architectural Association School, the largest in the Commonwealth, where he has revived planning and urban design studies and reformed its handling of technology. In 1962 he also re-entered private practice in partnership with two other architects, a quantity surveyor and a structural engineer. The firm, Associated Architects and Consultants practises in building science and engineering.*

Consider some of the factors involved in forming the 'idea' of the building. Is the noise great enough to demand a sealed and solidly-built structure? If so, we will need artificial ventilation. Will we need, and can we afford, cooling? How big will the plant room need to be then, and where will it best be placed? What sorts of construction techniques can we consider and what are their relative costs? These affect the number of storeys to use, and so do the circulation and mechanical equipment problems. Perhaps a radical, multi-storey solution will be more expensive in one respect and cheaper in several others. Which balance will be optimal in terms of human and mechanical functioning and cost?

One must be able, as an architect, freely to range over alternative solutions, weighing continuously such different factors as these, for design is only a semi-logical process; imagination and critical, informed judgment come into play alternately. The right idea is often to be caught on the wing. There is no time for laborious setting out of one scheme after another, each to be assessed accurately. Only as one begins to grasp key relationships should the design be allowed to begin to gel. Even so several probes will probably have to be made before the break-through is achieved to a sound design strategy. The tactics can mostly come later.

My point in recounting this is to underline that the proper approach to a modern architectural problem involves continuous integration and balancing among different kinds of factors, some engineering, some economic and some architectural, and all have to be held and knocked around at what I call strategic level, ie at the level of policy in design decisions. It is seldom right, in these circumstances, to worry about quantitative exactness, because one should be using judgment and some of the factors do not quantify, though they will be none the less important because of that. Any weighing-up may involve beauty or practicability or cost, or quietness or human comfort and delight. An essential skill of the good designer is to see his problem whole, get his priorities right, and get a concept formed in principle that is so complete and balanced that no major change is needed during development.

Imaginative partnership

Obviously not all of these factors can, or should, be left to the architect. He ought however to be able to have imaginative, easily flowing conversations and arguments with the other experts as ideas for the building get manipulated and begin to take shape. But the only kinds of people who are really useful at this stage are those who can and will speculate reliably, who will make both quantitative and non-quantitative judgments with equal willingness, and offer useful views and ideas directed to the common ultimate ends of architecture. In other words, engineers and quantity surveyors should comprehend and be concerned about, and be ready to offer views relating to all the qualities that good design should possess. Then their potential value in the design team could be fully realised and their own satisfaction would be at its maximum.

This may seem impractical to many engineers. If it is, it augurs ill for the development of high-grade modern building, because it would imply that our problems about cooperation in design will persist. It is no good an engineer saying to an architect 'you tell us what you want and we'll give it to you', (which is still an attitude one encounters) because it means there is no constructive feed-in of ideas that can improve the design. And besides, it implies a secondary or dependent role for the engineer, not that of a partner in imagination.

A good building is not just an abstract architectural concept to which structural support and mechanical aids are given; it is an organic embodiment of all its technical functions, its planning requirements and its aesthetic idea.

When the initial concept is sound, one can reasonably assume that detailed development of the design can take place successfully in every respect and lead to an

"—tied too tightly . . . to traditional ideas . . ."

economic, functionally sound and convincing piece of architecture. If the original idea has serious flaws, then you will find yourself going back to square one at a stage embarrassing and wasteful to the architect and to the other consultants. It must be correctly balanced when the detailed development of the design begins and this depends, as I hope I have made clear, on the easy cooperation of like-minded people with varied building expertise. This takes us back to the idea of the design team and the 'building professions'.

The practical problems

At the risk of over-emphasis I think this is a point at which I must try to summarise explicitly the present problems of collaboration between architects and engineers as I see them. Leaving aside the details and odd misunderstandings, the main problems are these: engineers too often find it difficult to be constructive or contributive about the project in hand. They are tied too tightly and uncritically to traditional ideas, a state of mind which has been described as 'hide-bound'. And they rely too much on aspects they can quantify and too little on judgment about unquantifiable factors. By these attitudes they are led either into the position of 'tell me what you want and I'll do it' or the opposite 'this is what is required and it must therefore be done this way'. Neither of these is a basis for collaboration between equals.

Two other problems are that it is very difficult to tell from services drawings whether things will work or not. And, in terms of selected equipment, engineers tend to work, not from what is needed, but from what is available, so that there is too easy acquiescence in the second-rate.

Ten years ago these strictures, which perhaps sound unkind but are not meant so, would have applied equally to structural and services engineers; today they mostly concern the latter.

I have wondered if I could usefully quote examples, but it is not easy to take actual instances without arousing defence mechanisms that only leave hard feelings. Perhaps one or two general matters could be mentioned however. One would be the noise of mechanical, and sometimes electrical, equipment. There did not seem to be much concern about it until recently and, as far as I know, there is little or nothing about it in normal undergraduate education. Because it affects factors such as comfort and intelligibility which are human, non-mechanical and not easily quantified, it seems difficult to arouse engineers to their importance.

Another example is the design of public address systems. The basis of high realism and intelligibility has been known since Haas's work with Meyer, shortly after the end of the war, but few engineers are aware of it, and if intelligibility is achieved, it is often without apparent concern for realism. Consider the public address systems in our railway and air termini, where the highest standards of intelligibility are needed to help foreign travellers; they are seldom intelligible even to natives.

In structural design the most common failing has been to excuse inelegance by assuming that certain things are merely utilitarian and therefore do not need to be 'aesthetically pleasing' (a phrase that makes architects' insides turn over).

"Consider some of the factors involved . . ."

Bridges built in out-of-the-way places, factories (which engineers often design *in toto*), even harbour installations are built with seeming unconcern for character, order or human pleasure. I wonder, too, about the overhead support frames for the wiring on electrified main-line railways. Were these really the best that could be done? They seem to me to compare poorly with motorway 'furniture' for example.

But it is all too easy to argue and criticise in this way and I am not trying to be pointlessly critical. I want to draw attention to those aspects of our work in which we are not achieving what we would all protest we want to achieve. I believe the roots of the difficulties lie, not in personal prejudice or shortcomings, but in omissions in engineers' education and a lack of concern in the engineering institutions for the reforms which would give its proper place to building engineering and design.

Engineers may say, of course, that at best only the odd man here and there could meet my criteria for true collaboration, but this is not an opinion I can accept, for it implies that a nationally important problem will have to be left to fate.

I will therefore try to make some positive suggestions.

What is design?

A training in design appears to be a necessary part of an engineer's education. In saying this I am not thinking of abstract aesthetics; I have in mind something which is much more akin to invention, in that it is concerned essentially with the formulation of new ideas and objects that are organically sound and practicable. In this sense every new building, every new railway bogey, every new piece of electrical gear, every new bridge requires inventiveness. But good design implies more than this, of course, for it means investing the object with the good sense of well-ordered relationships, and the taking into account of how it will be made, how used, and how the user will feel about it.

When one starts talking about feelings, most engineers start to squirm; but in an important sense feelings are the ultimate reality. Engineers do have feelings, like everyone else. But they seem to have insufficient confidence in their ability to formulate an object that will evoke any particular feelings, even supposing that they knew what feelings they wanted to arouse. Perhaps many do not even realise that particular feelings can be deliberately aroused, though they must have experienced different feelings when getting behind the wheel of a car that is well-designed as compared to their reactions to a poor design. They must realise that much the same feelings can be experienced by others.

It is not sufficient, then, to presume that a good design has been achieved when a merely utilitarian purpose has been adequately served. There should be some positive pleasure or other quality aroused in the user. But it cannot just be 'styled' in; it has to be inherent. Style is not applied to, but is inherent in an object, just as it is in a piece of writing.

Perhaps here I should take a moment to establish clearly this meaning of design, for among some engineers analytical examination tends to be thought of as the essence of the design process, whereas the vital part of it is the forming of the concept. This is the creative act; the other is a validating

"—seldom intelligible even to natives"

and refining business. As an architect sees it, the total process goes somewhat as follows. There is a preparatory study directed to the problem in hand. Then the imagination begins to function within the critical parameters established by instinct, education and experience. Eventually concepts emerge sufficiently clearly to be given conscious appraisal; and finally, when the balance of the key factors seems right in one of these, the design is ready for development.

Usually it is only during this last stage that exact analysis becomes a major activity. This should clarify the difference between design as an imaginative act rather than an analytical, confirmatory process.

'Industrial' design is often associated with sales appeal and carries overtones of fashion. What I am thinking about is, as I said, akin to invention. It is a constant, positive habit of mind, a discipline, an acquired, three-dimensional organisational skill applied to the solution of a problem, whether it be a building frame, an electrical circuit or a boiler house.

Educational training

It may be supposed, as I mentioned earlier, that one cannot be educated for this kind of approach, that it is a God-given talent; or, at the other end of the scale—as the Feilden Committee may have seen it—the assumption that it is an essentially logical process which any intelligent person can do who has obtained the requisite knowledge and that no special training is needed or can help. There is the barest trace of truth in each, but between them lies the much larger heart of the matter; which is that the ability to design can be developed to a very respectable degree by proper training, giving discipline to the occasional genius and a generous measure of capacity and confidence to others less favoured.

The vital point in this argument is that design is a skill which can be developed by training. Architects believe this best takes place during one's education, where one can be directly exercised in the use and combination of knowledge by doing creative problem-solving design under effective criticism. This is a question of learning to think easily in terms of complex three-dimensional arrangements and developing judgment about the values to be given to this or that factor, whether quantitative or not; and it is in such design situations that one discovers, often intangibly at first, then more confidently, how to invest objects not only with those orderly relations that make the

"—their reactions to a poor design"

whole greater than the sum of the parts, but with the sort of character that will arouse particular feelings in the user or onlooker—pride of use, or the desire for possession, or admiration, or confidence, or any of a dozen moods or emotions.

These are not superficial objectives but qualities of vital value. All objects and environments arouse feelings. Ugly things also do so. It is simply a question of recognising that feelings can be aroused by products of one's work and that human beings want and need positive feelings.

For an educational example of how this skill can be developed I have to fall back upon one aspect of architectural education which I believe has something to contribute here, for architects embody in their studies a series of what are now called open-ended projects. These are design problems without specific answers in which the student is challenged first to assess the problem itself in detail—to establish the so-called brief—and then to work out a piece of architecture which solves it. By this means he is forced to use his imagination and to exercise his skill and judgment.

These projects are done in a school where tutors are available for discussion and criticism; the results are usually presented periodically to panels of outside critics who debate with the student in public about his proposals. This is where a large part of an architect's outlook and many of his values are formulated, because a good panel of critics will look at his work from all the points of view of a real-life situation.

The standard answer to all this by engineers is that there is no time in a curriculum already full to overflowing. I think my reply must be that I (and now a growing number of engineers) believe it to be essential to achieve something of this sort somehow.

But, as an architect, I must, from this point forward, mind my own business with, perhaps, just this post-script, vital to my opening argument; that, when engineering education develops along lines that bring into play a broader set of values and a corresponding design outlook, I believe such barriers as inhibit our easy collaboration today will disappear. Engineering itself will become much more creative and perhaps more satisfying; and we in architectural education will benefit in turn from the better engineering content we can give to our courses with the help of engineers. In fact, collaboration in education, sought in the face of heavy odds, could then become a reality.

Let me add in

"Ugly things also do so"

"—a kind of total services engineer . . ."

parenthesis that I am well aware—none better—that architects, too, fall somewhat short of perfection in their relations with engineers. But this is not an article about architects; it is about engineers as an architect sees them.

In any case the situation is not symmetrical: in the engineering aspects of their education architects depend largely upon engineers and, of course, it is difficult for these to provide what is needed because their own education has not prepared them to give it.

Thus, for exactly the same reasons as we have difficult collaboration, we also find it difficult to find the right kind of engineering education for architects.

The time spent on engineering work of various kinds in architectural education generally amounts to 150 to 250 hours a year, for three or four years, which is perhaps not unworthy of the subject. But the results are not commensurate because, as I say, for the most part we get engineers' engineering.

British architects are at present very self-critical and reform-minded about their education, and probably have a world-lead in the development of improved techniques. So we are perhaps too easily led into assuming that others, too, could feel the need for radical reform and experiment.

What kinds of engineers?

As my last argument has been somewhat climactic I am now risking an anti-climax in wishing to offer one other matter for consideration.

I referred at the outset to the building professions, and tried to indicate that, in terms of national investment and social policy, we need a strong professional architectural and engineering design team dedicated to collaborating in the field of building.

Architects have been busily discussing what kinds of engineering team would most suit the evident needs of building design and I think that the most widely accepted conclusion at the moment is that we need two main groups, one structural and the other environmental. Since this last word is widely and loosely used today I must be more explicit. Architects (and some engineers) have in mind here a kind of total services engineer, covering (hopefully) mechanical, electrical and hydraulic services in buildings, and, of course, design-orientated in the sense I have described, interested in the whole social problem of giving people a good and efficient environment for their daily life.

Educationally this would imply a departure from present habits for engineers. It would not be a departure from trends in practice since, as I mentioned earlier, quite a number of consulting engineers already operate in this manner to the best of a sometimes considerable ability. One would be surprised if their principals were not as keen as architects to see the kind of education established which would facilitate the establishment of design organisations such as are now needed, both in the structural and environmental fields.

It may be argued that such an education would be more superficial than the present analytical approach, but need it be? I doubt it. I believe that if one looked carefully at creative design studies they would be discovered to have excellent attributes has a sector of engineering science, bringing a refreshing infusion of the human and economic sciences. In any case, the creative content would be a great merit.

In this context I am discussing engineers of university standard, educated at a professional and not at a vocational level. Whether they receive this kind of education at undergraduate or postgraduate level is another question. Architects hope that the bulk of such people will be produced in the course of undergraduate education. Whether it is reasonable to suppose that the necessary imaginative project work can be fitted into the three-year curriculum and still leave room for a thoroughly good theoretical education is a far more difficult question. In four years it would presumably be feasible, but three years might be too little. This is not something an architect can discuss; but we are glad that we have five years, like medicals, in which to develop our skill alongside our formal education.

Another possible way is to do the design projects as a postgraduate course. Whether this is a worth-while possibility is an open question, for it has the disadvantage of putting the two activities end-to-end, which is less satisfactory than side-by-side. And, of course, the total time is the same.

There is another mooted development, which is to take the graduates of any branch of engineering and offer them a postgraduate specialised course in building. This has the advantage of putting into the consultant field graduates expert in one engineering specialisation but with a broad view of building. Why not do building engineering for three years and then specialise; ie the reverse process? It is not for me to answer. But this would omit the design projects which, as seen by architects, are critical to the development of creative habits of mind and confidence in value judgment.

Architects want mechanical, electrical and civil engineers to prepare part of their student body deliberately to participate in building design. In terms of national need and expenditure, this seems the only socially responsible view.

We envisage two kinds of building engineer for normal practice—the structural engineer; and the environmental or equipment engineer who combines the technologies and sciences needed for this aspect of building.

We urgently want intimate collaboration but I believe it depends upon the development of design training in engineering education, not merely because this would be good in itself, but because it engenders valuable attitudes of mind. As a by-product, I believe it would develop far greater inventive capacity in engineers and thus provide one of the keys to the country's economic problems.

To close gaps and lower barriers between us,—as we must —is a question of attitudes, values, outlooks and that sort thing; not facts and figures.

There must be a firm foundation of concern about good design, and the kind of confidence about adventuring into new solutions that comes from the knowledge that one has been prepared for it in a balanced and whole manner and has tested oneself for the process. This I believe to be vital for the health and power of the engineering industry generally; but I am certain that it is the key to real success in the next stage of development in building, if we are to be able to build a good human environment at a price in resources that the people of the world can afford to pay.

And this is a joint responsibility.

REFERENCE

Creativity in Technical Education, pub. London and Home Counties Regional Advisory Council for Technological Education, 1964. 7s. 6d.

What's Wrong with the Technical Press?

by E. G. Semler, BSc, CEng, FIMechE

Engineers tend to expect a technical journal to keep them informed on all new developments relevant to their work and constantly express surprise at the inadequate manner in which the technical press carries out its ostensible function. But they rarely look into the reasons for this and what could be done to improve matters.

Before we get involved in the diversity of engineering publications, let us look at their readers—that is, ourselves. Ultimately—and I do mean ultimately—every public gets the kind of press it deserves.

We know, from a number of surveys, and from personal contacts, that the average engineer does not read. Perhaps that is a slight exaggeration; but while the pure scientist thinks it vital to his work to know what others in his field have published, engineers, on the whole, prefer to rely on the technical grapevine to tell them what's been done. We seem to read technical journals only in direct proportion to the amount of pure science involved in our work.

It is difficult to decide whether this is the cause or the result of the rather inadequate technical press we have; but, as with all chicken-and-egg situations, we have to make a start somewhere.

The fact that engineers are casual about their technical reading is, of course, reflected in the price they are willing to pay for it, the attention they pay to advertisements and, therefore, in the finance available to give them something better.

Do we need journals at all?

Perhaps the customer knows best: after all, if you scrapped all technical journals, would anyone (except their staffs) notice the difference? This question everybody must answer for himself. As I see it, much depends on what you demand of the technical press. Most journals try to fulfil four principal functions.

1. The reporting of new techniques in their particular field;
2. Reporting new techniques in other areas which the editor considers might be of use to his readers;
3. Serving as a buyer's guide by describing suitable equipment which is either new in itself or new in some aspect, such as price, availability, etc. Advertisements help here, though in a rather haphazard fashion;
4. Providing news and gossip about the industry concerned, which may not be directly relevant to the job but which gives useful background knowledge.

Some journals also seem to see themselves as textbooks in serial form but I do not consider this their proper function: most people prefer their textbooks in one piece.

"—the average engineer does not read"

Having listed these functions we can immediately see the inadequacy of most journals: very few indeed can claim to cover their chosen field even as well as the *Daily Tabloid* covers the world's general news. To understand the reasons why, it might help to look at the history of the technical press.

How did it start?

Technical books have of course been written for thousands of years but technical periodicals only started with the *Proceedings* of the Royal Society in the 18th century. As scientists became more numerous, they were no longer able to exchange enough information through meetings or personal letters and a regular publication had obvious advantages: inadvertent duplication of work could be avoided and intentional duplication could assist in checking the published results of others.

The commercial technical press—and the trade press from which it sprang—were 19th century developments, following upon the industrial revolution. As engineering ceased to be a craft and became first an industry and later a science-based professional activity, there clearly arose a need for journals serving the industry as a whole on the one hand; and learned publications for the experts on the other.

The Engineer was probably the first 'specialised' journal to serve the industry, and the learned society publications dealt with the more scientific aspects. (But a truly scientific approach in such Proceedings was, for a long time, rather rare; papers were mostly by practical men describing practical achievements.)

Throughout most of the 19th century, during the great days of British engineering, less than half a dozen journals of each kind served engineers. They derived their income largely from subscriptions and made a living by keeping their readers informed about what they wanted to know.

Two developments spoilt this idyllic picture: one was specialisation and the other was a related phenomenon, the information explosion. Specialisation has obvious advantages: you learn more by using a microscope than with the naked eye, even though you get a more restricted view. But this inevitably leads to departmentalisation and the so-called information explosion; which is really a misnomer because true information doesn't grow all that fast. What does grow at an exponential rate is printed paper.

If there was only one engineering journal you might read an article on, say, lasers and think about how to apply them in your own work. But if you have a specialist journal on chemical engineering you will obviously want an article on lasers applied to process work; and other specialist publications will cover applications to mechanical engineering, and so on. It is quite a thought that the Institution of Mech. Engineers was founded as a specialist body to deal in greater depth with mechanical engineering: today there are some 30 much more specialised bodies in this field alone; and an even larger number of specialist commercial publications trying to cash in on this departmental approach to engineering.

Commercial publications

And so we arrive at the technical press as we know it today, as prolific as it is inadequate to its task. It can best be described under three headings: commercial prestige journals, so-called controlled-circulation journals; and institution publications.

There is some overlapping, particularly between the first two categories, but by and large a prestige journal is an expensively produced periodical sold largely on subscription, which attempts to publish original articles, either contributed or staff-written and, in general, at least attempts to fulfil all four main functions of the technical press. Most engineers have seen such journals, though few of them buy them with their own money. Subscriptions are usually paid by firms or libraries and each copy may circulate among a dozen people, most of whom collect a pile in their in-tray and, having had time to skim through only one or two, pass the whole stack on to the next person on the list.

Curiously enough, while treating their reading matter in this casual way, most engineers seem to be under the illusion that every such journal is produced by a large, highly qualified staff, with correspondents all over the world, using all the devices of modern newspaper production; that it has a circulation of about 50 000 and relies on his subscription for its profits.

Nothing could be further from the truth. The economics of paper, print and postage (henceforth known as PPP) are such that even the rather pathetic resources devoted to the production of a technical journal could not be recovered from readers, even if they paid 50p for each copy. The advertiser pays 80 to 95 per cent of the cost of all commercial prestige journals and 100 per cent of others.

More and more, therefore, the advertiser is calling the tune. And so much for the wide-eyed innocent who writes to the editor, suggesting that he should keep down the advertisements and, furthermore, tell the advertisers off about the contents or presentation of their offerings. He might as well ask the editor to sack his boss.

Fortunately, there are among advertisers enlightened people who either believe that a quality publication will confer prestige upon the wares advertised in it; or that such a publication will at least advance the interests of the whole industry it serves, and therefore their own. Such an attitude, while far-sighted, is expensive and more and more advertisers want to see a direct return for their money.

However, it is very difficult to measure a direct return from advertising because so many intangibles enter into it and this explains why a number of reasonably good commercial journals still eke out a precarious livelihood. But the fiendish invention of the reader enquiry card during the last 20 years is

rapidly putting an end to this. Like all good things it came from America and, as there is one in this journal, I do not have to describe it. Its function is simple: while making it slightly easier for the reader to enquire about anything mentioned in the journal, it provides the advertiser with some measure of the 'pulling power' of

"—tell the advertisers off about the . . . presentation of their offerings"

one journal, compared with another. There are many ifs about this: still, as we know from our own work, if A is only one aspect of B but it is the only aspect we are actually able to measure, then, however poor the correlation between A and B, so long as it is positive, we grasp at A as a means of measuring B. Reader enquiries do not prove that an advertisement has paid for itself, nor is the converse true. But what else can an advertising agent show his client?

The tendency for the commercial technical press, therefore, has been to turn itself into an agency for promoting reader enquiries. There are a number of tricks known to editors which help to keep the advertiser happy, but not necessarily the reader. Every potential advertiser wants his products mentioned as often as possible, and if you don't mention them, how can you get reader enquiries? Short and bitty information is more likely to bring enquiries than long articles, and so on.

But the most obvious way to get enquiries is to have a large circulation. The average general-coverage prestige journal has a paid circulation of 5-10 000 and for specialist journals it may be as low as 500. So every publisher gives away a number of additional copies under various pretexts. But this is expensive.

'Controlled' circulation

The obvious answer to a publisher's prayer is the so-called controlled-circulation journal which is given away absolutely free to those expected to feed back enough suitable enquiries. I say 'so-called', because most publishers simply pick names from suitable lists and hope for the best. If you have ever received three copies of the same free publication it is because your name happens to appear on three lists. If they have no relevance to your work it is because the lists were inadequate.

Such journals have a comparatively large circulation because the publisher has sacrificed everything to pay for PPP. Fortunately (or do I mean unfortunately?), the kind of journal that brings in enquiries does not need much editing since most manufacturers kindly flood editors with information about products they wish to sell. All one has to do is send their puffs off to the printer.

Newspaper format is the cheapest and seems to convince a good many readers that what they get in this form must be news. And, of course, advertisements are so much mixed up with the text that only by a severe concentration of effort could you avoid reading them.

Alas, there seem to be many readers who actually prefer

this type of publication to the prestige journal because it makes fewer calls on their mental powers. This unmitigated rubbish is rapidly driving out the reasonably adequate. Gresham's law is still with us.

I do not wish to suggest that the controlled circulation idea in itself is deplorable. On the contrary, it is probably the only way in which technical publishing can survive. But, alas, it is not often combined with reasonable standards of editing and reporting.

Institution publications

This leaves us with learned societies and the more esoteric aspects of information dissemination. Though why they should concentrate on the highbrow stuff is not clear. We have seen that, while Proceedings started with those of the Royal Society, a century later they largely consisted of practical papers. With increasing specialisation, however, came jargonisation and Greek letters. Many people seemed to feel they couldn't really call themselves experts if an outsider was still able to understand what they were talking about. So they invented a secret language of their own.

There is only one snag about this kind of publication, which had its heyday in the first half of this century but is still very much with us: few people read it, though most consider it good form to display it conspicuously on their shelves. Latterly the cost of PPP has also hit institution publications so that many have begun to take advertising and otherwise to search their hearts (or committees, as the case may be). To keep their readers and to please their advertisers many had to turn themselves into something more like commercial prestige journals. Some have even gone further and are beginning to look like controlled-circulation sheets.

To my mind there is nothing wrong with this development, as long as it remains under control, for institution publications are in the happy position of having the best of both worlds; though few of them take advantage of this.

To begin with, their circulations are really controlled; not perhaps from the point of view of obtaining the maximum number of enquiries but, because of the institutions' qualifying function, the advertiser has a guarantee that the members are active in a reasonably defined field and qualified to a known standard. Most commercial publishers would be very happy to be in this position and they are always trying to take over good institution journals.

The other great advantage is that a learned society is not dependent on advertising. True, its journal could not possibly be produced to anything like a prestige quality without advertisements, but every society has to produce some sort of journal, even if it brings in no revenue at all. Therefore an institution editor need not take his orders from the advertisers; instead he should be in a positon to serve only the readers which, in the long run, should also help the advertisers. After all, they want their ads to be read.

Unfortunately, in many cases this is not how it works. Being in one sense 'house journals', such publications are exposed to severe pressures from committees and other staff departments to provide an excessively detailed and often boring coverage of institution act-

ivities, regardless of what the reader really wants to know.

There is, in addition, the feeling of having a tied reader who gets the journal for nothing and cannot give up his membership if he wants to keep the letters behind his name. Besides, he jolly well ought to read what his institution wants him to! Unfortunately (or fortunately?) this is not so: after all, he need not open his journal.

How it works in practice

It will be fairly clear from what I have said that a technical journal, in whichever category, and whether it appears weekly or monthly, is not the buzzing hive of technical and journalistic activity which the reader may imagine. Run on a shoe-string, its average complement is one editor plus one secretary for a monthly; with weeklies often boasting an even smaller staff in proportion.

The editor is frequently not a technical man for there are few engineers with the gift of the gab; and they can usually earn a great deal more in industrial management. Reporters are out and contributions must come from those willing to provide them for reasons of their own, in which payment must not play an important part.

The motive for technical writing ought to be a desire to share one's discoveries with others but commercial secrecy nowadays makes this largely impossible. The actual motives are therefore an enhancement of personal and company prestige or a desire to advertise one's products free of charge. And the cream goes to institution publications because they still confer the highest prestige.

The trouble is that, with this kind of contributor, what matters is to impress one's boss, one's learned colleagues, possibly one's customers. Which naturally makes for a great deal of jargon and mathematics, rather than an attempt at inter-industry and interdisciplinary communication.

Having looked at the structure of the technical press, we can now re-examine how it actually fulfils the four functions outlined at the beginning of this article.

Up to a point a specialist journal can be quite successful in covering new techniques in its own field; but only up to a point. It is bound to be parochial since it cannot afford proper foreign correspondents or even first-class abstractors of foreign literature. It cannot penetrate commercial security or, even if it got the information, it could not afford to publish it and quarrel with an important advertiser. But, since it largely relies on press releases, it can hardly ever get first-hand information while it is news. Press releases are prepared by PR men, who first have to find out about it themselves; and then reflect only the official views of a company about its own products or activities.

General engineering journals cannot even do that much. Although they are in a better position to jump the inter-disciplinary boundaries, their limited space and resources do not allow them sufficient depth of coverage. And while even a non-technical but specialised editor, after a few years in his job, can come to be much better informed on his subject than the average engineer in industry, the editor of a non-specialist journal can never learn much about any of the numerous activities he covers.

"—most publishers simply pick names from a list and hope for the best"

All this applies even more to the second function, the reporting of possibly relevant developments outside one's own field. Here nobody can be a specialist and only a large staff of experts could make a really satisfactory job of it. A general engineering newspaper might be thought best at this work for it would cover all specialities and perhaps it could be done by a daily like the *Financial Times*. A number of attempts at weeklies have failed for the sufficient reason that, to cover everything in a week means to cover nothing properly.

The buyer's guide function is carried out rather haphazardly by advertisements and notes on new equipment, none of which, obviously, can be at all comprehensive. Some editors publish more systematic reviews of equipment with comparative data sheets which would be extremely valuable, if only they were anywhere near complete. Alas, few manufacturers are willing to supply any but the most favourable data about their products and no engineering journal has time to carry out its own tests. There is a great need here for an engineering '*Which?*'

The buyer's guide function is, in any case, obsolete and is only retained because it appeals to advertisers. There are perfectly good buyer's guides in book-form and the new computerised and microfilm systems (when they finally get themselves sorted out) will obviously do this job much better than any periodical can attempt to.

Could the same be said about new techniques in general? By no means. To get information about techniques from a computer (or out of a conventional library, for that matter) you first have to know what you want to know. And, as we have seen, most engineers do not even know that they want to know anything, let alone what it might be. There is, therefore, no substitute for leafing through a reasonably attractive technical journal and getting a new idea (new in your job, that is) from something you would never have looked for in a library or buyer's guide.

Finally, industrial news and gossip: this is probably the function the technical press carries out best and it is a pity that many institution journals deprive themselves of this opportunity to interest the reader because they consider this sort of thing too 'commercial'. But here, too, a journal has to rely heavily on what the industry wants it to know and must not offend advertisers.

What can be done?

So where do we go from here? The present trend is quite clear in one respect: commercial controlled circulation journals will drive the prestige journals out of business in all but the most narrowly specialised fields. And quality will go down, not up, as the triple-P costs continue to rise.

At first sight one might think that the recent multiple mergers would counteract this. It seems logical to assume that a large publisher could merge several competitive journals, without sacking any of their staff; instead he would merge their revenues and circulations, make a first-class job of the editorial and save a good deal in overheads.

But not a bit of it. Three competing journals in a narrow field might seem too many but a single journal would barely earn 50 per cent more than one of the old ones, however good its editorial. Advertisers don't use two or three competitive journals at once because they like them but because they believe that only in this way can they be fairly sure to cover all potential purchasers. If they can do it in one journal, so much the better, but they will not pay three times as much per reader for the privilege. Again, improving the quality of the editorial is a hard slog and making an impression

with it on advertisers is even harder. Doubling the minimum editorial staff will probably yield a 30 per cent increase in quality with rapidly diminishing returns; and doubling the editorial quality will probably increase the advertising revenue by 20 per cent.

It is this sort of economics which has given the publishing giants a severe stomachache from trying to digest the smaller technical publishing groups they have swallowed; and which has hastened the demise of prestige journals.

Does this mean the extinction of the technical press as we know it? Despite its many shortcomings, this would leave an information gap much larger than most engineers imagine. Fortunately, there is another way.

As indicated above, the larger institutions at least are in a unique position which few of them have exploited so far. A highly qualified readership of exactly known size, operating in a fairly accurately defined field, is an enormous publishing asset. But institution journals need not be self-supporting, for their object of disseminating technical information is one of the principal functions of the institutions themselves. Indeed, since the foundation of CEI, it is possibly the primary function left to specialised institutions.

Think of the first-class technical journals they could produce if they really tried. If they rationalised their other services and concentrated their resources on those which involved the technical specialisation of their members, think of the money they would have for increased editorial coverage and promotional activities. Think of the wealth of specialised technical information which could be at their disposal if only someone had the time to tap it. Think of their international connections which could be exploited. Think of their educational activities which could be dovetailed with their publishing efforts. And, above all, think of the saving in overheads if all the commercial activities were undertaken by a single CEI publishing house.

It would seem, therefore, that in future the buyer's guide function will be carried out by computers (though there will be an increasing demand for it to be based on a prior unbiased product evaluation which could well be a profitable activity for engineering institutions). There will probably still be a market for tabloid-format, large-circulation, commercial, give-away sheets but the future of the quality technical press will increasingly be in the hands of the institutions, preferably under a central publishing house.

Mr E. G. Semler *was, until 1970, Editor of the CME and was responsible for its business management. He is now a Publications Consultant and freelance writer. Educated at Herne Bay College and University College London, he graduated in Mechanical Engineering in 1942 and served a college apprenticeship with the Metropolitan Vickers Electrical Co. at Trafford Park, where he also edited the house magazine, Rotor. For the next 12 years, he was a welding engineer with MetroVicks and other companies, his work ranging from technical development to sales promotion and technical writing. In 1956 he became editor of Automation Progress, a technical journal, and in 1960 he joined the staff of the Institution as Editor of the CME. He became a Student member in 1942 and was made a Fellow in 1970.*

The Place of the Engineer in the World

by the Rt Hon the Lord Snow, CBE*

Some of the reasons why engineers do not enjoy the status they deserve are to be found in their education. A wide and humane education will produce a climate in which engineering creativity can flower. If the profession adopts a defensive attitude to the status question, it is likely to become more rigid and so defeat its object.

I have said in public more than once that the status of engineers in Great Britain is lower than in any country that I know of. This statement I stick to. I believe this condition of things—if I am right about it—reflects a good deal that is wrong with our society. I should like to develop the point a little. I shall stray a good deal from our purely local problem, important as that is to us and ought to be. Right at the heart of this discussion there is the need, which all sensible people increasingly feel, for engineers all over the world to be close to all the decisions which are going to affect our human society.

Contrasts in prestige

To an extent, however, engineers get the rough end of the stick even in countries where they are more esteemed than in ours. You know the bitter American crack invented, I need hardly say, by an engineer. You fire off a rocket and a satellite moves successfully round the world. That is a scientific triumph. On the other hand, if it flops on the launching pad, that is an engineering failure. Engineers don't often win.

The only country I know where they really do have maximum prestige is the Soviet Union. It is hard to imagine an English or even an American engineer being given a State funeral—Korovolov got one in Moscow this winter, with 50 000 people lining the streets in the bitter cold. The present President of the Soviet Academy is an engineer. So, curiously enough, is the present Soviet Prime Minister. It is still difficult to imagine similar appointments to the Royal Society or to No. 10 Downing Street.

Our position here is far less favourable than in the Soviet Union, and a good deal less than in the United States or in Continental Europe. Somehow, and soon, this position has to be altered. But the first thing I want to tell you is that engineers themselves can't do much about it. Certainly not by administrative action, and probably not by expressing their own justifiable discontent. The protest and change have almost certainly got to come from others. In fact, that is why I am talking to you on this occasion.

My only personal connection with the engineering profession is that recently I was given an engineering medal from an American college, and that is an award which none of my acquaintances has yet begun to understand! My motive for being concerned with the status of engineers is that, unless we raise it to an extent commensurate with the importance of engineers to our society, our society cannot possibly exist in economic health, or in intellectual and moral health either.

As I said a moment ago, I don't believe that there is much that engineers can do for themselves. I would go further. I suspect the less engineers talk about the problem, the better for all that we want to attain. I am entirely in favour of the new Council of Engineering Institutions. But formal organisations of this kind usually don't affect the status of an activity by 1 per cent. I think that, with some reserve, I am in favour of the new designation of 'Chartered Engineer'. But again experience suggests that professions riding the crest of the wave tend to take this kind of designation rather lightly. Scientists in this country, for example, enjoy a very high status, probably as high as in any country in the world. Our pure science continues to be remarkably successful. After the United States, we come a clear second in the Nobel prize race, with all the rest of the world non-starters. Since 1945 our number of awards is greater than that of the Soviet Union, France, Germany, and the next three countries in the list, added together.

The unorthodox mind

I don't think it is an accident that, among our scientists, there is still a cheerful and relaxed attitude to the kind of qualifications some of our most creative scientists are equipped with. Cockcroft was a great physicist but had never taken a Physics course in his life. Einstein and Dirac, two of the most profound theoretical scientists, not only of this century but of all time, started life, believe it or not, as engineers. In both cases it seems to have been thought that, while they might not be specially good engineers, they both showed a certain mathematical talent. Crick has revolutionised modern biology; but he has had as much formal instruction in Hebrew as in biology.

The point I wish to stress is that the scientific climate is easygoing and encouraging. It would be a mistake and, to my mind, a suicidal mistake if engineers, feeling, and rightly feeling, that they were undervalued, went on the defensive as a result. And, because they were on the defensive, became not less rigid but more rigid. It seems to me probable that engineering has already suffered through the rigidity of its training. This is one factor, but an important factor, in the lack of appeal of the profession to men of original and unorthodox minds.

At all costs, we need to attract such unorthodox minds. I believe that constantly we ought to look over our shoulders and ask why pure science is so good at attracting them—and

*Twelfth Graham Clark Lecture delivered to the Institutions of Civil, Mechanical and Electrical Engineers in 1966.

what is wrong with both the education and the career of engineers that acts as a negative inducement.

Forgotten men

In a moment I will go back to the education of engineers, but I would like to throw out a few indications of how little engineers are understood or appreciated in this country. We were all grateful when the Royal Society seven years ago engineers.

Yet to anyone who has ever been interested in either the practical or aesthetic contribution of engineers, or ever even looked at an aeroplane or a computer, it seems extraordinary that this step wasn't taken years and years ago. Somehow, already in the 19th century, engineers became forgotten men, even though they were making the nation's wealth. That is still true.

I noticed just before the 1966 election was announced, an interesting piece of news. This is not in any way a political article. In fact, I thought that the subject I am writing about reflected credit on the good sense of the Conservative Party. It was proposed, so I read, that they should try to collect a kind of don's brains trust. That is, they should try to draw into Conservative councils a number of minds from universities in the country. This was, of course, in part a recognition of the very high repute of dons in our society.

There is nothing new in this. The Labour Party have been doing exactly the same thing ever since 1945. A great deal of economic and scientific thinking has flowed into politics from precisely this source. That is fine, so far as it goes. What I find depressing and lamentable is that no one in this country would even contemplate trying to collect the wisdom of engineers. On many of these same problems they would have much to offer: but no one thinks of them as he does of dons.

In just the same way it is significant, and once again depressing, how very rarely engineers figure in any kind of art. I have spent some time trying to get interesting fiction about engineers onto TV. I have had no success. How many engineers enter into the novels and plays of the last 100 years? The answer is, astonishingly few. In recent times the only creditable attempt to represent engineers at work occurs in the novels of Nevil Shute. He was, of course, an exceptionally popular novelist and, partly because of that, it hasn't been often recognised that at his best, he was a good and original one. He was also a good and original engineer.

These are small things that I have been talking of. But they are indicative. Pure scientists have been taken, although with some difficulty, into the main stream of our society. They can be depicted, with some hope of establishing communication, in films and plays and novels and in the mass media. With engineers the barriers are much more formidable.

Why has this happened? And how are we going to put it right?

Why it has happened seems mainly a matter of our social history, but a good many of the reasons we still do not begin to understand. It is at first sight puzzling that we, who began the first industrial revolution, one which was closely linked with the work of engineers, should during the 19th century come to think less of them and care less about their education: whereas in the United States and Germany, industrialising themselves about 20 or 30 years behind us, exactly the reverse was true. This is one of the oddest phenomena in English history. From it have followed many of our existing

Lord Snow was educated at Alderman Newton's School, Leicester, and at Christ's College, Cambridge. His early career was as a professional scientist; he was a Fellow of Christ's College from 1930 to 1950 and Tutor from 1935 to 1945. During this period of his life he began his series of novels under the general title 'Strangers and Brothers'. As a Civil Service Commissioner during the 1939–45 war, he was deeply involved with Britain's scientific war effort and from 1947 until 1964 he was a Director of English Electric. He has received innumerable honours and awards for his literary and scientific work from the USA, the USSR, France and this country. Lord Snow is the originator of that pregnant phrase 'the Two Cultures'. Knighted in 1957, he was created a Life Peer in 1964. In the 1964 Government he was Joint Parliamentary Secretary to the Ministry of Technology.

discontents. Relative to the United States and Germany, we were on the technological decline very early. One can even put a date to it. It was somewhere between 1851 and 1867. If we had put one-tenth of the effort into engineering that we put into the Indian Empire, we should now be a very prosperous country, individually more prosperous than Sweden and able to look American technology in the face. But we didn't.

It is no use repining about where our ancestors went wrong. The important thing is not to continue to go wrong.

A humane education

It is, however, going to take a long pull to get rid of our lack-of-insight syndrome, by which I mean the various symptoms which prevent us recognising what is necessary for us in terms of rudimentary social health. Features of this syndrome are lack of awareness of who makes the social wealth: absence of interest in, or even active hostility to, the technological basis of our society: nonsensical patterns in education. These malaises are very deep among us. Even with good fortune it is going to take ten or twenty years before we breathe a freer air and walk about with a more robust common sense.

We need a powerful educational effort, an educational effort which reaches the people who have to make the decisions, and also produces a new generation of those who will make the decisions in ten or 20 years. Above all, we need to produce new generations of engineers who have been taught engineering as a humane education.

Don't misunderstand me. I don't want all our bright boys and girls educated as engineers. God forbid! Any kind of talent is worth training to the full. Our society would be immensely impoverished without full scope for its artists and linguistic scholars. But I am more than ever convinced that our secondary education is too specialised; that we've got to search for and train—and probably specially train—every kind of really high ability; and that we have to make special efforts at the undergraduate level to make engineering education something which it is not at present, or at least has hardly begun to be.

It is not enough to think that someone who is a natural engineer will find his way into the profession. Very largely that is true of engineering, as it is true of pure science or music or painting. But the number of people with a deep and passionate vocation for any walk of life is quite small, a tiny fraction of 1 per cent. What is needed is to attract to valuable social functions such as engineering, people who could do many other things. Most of us respond to the pressures of our society. Where the esteem and the rewards appear to be, there able people will go.

Most of us graduates, if we had been born 800 years ago, would not have studied engineering or science, for the very good reason that there was no engineering or science to study. We should probably have become canon lawyers, however, just because that was the road to all the administrative jobs there were.

In the same way, the high-class professional English administrators in the 19th century studied Greek and Latin. Not because they could gain an intellectual training, which indeed they did, certainly not because of profound literary interests, but because that was the accepted training for anyone who aspired to run this country. Our task is to make engineering just as natural, just as accepted educationally—not the only way, but an important way—for young people of high ability who want to test themselves to the full. When we have done this, I believe the place of engineers in society will be much nearer the ideal we should all like.

I've often thought it surprising that engineering has not in fact become more of a humane education. The training is rigorous, and that is a good thing. No subject is really worth years of time for a good mind unless it is, in the old-fashioned academic sense, 'hard'. Engineering is certainly that. It is, or certainly ought to be, deeply concerned with human beings. That is, an engineer does not exist in a vacuum: his working life is going to be intimately concerned with people: the results of his activities are going to affect much larger numbers of people.

And yet, somehow, engineering, like medicine, with which it has much in common, has up to now been something of a failure as an educational discipline in its own right. I am not being abusive. I have spent much of my life among engineers and count some of them among my most intimate friends. By and large, engineers turn out to be very good citizens, honourable, responsible and upright. But I think we should most of us agree that they are not by and large as fully educated as we should like. It is partly for that reason that they don't play the prominent part in society that we should also like. There ought to be far more engineers where the decisions are really made—in Government, in Parliament and at the top levels of the Civil Service. This doesn't happen. In the United States it happens a good deal more, in the Soviet Union incomparably more. When we have good engineering education, I believe it will begin to happen with us.

The engineer as a conformist

I believe one can identify two defects—there are probably many more—in contemporary engineering education. One is, it does not encourage enough the speculative and rebellious intelligence. This is its great handicap, and its great lack of appeal, as opposed to pure science. Engineers, as I say, are good citizens. It is rare for engineering students—at least it was rare when I was a don at Cambridge and, so far as I can observe, is rare still—to question everything under heaven or earth in the way that good science students will.

Medical education also seems to repress the speculative intelligence. Engineering students and medical students are nearly always the most conservative-minded of any large student groups. This tells its own story. I am not asking the engineering profession or the medical profession to move as one man sharply to the left. What I am saying is that a uniform approach and a conformist approach to all the major problems of society suggests either that we are getting too homogeneous an entry in these professions or, alternatively, that the training itself has a rigidifying effect. I should like to see as part of all engineering education what one can already see at MIT and Caltech and at some colleges of technology here—a serious study of social history or economics, or any subject which, of its nature, is bound to arouse the speculative mind.

The second of these two defects seems to me to be the training's lack of verbalism. Engineers as engineers can get on very comfortably without words. Engineering students can make do with a very small vocabulary, and it doesn't matter if they misspell the vocabulary they use. But engineers, when they are not being engineers, when they are moving into management, board rooms, Government, are often handicapped just because they have neglected or despised the use of words. Language, verbal language, is a great invention. It is different from symbolic language, and a man is lucky if he is equally at home with both. But without some feeling for verbal language there is much of human wisdom shut off from one for ever. It ought not to be shut off from our engineering students. They will be less complete human beings without it.

On the practical side they will be less effective than they should. A surprising amount of authority still depends on the written and the spoken word. Perhaps too much. But we have got to live in the real world and not the world of desire. If our engineering students are going to hold their own in ten years' time with people of comparable intelligence but different training, they have got to be as good on paper and as good speaking on their feet.

I believe this widening of engineering training would of itself attract a great many students of high ability who at the moment are being drawn into pure science or economics. It is precisely such students that we need. Some of them will be among the decision makers of the 'eighties.

I should like to make one digression. I have been speaking of students of high analytical ability. It is those we want if we are going to produce our humane and humanely trained engineers. But there is another smaller group whom I want to make a special case for. These are people who almost invariably lack analytical verbal skill. They often lack any kind of analytical mathematical skill. Judged by IQ tests they come distinctly low. They are usually pretty bad at examinations. The result is that they get missed by our educational net.

Yet they may possess—all of us must have known some—a special ability which is most valuable in engineering proper. It is the ability to think concretely in three dimensions, to have an innate sense of design. Some of the most accomplished of all engineers have had this particular mental temperament. Our present education system is badly designed to cope with them. But somehow we have got to find another net which scoops them in.

The world is pretty easy with persons of high general analytical intelligence. That is a very desirable quality, and a lot of our future depends on training enough such persons. But one must not in the process forget about the odd fish who don't fit comfortably into our present forms or into any conceivable forms. If you doubt that—though I don't think

many engineers would doubt what I say about the odd fish—
look at what Dr Sommerhoff is doing at Sevenoaks school.

The question of reward

If we get our education right, then in the long run the
place of the engineer in our society will become right. But
I should be unduly timid if I didn't mention that what a
society recognises is ultimately expressed in concrete terms.
These terms are not only financial. Esteem to most sane
persons is more important than pay.

If we had the choice (which is unlikely to come to any of us)
most of us, I hope, would prefer to be President of the Royal
Society than the most successful pop singer in the world.
Yet the most successful pop singer in the world has been out
of proportion better rewarded financially than any of the
Presidents of the Royal Society I have had the privilege of
knowing. Nevertheless, pay, in a qualified sense, does matter
in a society like ours and is likely to continue to matter. I
don't regret this.

If that is agreed, then at present we certainly are not lavish
with our rewards to creative engineers. I have studied a lot
of evidence in the last 17 months. Take one of my theoretical
young people of high general ability who could become
reasonably accomplished as a pure scientist or economist or
engineer. At the moment, if he is thinking only of money,
engineering is not the first profession he would choose. His
starting salary as an engineer would be as good as, or better
than, the others. His terminal salary is likely to be appreciably
lower. This is another surprising fact, but the evidence is
overwhelming. Like Poincaré in an air raid, I take refuge
under the arch of probability.

I am sure this cannot be right. Once again, it will not be
adjusted overnight. The proper value placed on engineers
will only happen when we who are interested contrive to get
our side of the house right. And when we have persuaded the
rest of society what engineers are like, what they can do, and
how upon them depends a large part of whether our country
flourishes or dwindles imperceptibly into decline.

Hey Mac! — the Engineer in Modern English Fiction

by John Winton

While the other arts borrow increasingly from the engineer's vocabulary and use his tools and materials, modern English fiction continues to treat him as a non-character. Is this because very few engineers have ever taken up the novelist's typewriter — even though they may have designed it?

It is a modern paradox that while contemporary artists are making increasing use of engineering techniques and engineering materials, they still remain almost wholly ignorant of, and uninterested in, the engineer himself. The impact of engineering on art is evident in many modern paintings, modern furniture design, the sculptures of Anthony Caro and David Smith, the photomontages of John Heartfield, films such as *2001: A Space Odyssey*, the buildings of Frank Lloyd Wright and Alvar Aalto and in the designs of Walter Gropius, Buckminster Fuller and many others.

Some of the latest art-forms are kinetic playthings, attractive assemblies of mobile objects, shapes, lights and colours. They are admittedly ephemeral, expendable 'fun-pieces', but their authors nevertheless display a quite sophisticated knowledge of engineering possibilities—the use of machine tools to create contrasting surfaces, the different textures of paints and aerosols, and the various properties of perspex, plastics and metals. But although modern poets record 'sound poetry' on tape and modern composers write scores for electronic music, vacuum cleaners and pneumatic drills, few anthologies include an *Ode To An Engineer*, few repertoires a *Symphonie Mechanique*.

Initially, these omissions may seem surprising. In a world where his handiwork lies visible on all sides, in motor cars and motorways, aircraft and airfields, ships and shipyards, television and radio, new machinery, new buildings, dams, harbours, bridges, pipe-lines and pylons, the engineer might have become a folk-hero, widely celebrated in word, song and dance, like the early pioneers of the American West. Those who have so radically changed the shape of life and landscape might expect at least a nodding recognition in contemporary folk-art. Such celebration has not occurred, indeed the mere notion of it seems faintly ludicrous. Art gladly utilises engineering but consistently rejects the engineer as a character or subject.

The rejection is almost total in modern creative fiction, where the engineer is a non-hero, often ignored, sometimes caricatured, almost always a Laertes-figure in the plot—ground-crew for the hero-pilot, pit-crew for the hero-racing-driver, engine-room crew for the hero-captain. To borrow Danny Blanchflower's adroit analogy in a *Sunday Express* article written after Donald Campbell's death, the chief mechanic is like the faithful Indian who holds the horses. He is always Tonto, never the Lone Ranger.

For some reason which is worth discussion, the authentic, immediately recognisable novel about engineers and engineering has not yet been published (at least in English; nobody knows what might be stirring inside Russia). Any phrase such as 'The Great Engineering Novel' somehow seems absurd.

Perhaps the paradox is best approached by taking examples of modern fiction and examining how engineers and engineering are handled by the authors.

The engineer caricatured

The engineer as caricature has a long literary tradition, particularly in sea stories, in which a Chief Engineer only rarely achieves the metaphysical dignity of Kipling's archetypal McAndrew—the man who "saw from coupler flange to spindle guide the hand of God and predestination in the stride of yon connecting-rod". More generally, engineers are represented as 'Hoots mon, ye ken they big-ends will dae for anither twa hoors, och aye' characters; Harry Lauders in boiler-suits, with their inspiration presumably in the time-honoured premise that one can shout 'Are ye there, Mac?' down the engine-room hatch of any merchant ship anywhere in the world and up will come the Chief Engineer.

Yet, curiously, from a caricature in a work of gentle and genial satire, *England, Their England*,[1] there emerges one of the most credible and endearing engineers in fiction—the ineffable Mr William Rhodes. A Yorkshireman from Leeds who happens, by an accident of parentage, to be bilingual in English and German, he travels the world on behalf of his firm, delivering, erecting and demonstrating ever more exotic items of machinery: machines for digging irrigation canals in Spain, for boring oil-holes in Rumania, weeding between fruit trees in Hungary and for pumping out the sewers of the city of Warsaw. He is one of nature's aristocrats and, in his own way, a poet; he derives the same satisfaction from machinery as other men get from books, music or poetry, and he will go on designing and making machines until his dying day, even if they are only toy engines for his grandchildren.

It is one of this country's remaining strengths that one can still see William Rhodes, or someone very like him, in the waiting room at Heathrow most weeks in the year.

The concept of engineering as a wholly satisfying way of life is completely turned on end in the work of Alan Sillitoe. *Lady Chatterley's Lover* was once described as 'a day in the life of an English gamekeeper'; similarly, Alan Sillitoe's first novel *Saturday Night and Sunday Morning*[2] could be described as 'the life and hard times of a Nottingham capstan-lathe operator'. Except for the limited satisfaction in the perfect co-ordination between hand and eye demanded of the skilled lathe operator, Arthur Seaton looks upon his work at the bicycle factory as no more than remunerated slavery; the lathe, the rate-fixing, the foreman, the factory itself, are all segments of a huge pattern of imposed and resented authority. The job is a means of filling the time between Monday morning and Friday evening; so many hours of servitude exchanged for a pay-packet. Life, real life, begins at the weekends.

The background is authentic and, unless factory attitudes and working conditions have greatly changed since the 1950s, every employer or manager of assembly-line labour would do well to study the shop-floor episodes in this book.

On tap, not on top

Engineers are often conscious and resentful of the tendency to place them at a subordinate level in any group or organisation. 'Engineers on tap but not on top' is a familiar and hurtful saying.

Nowhere is this subordination more clearly demonstrated than in novels describing life in a particular closed society which includes engineers. A good, if unintentional, example is C. S. Forester's *The Ship*.[3]

Published in 1943, the novel contains a strong propaganda element, its theme being the old military maxim that morale is to material as three is to one. The story strikingly resembles a famous naval action in the Gulf of Sirte in March 1942, when a force of cruisers and destroyers under Admiral Vian successfully defended a Malta convoy against an Italian battle fleet. The cruiser *Artemis*, the ship of the title, is popularly supposed to be HMS *Penelope*.

One whole chapter is devoted to the Engine-room Branch, with an admirable and evocative description of the cruiser's Commander(E) and his men and their stoical composure under stress, isolated as they were at their action stations far below the water-line and knowing little of the violent events taking place above. The chapter is undoubtedly C. S. Forester's own sincere tribute to the Engine-room Branch of the Royal Navy. But it adds nothing to his story. If it were deleted, the novel's structure would hardly be impaired. In *Artemis*, as in so many other situations, the engineers are auxiliaries—vital to the ship it is true but irrelevant to the story.

It is arguable that a warship in the presence of the enemy is a special case; in such circumstances, all on board are the captain's auxiliaries. If other examples were chosen, nearer to the everyday norm of life, surely an engineer could expect to find his profession shown in sharper relief?

It does not seem to be so. Certainly many 'closed society' or 'inside story' novels are published every year and certainly a fair proportion of them rely upon an underpinning of engineering technology. But the supporting structure is firmly kept out of sight. One excellent example of the *genre* will demonstrate this: Arthur Hailey's *Airport*,[4] published in this country in 1968.

The author has obviously carried out research in depth into the workings of a modern airport and he describes the solutions to problems of passenger relations, staff management, baggage handling, air traffic control, security, car parking, customs and excise, catering, hygiene, noise abatement and publicity, in exhaustive detail. The manager of a large international airport clearly needs to have administrative ability of a very high order. The author does not lay the same

John Winton *is the usual pen-name of Lt-Cdr John Pratt, RN, CEng, MIMechE: He was educated at St Paul's School, the Royal Naval College, Dartmouth, and the RN Engineering College, Manadon. He joined the Royal Navy in 1949 and served as an engineer officer in the Korean war, at Suez and in submarines, before retiring in 1963 to become a full-time writer. He has published six novels, including 'We Joined the Navy', 'HMS Leviathan', 'The Fighting Téméraire', and has written many technical handbooks. He is 40, married, has two children and lives in Cheshire.*

emphasis on technical matters, although an airport, of all complex modern installations, depends upon the efficient working of a great range of machinery.

The only engineer among the principal *dramatis personae* is notable for his sexual prowess and his expertise as a 'trouble-shooter' when, for example, he extricates a bogged-down airliner which is obstructing a desperately needed runway. Once more the engineer is ancillary to the main plot. The reason for this is important and crucial to any examination of the engineer's place in fiction: people like to read about people and, understandably, the novelist calculates on a better livelihood from love affairs amongst the aircrew than from engine maintenance routines.

Science-fiction

None of the books so far mentioned could, by the largest stretch of imagination, be called engineering novels and indeed their authors would have been dumbfounded by any such suggestion; they do touch upon engineers and engineering but they do so obliquely, by inference or by accident. A more direct concern might be looked for in science fiction novels, more especially today when so much of sci-fic is fast becoming sci-fac and the authors must write while looking over their shoulders at the real world rapidly overtaking their wildest fancies. After all, the very words 'science fiction' presuppose an imaginative interest in scientific and technological affairs.

But, once again, expectations are disappointed. At their crudest level, science fiction novels are what is known as 'bug-eyed Westerns': the spaceship is a kind of galactic stage-coach; meteorites whine and ricochet like lead slugs around the heads of the unwary; death-ray guns replace six-shooters; and for Injuns read Martians or Alpha Centaurians.

Although these novels blandly assume, and their *denouements* often depend upon, the accomplishment of marvellous feats of engineering, the engineer generally pays no more part in them than does the saddler or the wheelwright in the average pulp Western.

Fortunately, there is a more subtle and rewarding type of science fiction novel, in which the author takes as his starting point one or more facets of modern existence—of technology, philosophy, religion, medicine, social habits or morals—and projects that concept forwards and outwards to its furthest logical, and sometimes illogical, conclusion. These novels are often valid and revealing commentaries on present-day life; by postulating what might be, they illuminate what is. Of several first-class science fiction novels which have attempted an answer to the question 'Where is our present technological race leading us?', one of the most brilliant, and blood-chilling, is *A Canticle For Leibowitz*.[5]

The story opens in America, some hundreds of years after a nuclear war which devastated much of the earth's surface. In a great frenzy of 'Purification', the survivors of the 'Flame Deluge' have expunged all traces of science and technology. Books and documents were burnt, instruments and machines destroyed, and all who created them or were in any way connected with them, massacred. It must never happen again.

However, in a remote desert monastery an order of monks have preserved and now venerate their *Memorabilia*—a collection of miscellaneous technical and household documents such as blueprints and shopping lists, once the property of a long-dead man called Leibowitz, which survived the fire. The *Memorabilia* give tantalising glimpses of a more enlightened and seemingly happier existence before the Flame Deluge, and generations of monks spend their lives studying them.

Eventually, with the help of the *Memorabilia* and a gigantic leap of intuition, one monk succeeds in making a primitive form of electric generator.

At once, the way is open to a second technological revolution and in due, inevitable course, to a second nuclear holocaust. This time, the monks escape by starship to a distant galaxy. But, ominously, they take with them the *Memorabilia*. Technological man carries with him, willy-nilly, the seeds of his own destruction.

The author, Walter Miller Jr, deploys an enviable literary skill and a keen intellect in unfolding this appalling train of events and its implications for mankind. There is much food for reflection in his book for theologians, politicians, historians, nuclear physicists, sociologists, even for anthropologists. But there is nothing for engineers. The novel's timescale covers colossal strides in engineering theory and practice, but not one of the richly-drawn and memorable characters is an engineer. It is simply assumed that engineers will continue to invent, build, maintain and repair the necessary machinery to fulfil the story.

Not so Jules Verne

It was not always thus. Jules Verne, the 'father' of science fiction, held engineers in the highest regard and allowed them major roles in his novels. The technical flair of the redoubtable Captain Nemo in *20 000 Leagues Under The Sea* is well known. Not so famous, perhaps, is Captain Cyrus Harding of *The Mysterious Island*,[6] a novel set at the time of the American Civil War. With four other Northern prisoners-of-war and his dog 'Top', Harding escapes from Confederate captivity in a balloon which is then driven by storms across to the Pacific and wrecked upon a small, remote island, where Harding and his companions are left, literally, to their own resources.

Harding, or simply "the engineer" as he is often called, is described as being "courage personified", "learned, clear-headed, and practical", with

one of those finely-developed heads which appear to be struck on a medal, piercing eyes, a serious mouth, the physiognomy of a clever man of the military school.

He is the natural leader of the party and is capable of turning his hand to anything, from repelling pirates to making soap; according to the author, Harding might have taken for his motto that of William of Orange: "I can undertake and persevere even without hope of success".

The Mysterious Island, published in 1875, is one of the very few novels in which an engineer is acknowledged as a genuine hero, using that word in its unqualified heroic sense.

If engineers could write

The scarcity of engineer heroes is, of course, directly related to the scarcity of engineers who can write (writing, in this context, means writing for a lay public). There is clearly some inhibiting quality in machinery, for engineers in general have remained grimly mute about their working lives. Meanwhile, doctors, teachers, politicians, lawyers, people whose business is people, write prolifically and indeed seem to look upon publishing books as a necessary prerequisite of their professions.

Anthropomorphism, even towards such a mechanism as a motor-car, is evidently no substitute for a real live breathing client. Professional writers engage in all manner of esoteric activities, working as waiters, attending murder trials, going to sea in trawlers, climbing the Matterhorn, to gain material for books.

So far it does not appear that any author has published a novel as a result of spending time in an engineering firm or at a work-bench. The best indeed the only, possible source of a modern novel which engineers would recognise as having any relevance to their profession would seem to lie within the profession itself.

It is therefore not a coincidence that the two writers who have dealt most fully with engineering matters in their novels both had technical or scientific backgrounds. One is the late Nevil Shute, a master storyteller, and the other is Nigel Balchin, a novelist of genius.

After taking an engineering degree at Oxford, Nevil Shute worked with Barnes Wallis on the airship R100 and later founded Airspeed Ltd; in the 1939-45 war he was a Lieutenant Commander RNVR serving in the Department of Miscellaneous Weapon Development, better known as 'The Wheezers and Dodgers'. In a long writing career he published several novels with an aero-engineering background, of which *No Highway* may be taken as representative.

Nature follows art, and in at least one respect *No Highway*,[7] published in 1948, was disturbingly prophetic: the metal fatigue failures in the tailplanes of Shute's *Reindeer* aircraft were uncomfortably close to the actual circumstances of the *Comet* disasters only a few years later. No doubt, too, there are or were at Farnborough men such as Mr Theodore Honey, the boffin whose research work first forecast the possibility of a disaster.

Mr Honey was probably one of Nevil Shute's favourite creations; a middle-aged widower with a wide, if somewhat eccentric, range of interests, he emerges as a man of unexpected personal charm and great moral courage, sticking to his theories in the face of disbelief and ridicule.

In one sense, *No Highway* broke new literary ground. There have been many novels about human beings under stress. But this was probably the first to show that there could be drama in a piece of metal under stress. Three years later, in *Round The Bend*,[8] Shute developed the same broad theme to the point where he seemed to attach to aero-engineering the significance of a religion, through a main character who was both an engineer and a mystic—a combination of talents to make the unprejudiced mind boggle. Not many engineers would follow Mr Shute that far.

Nevertheless, his novels did contain engineers as convincingly-drawn major characters, and his place as a pioneer is therefore secure.

Shute's 'Wheezers and Dodgers' oddly resembled the anonymous weapons research establishment of Nigel Balchin's *The Small Back Room*—one of the half-dozen finest novels of the 1939-45 war. Nigel Balchin first began to write scientific stories when he was at Cambridge and later led a 'double life' as an author and playwright while he was a consultant at the National Institute of Industrial Psychology; in the war he was Scientific Adviser to the Army Council, with the rank of Brigadier.

Sammy Rice, the boffin of *The Small Back Room*, is a superb creation. A man of complexes and contradictions, he is possessed of a beautiful mistress but is haunted by fears of his own inadequacy; crippled with an artificial leg, he is, when the test comes, as brave as a lion. No office politician himself, he is exasperated by departmental intrigues and many scientists and engineers will sympathise with his feeling of helpless resentment that, while he in the small back room is doing the effective work, some smoother operator in the large front office is getting the kudos and the preferment. He suffers from a huge inferiority complex which is aggravated when he compares himself with his glamorous younger brother, a much-decorated RAF pilot, and which reaches a climax in the famous scene on the beach—one of the most exciting passages in all fiction—where he attempts to dismantle a new type of anti-personnel bomb. Here, characteristically, an achievement which most men would regard as the finest moment of their lives, Sammy himself looks upon as a failure. Any engineer with fears of futility, working in a large organisation, can read *The Small Back Room*[9] and marvel that a novelist has already passed that way before, in imagination.

The Small Back Room was published in 1943 in what now, looking back, seems almost to be engineering's age of innocence. Nearly a quarter of a century later Nigel Balchin returned to the subject when, in 1965, he was invited to go to America and was given every facility by the National Aeronautical and Space Administration. The result was *Kings Of Infinite Space*,[10] a novel about the exploration of space.

Except for the sound and fury of an actual rocket launch, a moment of hope and pure poetry, Mr Balchin evidently had mixed feelings about what he saw at NASA. While the scientist in him was thrilled, the human being was chilled. It was all too, too sane. Man's technical resources have now outrun his imagination—the opposite of Icarus. Plucked from his Cambridge college and hurtled into the NASA training

programme, Balchin's physiologist-astronaut at last concludes that a man needs more than an engineering degree to be a king of infinite space. For space, a poet is really the only man. Mr Balchin points his scientist-astronaut's final view of the space race with an apt quotation from Tom O'Bedlam; it is a nice irony that a madman can best define the sanity of NASA:

> With a host of furious fancies
> Whereof I am commander,
> With a spear of fire,
> And a horse of air
> To the wilderness I wander.
> By a Knight of ghosts and shadows
> I challenged am to tourney,
> Twelve leagues beyond the wide world's end
> Methinks it is no journey.

Engineers need not be offended when an author of Balchin's stature implies that, by itself, engineering is empty and not worth one moment of a novelist's time. A novelist is concerned with people, their relationships with each other, and their reactions to their environment. However, while conceding the point, the engineer might then retort, ungrammatically but truly, that he *is* people. Prick him, does he not bleed? Tickle him, does he not laugh? Admittedly, the novelist can disdain the machine, but why should he also ignore the man?

Strangely, no engineer has yet made this riposte. Although occasionally exercising themselves over the question of better publicity for their mysteries, engineers do not apparently think it extraordinary that they should have no place in their community's literary culture, except for rare instances such as those already mentioned.

It hardly seems likely, but it is still just possible that at some time in the future a writer of genius who may himself be an engineer will summon up a host of furious fancies and write the definitive novel on the modern engineer and his role in modern society. Methinks it will be some journey.

BIBLIOGRAPHY

1. MACDONNELL, A. G., *England, Their England*. 1933, Macmillan, London.
2. SILLITOE, ALAN, *Saturday Night and Sunday Morning*. 1958, W. H. Allen, London.
3. FORRESTER, C. S., *The Ship*. 1943, Michael Jsoeph, London.
4. HAILEY, ARTHUR, *Airport*. 1968, Michael Joseph, London.
5. MILLER, JR, WALTER M., *A Canticle for Leibowitz*. 1959, J. B. Lippincott, Philadelphia.
6. VERNE, JULES, *The Mysterious Island*. 1875, Sampson Low, Loadon.
7. SHUTE, NEVIL, *No Highway*. 1948, Heinemann, London.
8. SHUTE, NEVIL, *Round the Bend*. 1951, Heinemann, London.
9. BALCHIN, NIGEL, *The Small Back Room*. 1943, Collins, London.
10. BALCHIN, NIGEL, *Kings of Infinite Space*. 1967, Collins, London.

The Status of the Professional Engineer

by F. H. Towler CEng, FIMechE

There is little doubt that the status of the engineer in this country is lower than desirable in the interests of the nation as a whole and of the profession itself. High status is both a cause and an effect of high earnings but it is also linked with educational and professional matters, as well as with recruitment. The ultimate solution must lie with a better organisation of the profession. This article was written in 1965 before the advent of UKAPE and the Industrial Relations Act; but those events have made its message, if anything, even more urgent.

The status of the professional engineer is low because he is underpaid and he is underpaid because his status is low. That is known as a vicious circle. It is particularly vicious because no one has done anything about it. The engineering institutions feel they are precluded by their charters and constitutions from taking any active interest in the remuneration of their members. And the bulk of the members have taken no interest in the Engineers' Guild which was formed specifically to fill this deficiency. The Guild has a membership of less than 6000 or 4·5 per cent of the corporate members of the institutions affiliated to CEI.

I have spent my life thinking, designing, writing, lecturing and describing machines and it is only in the last 18 months that the status of the engineer has even entered my mind. I joined the Guild long ago but I was so immersed in engineering that I had no time to give to it. I left it to the Civils, as so many others have done; and that is where it is today, mainly supported by the Civils. And why? Because their members are more individually professional than any other of the engineering institutions—more like the architects.

I have mentioned my own shortcomings because I believe that they are similar to those of most of my fellow engineers. We are not interested in status; for we are almost wholly engrossed in engineering!

What effect has this apathy (or unworldliness, if you wish) on our profession? Does status matter? It does, for it means that engineers are not being given posts of responsibility in government and industry or in our national life generally. If they are given such posts in the Civil Service, it is not as engineers; they are called Scientific Officers or Experimental Officers, anything but engineers, for otherwise they would suffer an immediate reduction in salary.

That is what our Civil Service thinks of engineers, but this view is not peculiar to it. The same thing is happening in industry. Why do the highest posts often go to Arts men? Because they have the ability to write, talk and administer what they do not understand. They make wonderful salesmen and, believe it or not, the highest posts go to the man with sales ability, the gift of being able to 'sell it' to the Board, to the shareholders, to the workers, to Government and public. And, of course, he also has to 'sell' his own personality.

Salesmanship

It is all the result of the 'two cultures': the Arts man can explain what he does not understand, he has the command of words and he only needs to receive the 'gen' from some lowly engineer, scientist, back-room boy. The frontroom boy can do the rest and he does it with disastrous success.

As for the scientist or technologist, he spends his life groping for fresh knowledge or new techniques. He would never claim that he wholly understands anything or that he has more than a hypothesis which he may have inherited from a colleague or predecessor. Nothing is certain and what he barely understands he cannot explain. He is engrossed in thought, supposition and qualification. He cannot even be sure what time he is going to lunch; it all turns upon so many independent variables.

But industry, commerce, government and public cannot be told about all those independent variables. They want certainties and they are prepared to pay high salaries to men who are capable of convincing them of certainty: even though it does not exist.

My view is that we need a broader culture, embracing science and the arts; and, in the early stages, learning should not be specialised. Our learning should be nearer to that of Goethe, Pascal, Newton—but with the corrections and short cuts which we have learned from those who followed them. A little learning is a dangerous thing; but too much specialised learning stifles the imagination.

It is just as important for scientists to learn engineering, as it is for engineers to learn science; but they must both know something of art, painting, music, the poetry of life. The world cannot be safe in the hands of lop-sided, incomplete beings: all brain and no mouth, or vice versa.

If, in the course of this introduction, I have seemed to stray from my subject, the professional engineer, it is because I believe the engineer must learn to be more than an engineer. All great engineers have had that quality; they were able to explain their ideas to a large body of people who were not engineers. They had both sagacity and daring.

Imagine for one moment that scene when Charles Parsons steamed the *Turbinia* through the British fleet, outpacing its fastest ships. That was salesmanship of the highest order.

In the past few decades the engineering institutions have become a trifle pompous; 'selling' has become a dirty word; most unprofessional. Doctors and dentists don't sell. No, but Great Britain must sell or starve. Engineering products represent nearly 50 per cent of our visible exports and the rest are mined, manufactured or transported by machine. As for our hidden exports: banking and insurance; they are becoming increasingly dependent upon the computer.

Our whole civilisation may be destroyed by pressing a button. Is it not important that those in places of responsibility should include first class engineers—be it, in the Cabinet or board room—but not tucked away in a back-room? Civilisation has become an engineering product and it needs engineers to control it; but engineers who understand machines and people.

The problem

In December 1964 Lord Snow said in the House of Lords[1] "Of all countries I know, this country respects engineers least;" and Lord Bowden in the same debate said that engineering places at universities were not being filled and that some engineers made their career choices at school "in a total fog, a complete absence of understanding". Perhaps the two statements are not unrelated.

In spite of his comparatively lowly condition, the professional engineer in Great Britain still contributes some of the finest engineering and some of the foremost inventions in the world. Affluence does not necessarily produce excellence nor does it always favour creativity. But must the artist and inventor always starve?

Is there a status problem? A survey by the Oxford University Department of Education[2] showed clearly that elsewhere in Europe engineering attracts the best brains. In some countries, notably France and Sweden, the elite of university entrants on the science side prefer engineering to pure science. It no longer needs stressing that the position is very different in Britain.

Prof. H. A. Prime of Birmingham University[3] believes that this bias against engineering as a career originates in our educational system in which industry is largely ignored because the teaching staff has negligible industrial experience; and that it is also fostered by the press, which credits the scientist with the achievements of the engineer. Somewhere along the line something did go wrong with the engineer's image but it is doubtful whether the position is quite as black as the professor painted in his lecture when he said, "Our social structure still regards engineering, like profit, as a dirty word".

Is it possible that professional engineers suffer from the same handicap as school-masters? that they are unwordly? I submit that most engineers have been so busy with engineering that they have neglected their profession. Thus, the Royal Commission on Doctors' and Dentists' Remuneration 1957–60[4] surveyed other professions: accountants, barristers, architects, engineers, university teachers, etc. About engineers it had this to say:

> This profession is not organised in a single association and the three institutions, respectively of Civil, Mechanical and Electrical Engineers, who speak for it on most matters, are prohibited from dealing with questions of remuneration. At their suggestion we accordingly approached the Engineers' Guild . . . .

How very unwordly engineers have been! If all Chartered Engineers joined the Guild, it would have a membership of over 100 000. But, in spite of the efforts of many worthy and distinguished engineers, including Presidents and Council Members of this and kindred institutions, the Guild has been treated with apathy by the vast bulk of professional engineers. They do not appear to see the basic need to be organised on an equal footing with other professions, such as doctors, dentists, teachers. Why this apathy? Are we afraid of losing our dignity—to be classed as trade unionists?

As regards our other professional organisations, there are any number of them. Thirteen at last got together three years ago to form a common organisation, now called the Council of Engineering Institutions and granted a Royal Charter of its own. Thus, for the first time since 1847, engineers in Great Britain will be able to speak with a common voice—about some things at least. One hopes they will use it.

Remuneration

Every member of the public has at some time suffered from irresponsible action official or unofficial, which has done damage to the trade union movement. But one only has to compare the rising standards of manual workers with those of technologists to decide that the trade unions have done the job they were set up for. Have we professional engineers counted the cost of our own lack of organisation? Where would the doctors be without the BMA?

With reference to the relationship between status and salary, Mr J. G. Orr, Secretary of the Engineers' Guild, says[5]

> There is no doubt that the engineering profession is much lower in status than many other professions, and lower than it is abroad. How far status determines salary, or salary status, is a matter for argument, but at least in the commercial and industrial organisations, where most engineers are employed, it is reasonably certain that status and salary are related . . . .

Consider the effect of inflation: the pound of 1938 was worth 10s in 1950, and its 'value' is about 4s in 1965. During this period the trades unions have been able to protect their members by continual demands for increased wages. Those demands had to be met; the average wage for men over 21 is now[6] £18 18s 2d, 60·5 per cent more than in 1956. In common justice, most employers have tried to keep salaries in step with wages. But where there is no pressure, no collective representation of the employee, the good employer suffers unfair competition from the bad one. Also, he is often as ignorant of the 'rate for the job' as the employee himself. This situation, in which employers are acting blindly, reluctantly, to increase salaries, must ultimately result in a brain drain to other professions, industries or countries, where salaries are more in step with wages.

For instance, a survey of almost 2000 Graduates of the Institution of Electrical Engineers by Research Services Ltd[7] in 1964 revealed widespread dissatisfaction among engineers—9 per cent of those under 44 said they would be interested in emigrating permanently from Britain now; and 38 per cent "some time". Asked "What do you think the surest road to top management tends to be?", their answers ranged as follows:

36 % accountancy
29 % sales
17 % family connections
14 % production
 3 % design
 4 % research and development.

That surely must affect the pattern of recruitment in a manner not at all in conformance with the industry's real needs.

I wonder how many industrial employers know that the Government is beginning to offer more than private industry, and that leading executives are also being recruited by the universities. We have a Professor of Marketing at Lancaster—a very healthy move which should benefit industry in the long run. No one would suggest that such a brain drain is always against the national interest but if present trends continue, too little talent may be left in industry.

An exchange between the UK and the USA can also be of benefit; but it must not be all one way. We cannot afford to spend huge sums giving free education in our schools and

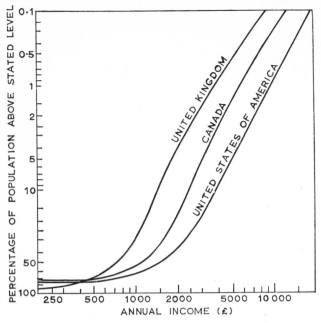

Fig. 1. A horizontal line across this graph compares the earnings required for the same social position in the USA, Canada and the UK

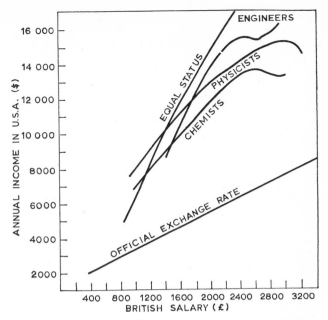

Fig. 2. What an engineer could earn in the USA compared with the salary conferring equal status. The official exchange rate is misleading

universities to students who emigrate in large numbers because we are unable or unwilling to offer comparable salaries and positions of responsibility.

On the other hand, salaries which look magnificent at $2·8 to the £ are not all they seem to be. We need some way of comparing salaries, status, cost of living, terms of employment, etc, as between America and Britain. For instance, there is no free health service, medical care can be very expensive; one can be out of a job at a week's notice; there are fewer pension or superannuation schemes. We take many of our benefits for granted, they have become a part of our way of life.

The brain drain and the pecking order

Supposing that 2·5 per cent of the population of the UK have an income of over £2,600 and 2·5 per cent of the population of the USA have an income of over $18 800: then, as

regards income status, the two are equal; though not, of course, as regards purchasing power. What this means is that 'keeping up with the Jones' will cost $18 300 (£6,700) in the USA as compared with £2,600 in UK. But the tastes and way of living of the Jones in the two countries are as different as baseball and cricket. Fig. 1 gives curves[7] of the respective salaries plotted against per cent of population in UK, Canada and USA. A horizontal line drawn across the graph will indicate equal income status in each of the three countries, expressed in £/year (this is very different from the official exchange rate or even the equivalent cost of living, as can be seen from Fig. 2).[8]

On the other hand, incomes within a profession do not necessarily follow the pecking order of the community as a whole. Clearly, before moving to the USA, a young man should decide what he is after: more prestige or more purchasing power; otherwise he might be disappointed.

Table 1—Earnings of graduate engineers in the USA 21–25 years after graduation or at 43–47 years age

Page of Report	15	17	23	33	37	39	51	57	67
Industry	All Activities	All Industries	Aerospace	Electrical Machinery & Electronics	Instrument Mfg.	Machinery Mgf.	Research & Development Activities	All Government Levels	Engineering Colleges
21–25									
Supervisors									
Median $	16,162	16,326	18,217	17,460	19,939	14,521	18,793	15,168	14,932
*£	2180	2200	2480	2380	2750	1930	2500	2020	1989
Total number	7522	6406	920	753	128	323	349	1093	734
Percentage of total	50%	50%	38%	45%	53%	57%	52%	41%	100%
Non-Supervision									
Median $	12,741	12,869	13,853	13,506	15,250	10,654	13,358	12,313	
*£	1680	1620	1840	1800	2040	1400	1760	1620	
Total number	7915	6351	1520	950	112	230	320	1539	

* Conversion $ to £ on the basis of income status or 'pecking order', as between USA and UK.

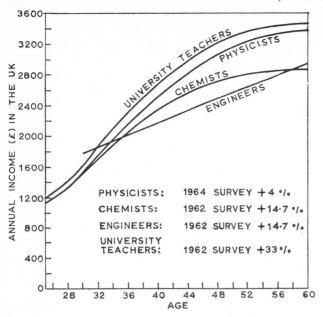

Fig. 3. Engineers in the UK earn about as much as other professions at 35 but are worse off in later years

Fig. 3, taken from the same source,[9] makes another interesting point: that in the UK engineers earn about as much as similar professions at around the age of 35 but that their further increments are much smaller. By retiring age the average physicist is £1,000/year better off. Is this because there are more engineers available? because they are more reluctant to change their jobs or because their status is lower? Or do physicists with their rather broader education, tend to 'grow with the job' more than engineers? We don't know the answers to these questions but we ought to.

We are severely hampered by the paucity of information about professional salaries in UK as compared with the magnificent biennial survey of professional incomes carried out by the Engineering Manpower Commission under the direction of the Engineers Joint Council of the USA.[10] Their survey for 1964 covers 231 613 graduate engineers in a large variety of industries, colleges and government depart-

ments. Here, the Engineers' Guild has done some spade work but it is greatly handicapped by its small membership and lack of reserves. I hope that the CEI will make a comparable survey in the UK; it is of the utmost urgency.

The American survey has been made possible by the cooperation of employers, heads of colleges and government departments. Its aims were:

1. Aid in establishing the importance of engineering to the national economy.
2. Aid in maintaining an adequate supply of engineers.
3. To promote the most effective utilisation of engineers in support of the national health, safety and interest.

It would be impossible to do justice to the results of the 1964 US Manpower Survey[10] within the space available. Figure 4 shows the trends of earnings of engineering graduates from 1953–1964. The total increase in the median earnings over the 11-year period has been about 62 per cent, compared with an increase of 16·2 per cent in the consumer price index and 43·6 per cent in the average weekly wages of production workers. The dotted lines connecting the various curves represent increases which would have been received by individuals (represented by this median) as they accumulated experience since 1953.

The two charts in Fig. 5 show the range of earnings of graduates in all activities. The thick lines represent median salaries; thin lines the salaries of the upper and lower quarters; broken lines the salaries of the upper and lower tenths. The median and upper tenth of all graduates is repeated on the charts for convenient comparison.

There are 55 pages of charts in this survey; and I have summarised some of them in Table 1.

This US survey should serve as an example to CEI and the Ministry of Technology, showing the sort of statistics we require to improve the income status of professional engineers; so that employers and employees may no longer have to grope blindfold in search of a 'rate for the job'.

In the meantime, when considering the statistics which follow, we must remember that we are not comparing like with like: many of our Chartered Engineers are not graduates. In the USA, according to the Manpower Commision, there were 935 000 engineers in 1963, of whom about 45 per cent, or 514 000, were graduates and the survey covered about half of these.

Professional engineers are so widely spaced in status among the community that I have thought it best to convert the

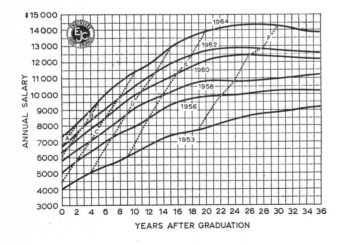

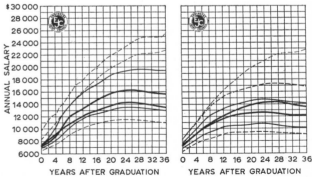

▲ Fig. 5. Graduate earnings (supervisors on the left, others on the right) in all activities in the USA. For details see text

◄   Fig. 4. The earnings of engineering graduates in the USA

Mr F. H. Towler, *born in Leeds in 1897, was educated at Ilkley Grammar School and Leeds Central Technical College. He was apprenticed at the age of 16 to the Leeds Engineering and Hydraulic Co., and later to Joshua Buckton and Co. The first world war saw him in the Royal Flying Corps in Egypt. In 1919 he returned to the Leeds firm, starting as a draftsman and eventually becoming assistant manager. He designed hydraulic presses for Vestey Bros and in 1926 became manager of their subsidiary, Corchera Espanola in Spain. He returned to England in 1931 where, with industry paralysed by the slump, he could find no work. Eventually he and his brother formed a small company, Electraulic Presses Ltd, to exploit their ideas. Later they bought control of the Leeds Engineering and Hydraulic Co. and finally formed Towler Bros (Patents) Ltd in 1935. Mr Towler shared a James Clayton Prize for "his lifelong work on the design, development and manufacture of heavy hydraulic equipment and related controls". He is a member of the ASME, the Institution of Production Engineers and the Iron and Steel Institute. He serves on various other bodies concerned with hydraulics. He formed a new Company, Pressure Dynamics Ltd, in 1965 at the age of 70.*

dollar earnings in Table 1 in accordance with Metcalf's equal status' data (Fig. 1). Thus, Table 1 shows dollar earnings in various key industries and functions, as well as their equivalent 'money-status' in the UK.[10] Thus, we may assume that the financial status of engineers does not differ greatly in the two counties. But while we in this country may enjoy roughly the same rank in the pecking order we have a great deal less to peck!

The engineer in the boardroom

Money certainly improves status but it is not the only criterion; every profession and vocation has a status of function, eg, a married woman as such, has a higher status than a single woman, and doctors have a higher status than dentists, possibly because of their greater responsibility.

It may be argued that the status of the engineer is comparatively low because he does not come into such intimate contact with the public as the doctor and dentist. The public does not realise how much its comfort and safety, even survival, depend upon the engineer. If the doctors and dentists went on strike for a month, there would be much suffering and a number of additional deaths; but if all engineering stopped for a similar period, half the population would starve. Just imagine the effect of a prolonged power cut and complete paralysis of communications by sea, rail and air.

If the general public does not realise this, it is our job, individually and through our organisations, to tell them so. There is no doubt that publicity is one of the things that determines status. This matter has been dealt with repeatedly in this journal so that I shall not go into it further, except to express a hope that, whatever the final structure of CEI and the Engineers Guild are to be, they will not neglect this aspect of status.

The engineer thinks in three dimensions, not primarily in words. This is why he may feel like a fish out of water in many places where his status needs to be improved—not least in the boardroom. To begin with, he will almost certainly be in a minority, surrounded by accountants, lawyers, administrators, salesmen and large shareholders. I do not suppose that any company director would care to admit that he could not read a balance sheet; but how many of them can read an engineering drawing?

From many points of view there ought to be far more engineers in managerial and directorial jobs. But how many try to fit themselves for such work? The engineer tends to concentrate on things, to the exclusion of people. He is always afraid that people may introduce an unpredictable element of anarchy which will result in disaster. That is why he tries to make his machines fool-proof, so that they can be operated by anybody, the lowest common denominator, a complete nonentity.

Consequently, when the engineer is thrust into association with people—particularly non-technical people—at a committee or a board meeting, he is appalled to discover that he has to deal with individuals who obey few known laws of thinking. There is no lowest common denominator, nothing is standard. His colleagues are not fool-proof and there is not a hope of making them so. The result is that the engineer tends to withdraw into himself. He restricts his thoughts and observations to the mechanical things which he understands and which he feels he can control.

This shyness of the engineer and his inability to express himself in words—and to explain his work to non-engineers—is like speaking in a foreign language. These factors militate against his reaching boardroom level, as they do against the profession as a whole obtaining higher status via a better public image.

However, it is my experience that, when engineers are given responsibility as executives or directors, and have to meet customers and users of their products, they quickly become articulate and presentable. And their testimony is often more convincing because more obviously honest, than that of the glib salesman.

It would be beyond the scope of this article to survey the boards of our large corporations and analyse the qualifications of the directors. Still, one wonders how much they have been helped by the retired politicians, civil servants and Service chiefs. Can we be surprised that there is little room for engineers on the board? Many of these extraneous additions to the business world may serve industry well, either because of their administrative ability or because their experience in dealing with Government or Service departments; but they are often lacking in business sense and few non-engineers understand production. The following letter appeared in the *Financial Times* in July:

DIRECTOR POLITICIANS
Sir,—Mr Joe Hyman (The Financial Times, July 26) may think it is "wrong to make retired politicians directors." Perhaps it would make better sense to make retired directors into politicians, and to retire them especially early for the purpose!
W. Crossland.

Civil servants are trained to be civil; politely to frustrate but never to please. But it is sound business practice to please the customer, the wage and salary-earners and the shareholders—in that order. If the order is reversed, so that dividends take precedence, the business will surely perish.

But the dice are heavily loaded against the engineer pleasing the board. Directors are elected by the shareholders who almost always accept those nominated by the existing board. Consequently, there is a tendency for like to choose like; for cliques to perpetuate themselves. And the engineer is

Table 2—Professional Qualifications of Company Directors

	MA BA	MSc BSc	Eng.	Acc.	Law	Others	All qual.*	No qualif.	Total*
No.	145	134	182	308	37	160	811	3193	4004
%	3·61	3·35	4·55	7·70	0·92	4·00	20·0 approx	80·0 approx	

* those with more than one qualification are listed in several columns.

regarded as a queer fish. He does not speak the same language.

We have far too many amateurs who seem to think that they are equally competent to play a part in politics or business: the 'gentlemen versus players' problem. Some boards are overloaded with nice chaps who can get on with one another; but can they get on with the business? Do they understand it?

In no country in the world is the amateur so highly regarded and the professional of so little account. However, we have been fortunate in our exceptions: our profession has recruited some highly gifted amateurs in the past—for instance Lord Armstrong was a lawyer. There may even have been some great accountants! Indeed, I can claim many exceptions among my friends and colleagues, otherwise I should have found it difficult to write the above strictures.

To get some idea of the proportion of professional engineers on the board of companies, large and small, I have made a rough survey of the *Directory of Directors 1965*, taking every tenth page. The result is as shown in Table 2.

Out of a sample total of 4004 directors, 3193 or 80 per cent have no declared qualification. They may be large shareholders; able businessmen; active salesmen; politicians; promoters; negotiators. Also, there are some whom I know personally, who have not disclosed their engineering qualifications; though titles, medals and decorations do not seem to have been overlooked! Perhaps titles ought to be counted as qualifications, in so far as they must impress the shareholder; and they are some indication of past service rendered to the nation, though not necessarily in matters bearing on the Company's business.

The analysis of those with declared qualifications is not so very depressing: engineers 4·55 per cent, as against accountants 7·7 per cent, Arts men 3·6 per cent, scientists 3·35 per cent. But many of those with science degrees were also Chartered Engineers, so we cannot fairly add them together to beat the accountants.

Again, for my own part, I must confess that I do not appear in the *Directory of Directors* and I made no attempt to get in, even when I was chairman of my company. I would rather be known as an engineer than a director! I wonder whether that is professional pride, arrogance, or unworldliness? Unworldly engineers are not likely to improve the nation's economy.

The following appeared in *The Times*,[11] some time ago:

SNOBBERY IMPAIRS BRITISH ECONOMY, SAYS DANE: Copenhagen, Dec. 29. British employers do not work hard enough, and Britain's economy is weakened by the snobbish title system and the urge to enter high society, Mr A. P. Moeller, the Danish shipowner, who has been described as Denmark's richest citizen, said here today.

It may be salutary to see ourselves as others see us; but I wonder if he is right about "the urge to enter high society". I should have thought it was the other way round; "the urge of high society to enter business". But how can we prove it? Does anyone wish to finance a survey?

I know many engineering employers who work very hard indeed; and I know a few with titles; they work hard too. And there are poor people with titles who have to work for a living. As for high society, where is it in England now? The urge today is to be, or to appear equal to, a Grammar School boy! Snobbery is a thing we catch at school like measles; but some people grow out of it.

Professional rights and duties

In an address to the Engineers' Guild in May 1964, Sir Thomas Lund, CBE, Secretary of the Law Society said:[12]

> ... an inherent requirement of a profession is that the service rendered to the public is mainly supplied through the use of brain rather than hands. ... It is a peculiar feature of a profession that primarily what they have to sell to the public is their ingenuity, their thought and their wisdom.

Sir Thomas' view is worth quoting at some length, not only because he administers the professional affairs of one of the senior professions but because his is a legal opinion. He goes on to say:

> The acquisition of professional status involves the acceptance of certain rights and of certain duties and responsibilities. The rights I think are of two kinds, rights of the profession collectively and of members of the profession individually.
> One of the rights of the profession is normally in the course of time that of the exclusive (or nearly exclusive) privilege of using some special description entitling its members to perform some function in which their training and education make them especially skilled. In your case, I would suppose it might be some such title as 'Chartered Engineer', carrying with it the right to perform for others certain specific engineering tasks which would be forbidden to persons not in the profession. ... The obligation to provide a public service goes to the root of every profession and means that the public interest must always take priority over the personal interests of members of the profession. In other words, the thought of striking for better pay or better conditions can never be accepted because that would mean depriving the public of your expert services in order to advance your personal interest. ...

Sir Thomas Lund also said that a profession must lay down standards of correct behaviour for its members, that is to say, standards of ethical conduct, and when they are laid down they must be enforced with absolute strictness.

Surely professional status cannot have much meaning unless its standards of conduct can be enforced by law, and loss of professional status must constitute a severe penalty. But this can only be practicable if we make the professional status of the engineer really worth having, both socially and by way of remuneration. And, in turn, a professional code of ethics would raise the status of engineers, as it has done for doctors.

We have to remember that, above the subsistence level, no one works for money alone. We work for power; the power to create, persuade, or dominate. Indeed, there are some who will starve in order to create; and some of these are engineers.

Nevertheless, we cannot separate status and remuneration. The relative remuneration and the status of one profession as compared with another must depend to a great extent on how well it is organised as a body, the way in which its public image is presented and its manifest services to the community.

And all these factors will affect the rate of recruitment and the quality of those who choose engineering as a profession in preference to any other, so that there is a strong element of positive feedback!

Qualifications and education

The title 'Chartered Engineer' covers an enormous variety and diversity of engineers. Can all of them be said to have the same standard of learning? do they need it? Some machines are more sophisticated than others, but with the advent of automation, all machines will become highly sophisticated. Therefore, it looks as if the standard of learning of all Chartered Engineers in the very near future will have to be of an equally high level. They will all have to be graduate engineers.

What sort of university education should they receive to enable them to qualify for the profession? What is an engineer? What part does science play in the profession? The question of educational qualifications is of obvious relevance to that of status.

Many recent papers in this journal and elsewhere suggest strongly that engineering education and training is not meeting engineers' need, particularly because it is too narrow and cuts us off from other professions. In so far as the status accorded to us depends on a mutual understanding between the engineers and society as a whole, this is bound to have an adverse effect on status. Here again I shall not go into details since educational problems have been amply discussed in these columns.

Supply and demand

In the last 50 years, there have been many fluctuations in the fortune of the engineer. At one time he is in great demand and at other times becomes a drug on the market. This is plainly due to our policy of *laisser faire*, reliance upon what are termed the 'natural laws' of supply and demand. But anyone who has studied nature must know that its laws inevitably result in alternate glut and famine, fantastic waste and redundancy. Thus, if we are to live civilised lives, we must develop some form of homeostatic control by which the supply of engineers may be made to anticipate the demand.

This is where the Robbins Report, for instance, failed to meet the national need.

From the age of nine, when a boy begins to show some signs of interest in engineering, we ought to take steps to encourage or discourage him from entering that hazardous occupation, bearing in mind that it will take 20 years to convert him into a qualified engineer; by which time his services may not be in such demand as they are today. The potentially great engineer will not be discouraged, no matter what we say, because he is born for the job. But others, who are not so dedicated might make better citizens in other occupations. They might even be able to understand engineers better if they had, after taking an early interest, found that this was not their vocation. In the present state of society it would be a distinct advantage if some lawyers and accountants had even a superficial understanding of engineering.

Action now

What action must we take to improve the situation? Status does not fall like manna from heaven; it is not even distributed at the command of politicians or Ministers of the Crown.

A common title will help but at present it means nothing at all in the world at large.

/Status has to be earned by every member of the profession. First we must improve the quality of the engineer; improve his training, give him a broader background, of arts as well as science; but also more practical manual work. Mechanical engineers learn engineering with their hands. There is very little engineering which is not mechanical when you come down to detail; and it is details which lead to success or disaster. /

The next important point which emerges from this review is the serious lack of information about the actual facts connected with status and remuneration of engineers in this country and how they affect both the performance and satisfaction of engineers. A survey on a large scale is needed here—yet another job, perhaps, for CEI.

/The engineer must take more interest in politics, economics and public life in the widest sense. We must do our best to improve our public image which mainly means explaining what we do in non-technical terms. Visit the old school, take a few films, tell them what engineering is about. Tell them that the affluence they enjoy today, little though they may appreciate it, is largely the product of engineering. That the civilisation which we have built up is so complicated, so fragile, that it can be destroyed by pressing a button—or by throwing a spanner, a tiny spanner, into the computer. Civilisation depends upon men who understand machines, but, in the end, our survival depends upon men understanding men./

I believe that the status of the professional engineer will not be improved until we are represented by a professional union, such as the Engineers' Guild.

I have been an employer for most of my working life, so that this article could be a trifle biased. Nevertheless, there is one thing of which I am convinced: a union, to have any real negotiating value, must be represented by, and be representative of, the rank and file who comprise the bulk of the members. It must not be sponsored or controlled by the employers or by the upper tenth of the members.

Our greatest problem today in any organisation is the apathy of the rank and file, which must inevitably end in a depletion of material for leadership and its arrogation by a self-perpetuating minority. Result: no leaders or the wrong leaders.

The membership of the chartered institutions has increased enormously in the past 50 years; to such an extent that their administrations are almost like government departments; and filled with the same sort of people, ie: civil service types. These corporate bodies have become large, slow-moving and ungainly like brontosauri. Communication between the feet and the head has become somewhat attenuated and only a great and continuous interest by the rank and file will cure this. We need an Ombudsman.

And, finally, engineering is so diverse that it seems unlikely the title CEng will adequately describe all members of the 13 chartered institutions. For instance, would it be proper for a mechanical engineer to practise as an electronics engineer? Consequently, a great deal more thinking will have to be done about the new structure of the profession and I hope that its effects upon our status will not be overlooked.

REFERENCES

1. 1964, 6th December, *The Observer*.
2. 1965, 5th August, *The Times*.
3. ibid.
4. 1960, February, Royal Commission Report, Cmd 939. Her Majesty's Stationery Office.
5. 1964, *The Professional Engineer*
6. 1965, 31st August, Ministry of Labour Press Notice.
7. 1964, 6th December, *The Observer*.
8. METCALF, W. S., 'Transatlantic Salaries'. 1965, 17th June. *The New Scientist*, p. 802.
9. ibid.
10. 1964 Survey, Engineering Manpower Commission, directed by the Engineers Joint Council, USA.
11. 1964, 29th December, *The Times*.
12. LUND, SIR THOMAS, 'Professional Status'. 1964, *The Professional Engineer*. Address delivered on 22nd May.

Our Public Image

by F. B. Roberts, MBE, FIMechE

Do engineers and engineering in Britain get a bad press? The question is: what kind of information do they wish to communicate to the public?

Among engineers in Britain there is a widespread feeling that their work is not adequately recognised by society. Many believe that the standing of engineering as a profession is low. Inevitably, the Press is blamed. What is the truth of the situation? What is being done that will improve it? And what still needs to be done?

Before these questions can be answered, it is essential for engineers to be sure they understand the nature and place of engineering. Few do.

To appreciate the significance of engineering today, we must try to stand outside ourselves and outside our time. Such an effort is neither natural nor easy. We are by nature highly self-centred. Even if we succeed in stepping outside our individual selves, it is still difficult to get outside our generation.

The extent to which men think outside their own personal, present circumstances is very small. It could hardly be otherwise. It is, after all, self-interest (or family-interest), enlightened or softened by religious or humanist concepts, which has provided most of the emotive power of civilisation.

But one of the extraordinary consequences of this self-centredness is that we are half blind to the explosion of man's creative powers—an explosion into which we have been born. Because a man was born in England in, say, 1925, and because, therefore, he is familiar with aircraft, ships, railways and cars as means of travelling; and because the previous 150 000 years of man's development mean little or nothing to him—because, in short, of his remarkable self-centredness, he is curiously insensitive to the achievements of his generation when considered—as they should be—on a base of historical time.

Where are we going?

Engineers are often just as blind as laymen; perhaps more so, because engineering development is more familiar to them. Thus an engineer might spend all his working life on the design of diesel engines or electric motors. On retiring he might think of engine development or motor development much as a horticulturalist would think of the new strains of broad beans that he had introduced during his career. In both cases they would enjoy a certain pride in their achievements but they would both feel that development was, after all, slow, often marginal. The engineer had finished his career working on the same products on which he had started as a young man. And so had the horticulturalist.

Some readers may be asking, at this stage, why I am pursuing this line of argument. They agree that modern science and technology are wonderful; so what? There are at least two answers to that question.

First, an enquiry into the place of engineering in society leads unavoidably to an examination of engineering past, present and future. It is not sufficient to look at engineering today. Tomorrow it will be different; and the day after tomorrow it will be different again. In fact, the rate of development is accelerating. It started from near zero a mere 150 or 200 years ago, after ages of comparatively static, primitive technology, and now it is gathering speed all around us. In what directions, then, is it heading?

The second answer has to do with the fact that man himself has changed very little—some would say not at all—since, say, ancient Greece and Rome. If man is thus static in his nature, it must surely affect his attitude to this scientific and technological explosion. And if, as I have suggested, he is half-blind to the very changes and developments of which he is the agent, this blindness must obviously affect his estimate of the engineering profession.

The engineer who was born in 1925 is like most of his fellow men, therefore, in his self-centredness and blindness. Edwardian and Victorian times mean something to him only because his father may have lived then. He may have known his grandfather, and tried to picture the days of his grandfather's youth, but beyond that everything is history. The only 'real' life is the life he knows.

As it happens, those who study the past are usually arts men, and most of them turn their backs on the technological present.

Is it so surprising, then, that the place of engineering in the history of man is largely unrecognised? Is it surprising that most people are quite unaware of the near miraculous quality of the developments of the past few generations?

In making this generalised criticism of our blindness, I am of course aware that there are engineers and laymen who stand in awe of the creative explosion, but they are a minority. We are concerned in this article with the general ignorance of engineering and therefore I must necessarily speak of the general blindness.

Creative explosion

At the same time I must, in fairness, try to define what I believe to be the significance of the creative explosion. If it is so remarkable, what does it mean?

To answer that I must return to the historical perspective which I used earlier. I hope that when engineers see the word 'historical' they will not think of school history but of a time base such as they are accustomed to using for a graph which illustrates some change or phenomenon. It is necessary to stand outside ourselves and outside our generation, and think of the creative engineering explosion as a remarkable phenomenon which has suddenly peaked after an incredibly long time of almost no change.

Mr F. B. Roberts, MBE, was the Editor of the weekly journal, Engineering. He was born in 1917 and educated at Chester City Grammar School. Trained as an engineering apprentice at LMS Locomotive Works, Crewe, followed by a period in their drawing office, he was commissioned in the RAOC as Ordnance Mechanical Engineer in 1941. He served in the Middle East, Italy and the Far East and was mentioned in dispatches. Demobilised with the rank of Lieutenant Colonel he joined Engineering in 1946, became Chief Assistant Editor and a Director in 1950 and Editor in 1954. Mr Roberts has served on Institution committees, including those of the Education and Training Group, Promotion of Interest in Mechanical Engineering and Design Memoranda. In 1971 he relinquished his editorial duties to set up Technical Evaluations Ltd, a company specialising in finding engineers for urgent assignments and appointments.

The development of engineering on a time base is well illustrated by the foundation of professional engineering institutions in Britain during the past 140 years. Leaving out of consideration the minor branches of development, I would say that the main stream has been: civil, mechanical, electrical, chemical, control/electronic. First, stationary structures; then mechanisms and the first use of power; then the discovery of electricity and its potential in power and other applications; the production of new materials; and, more recently, the parallel development of control and information engineering. None of these primary growths has been superseded by subsequent growths: on the contrary, civil engineering has been given a fresh spur to development by the later engineerings.

This line of development is familiar enough to engineers but I have recalled it because I wish to point out the similarity between it and the evolution of the natural world. First the 'civil engineering' of the universe—the planets and stars; then the simple mechanisms of the animal kingdom; and finally the 'control engineering' of the human species.

Compare, now, the two lines of development (the natural world and engineering): the astounding difference is in the time scale. Whereas the beginning of the earth is put at 4500 m years ago and the age of *homo sapiens* at 150 000 years, almost the whole of the development of engineering has taken place within the past 150 years. There is no suggestion that computer engineering, for example, has reached the refinement of development that we see in man (we are, in any case, rather ignorant of our own nature), but even allowing for a big difference here, the rate of engineering development is so extraordinarily high that it should be recognised as the greatest mystery of our time, and one of the greatest of all times. Not the mystery of how sophisticated machines work, but the mystery of why man has so suddenly become creative after thousands of years of primitive existence. In the amount of his creation, man of the Middle Ages differed less from animals (ants, moles, beavers, etc.) than from modern man.

We are blasé

In the whole long history of the human race, only twentieth century man could be so dulled, so self-centred, as to go about his daily work as though nothing had happened. Only twentieth century man, after seeing the Olympic Games 'live' on television from the other side of the world for the first time, would soon take the service for granted and become irritated if there were a breakdown. Only twentieth century man, watching the announcement of election results, would barely pause to think that an electronic 'brain' was doing the really hard thinking for him by computing the swing and the probable result.

Why are we so blasé? Why are we so nonchalant about the engineering explosion? I have already suggested that it is partly because the individual man is centred on his own life and times. But there is another reason. It is that the total work involved in advanced engineering is broken down into individual tasks which are in themselves quite simple. The girl in the electronics factory, the man on the car assembly line, even the man in the toolroom who is working to the nearest 0·0001 inch—all these have been given work which is well within their mental and manual capacities. In many cases they do their work without understanding the end product to which they are contributing. Yet they are the people who physically make the astounding products of modern engineering.

Behind them, of course, is the engineer whose brain has done the thinking and planning. His is the truly creative work. But he, too, tends to feel that his work is quite unremarkable because he is usually one of a team, and because so much of his work is an extension of other engineers' earlier work.

Engineering is evolutionary—though, as we have seen, it is evolving at an unprecedented rate.

I am thus led, through this examination of the place of engineering in society, to formulate a generalised statement: that we who live today are in the midst of an explosion of creative powers which is without parallel in the long history of man; that there is a faint but unmistakable parallel between this explosion of creative powers and the sequence of events which the first chapter of Genesis so simply and dramatically summarises; that the time scale of the creative explosion is short compared with the total time scale; that the growth of creative power has been made possible by a new division of labour—a division of greater complexity and greater specialisation than ever before; that this creative explosion has happened to men without their being entirely aware of it; and, finally, that they are still for the most part blind to it, or at any rate to its consequences.

And what will be the consequences? This article is not the place to discuss that question. Nevertheless, it is a question which urgently needs discussing by engineers, and if they were to make the attempt, I believe it would go a long way to communicate the message of engineering to the public. It would enlarge engineering beyond the confines of specialist papers to institutions. And it would lead to a fruitful meeting of minds from within and without engineering. Two 20th century authors whose writings one would need to study are, in my opinion, Pierre Teilhard de Chardin (*The Phenomenon of Man* and *The Future of Man*), and Arthur Koestler (*The Act of Creation*). It would help also to read Sir Leon Bagrit's Reith Lectures, with particular reference to his notion of the extension of man and "that the direction of modern science and technology is towards the creation of a series of machine-systems based on man as a model".

Here I have reached the watershed of the argument: beyond the ridge is a vision of creativity far more astounding than anything man has achieved to date. When men speak of engineering, this is what they mean, or should mean. In practice, on this side of the ridge there is a thick mist of ignorance. Men are afraid of machines, or they are indifferent, simply because they do not understand them. At one moment they welcome machines (for the old, self-centred reasons)—television sets, washing machines and motor-cars. At another they actively resist them.

What interests the public?

With this background of indifference and fear of the unknown, what in fact interests the public, as shown by the typical contents of the Press? Table 1 shows a list of the headings on the front page of a popular newspaper of 27th January 1965, with brief annotations. In the case of short news items the length in lines is given.

Out of these sixteen items, at least ten give news which would have been appropriate in ancient Rome. Our interests, outside ourselves, have changed little in spite of the engineering explosion.

What of the other items? The *Spitfire*—one of the most famous machines of modern times, but mentioned in the context of Sir Winston's funeral. *Valiants*—the taxpayers' money; a phenomenon (fatigue) which is popularly known because of earlier disasters and the writings of Nevil Shute; coupled with a hint of risk of disaster. A £200,000 stamp—fascination of (to the reader) outrageous values; the stamp was once sold by a schoolboy for 6*d*. Ferries halted—not a disaster but a major inconvenience to the British (N.B. the headline brings out this point, not the basic news that there is a French rail strike). Help for exports—"The Government's export-boosting plans will be revealed today by the Chancellor . . ." This at a time when the Labour Government, shaken by recent by-elections and economic difficulties, is seen to be on trial.

In considering these items of front page news it is as well to remember that they must immediately stimulate the interests of men and women, young and old, and they must do this at a time of day when most of us are only half awake and not very sociable. It is no good blaming the Press for the primitive and basic nature of our minds! Why, indeed, complain about it? Is it not human and humanitarian to be concerned that one Premier has died from wounds, that another is ill, and that a third, one of the greatest of all time, is to be laid to rest? Is it not, in fact, perfectly natural and reasonable for us to be interested in all those news items?

This one page of one daily newspaper on one day is only a small sample on which to base conclusions, but it is worth observing several other points. There is no mention of engineers; but then there is no mention of doctors, lawyers, soldiers, architects, priests, butchers, bakers, or candlestick makers. In fact the only professions which get a look-in, after statesmen and diplomats, are professional footballers and impresarios. This is not because professional footballers or impresarios are more 'important' than engineers, doctors, etc., but because the majority of readers of newspapers are interested in a footballer being sent to prison or a famous impresario going into hospital. The names of the two men concerned were household names.

If these exacting requirements, imposed quite properly on the Press by ordinary men and women, are appreciated, one can begin to understand that whether engineers and engineering get into the newspapers has little to do with their importance to the country or the economy, but depends on their news value to those same ordinary men and women; and news value measured over the bacon and coffee, or during the five minutes at the bench before work starts in the

Table 1—Front Page of a Daily Paper, 27th January, 1965

Headline	Notes	Lines
INTO THE HALL OF KINGS	Churchill's body lying in state	
Four years for Gauld	prison for a footballer	
Spitfire ban in fly-past	"on safety grounds"	
Valiants are scrapped as unsafe	due to fatigue (cost £60m.)	
Rising Hill school is to close	comprehensive schools are in the news	
£200,000 stamp flown in	most valuable in world	10
Ferries halted	due to French rail strike	5
Man stabbed	—	4
Tremor town	earthquake	4
Envoy expelled	Russia accuses an American	6
Help for exports	Big current political issue	6
Bustamante ill	Prime Minister, Jamaica	6
Shot Premier dies	Prime Minister, Persia	4
13 handshakes	New UK Foreign Secretary receives ambassadors	4
Jack Hylton ill		7
Blackout chaos	caused traffic jams in London	6

morning. The popular Press mirrors the instinctive interests of the public, and the public is no respecter of professions as such.

It is unnecessary here to discuss the role of those newspapers which appeal to particular groups—misleadingly called the 'serious' Press. They devote a considerable amount of space to engineering events and developments—as indeed they are bound to, since all the engineering industries, taken together, constitute the main part of the economy. Nor can anyone justifiably criticise the BBC and other media of broadcasting. The BBC, especially, caters for a wide range of interests, including engineering and science, and almost certainly allots more time and energy to them than any other broadcasting service in the West. But the BBC cannot ignore the taste of the millions—why should it?—and therefore has to take care in allotting peak watching and listening times.

Professional status

There are two other obstacles to effective communication and public relations. One arises from the vague meaning of the word 'engineer', and the other from the essential difference between engineering and science.

In Britain an engineer can be the designer of the Forth Bridge or the man who calls to repair a domestic vacuum cleaner. Before we too quickly deplore this ambiguity, however, it is worth reflecting that, forgetting personal considerations, this is not wholly a loss. Generals are usually proud to call themselves soldiers, and so are privates. Similarly, most Chartered Mechanical Engineers would heartily agree that there are many humble craftsmen and semi-skilled workers whose work is every bit as essential as their own.

It is pointless to argue that we couldn't do without, say, professional engineers. We couldn't do without any class of worker, so interdependent have we become—largely as the result of the development of engineering. We are all members one of another: if that was true 2000 years ago, it is a startlingly apposite truth in the twentieth century.

Chartered Engineers (of whatever specialism) are thus placed in a peculiarly delicate position: they naturally desire some recognition—which inevitably means some class distinction—but they are reluctant to achieve it by a move which might appear to separate them, more than is necessary, from those whose hand-work is so closely allied to their brain-work.

The group psychology of the situation is further entangled by the fact that Chartered Engineers are nervous of any movement within their ranks which smacks of the popular image of trade unionism. They are already confused with shop-floor workers; they do not wish to make for worse confusion.

I believe that this farcical situation is likely to be resolved soon. The Engineering Institutions' Joint Council has come at the right moment and, if a Royal Charter is granted, the profession will become united on a firm foundation. Before EIJC was founded in 1962, the Engineers' Guild had laboured single-handed, and with little support, to serve the profession as a whole. No doubt many engineers who have been indifferent to the activities of the Guild have indeed been indifferent precisely because they feared personal contact with 'trade unionism'. In practice, the Guild has been more a professional organisation than a trade union. It has been led more often than not by members of the senior institution—the Civils, i.e., by those engineers who, because of their seniority, were the first to recognise the need for a unified profession. That long-overdue development is almost within our grasp. If it comes, much of the difficulty of communication, public image and the like will vanish.

In Britain, the tide of public opinion is turning. Technology (or engineering) is moving up to join science in popular esteem. The Ministry of Technology was only one such sign. The Royal Society is taking steps to correct the balance. Academic institutions are adding their support. The schools are waking up to Britain's predicament. Our own institution is conducting a survey of the teaching of applied science in schools, on the results of which positive action may be based.

Understanding one's audience

So much for the vague meaning of 'engineer'. What of the essential difference between engineering and science? There are two prime factors at work and they need to be seen in parallel. The first is that science is knowing and understanding: whereas engineering is designing and making. The second is that teachers are experts in knowing and understanding but not generally in designing and making.

Now engineering itself requires science, i.e., knowledge, and this part, though no more important than the practical making, is the part that is most easily taught and examined.

The result of these twin factors is that there has been an unreasonable bias towards science and an unreasonable neglect of engineering—in our schools, in the minds of parents and the public generally. For reasons I have mentioned earlier, I believe that our democracy has a built-in mechanism for correcting such errors.

How easily are false values venerated! There is nothing inferior in the practical aspects of engineering—what could be more practical than Greek sculpture and architecture? Nor are the contents of our daily newspapers as inferior as is sometimes alleged—were not those front-page headlines closer to Greek tragedy than a separation of 2000 years might lead us to suppose?

Of writing and the other arts of communicating—as essential to the engineer as to other members of the community—I have not much to say that has not been said many times before. Technical writing and writing for laymen—these are skills which the engineer must develop, but no amount of mere verbal juggling will make up for a lack of understanding of one's audience.

It is useless for the engineer to be annoyed when the world won't listen to him, whether he has tried through a public relations officer or directly himself. He must have something to say that is of personal concern to other men, and if he doesn't know what interests other men, he had better keep quiet—unless, of course, he is a genius.

I am thus brought back full circle to the theme of the opening of this review: where is engineering leading men? Is it not time we grasped the nettle and worked out our philosophy? We have inherited and developed the ability to design and make those magical machines which generate indifference or fear in other men (and often, too, in ourselves). Why? To what ultimate purpose? What is our goal?

This is a challenge which we have evaded far too long. I believe it is a challenge that confronts engineers more than scientists. And when we take up that challenge and begin to speak in the forum, addressing ourselves to other men and forgetting ourselves, we shall be heard.

The Effects of Background and Training on Success in Engineering*

by S. P. Hutton, PhD, DEng, FIMechE and J. E. Gerstl, MA, PhD

Considering the shortage of engineers and the wide interest in better educational methods, very little is known about such basic facts as what professional engineers do with their time, and what kinds of social background, education and training actually correlate with success in later life. The detailed survey reported here, unique among mechanical engineers in this country, answers a host of questions on these and related subjects; but raises a great many more. This is a sociological document second to none and essential reading for anyone who wishes to form a valid opinion on the education, training and employment of engineers.

It is surprising how unscientific engineers have shown themselves to be when discussing the important problems of engineering education and the shortage of technical manpower. These subjects have been widely debated, mainly in subjective terms, and many recommendations made on little factual evidence. With all the emphasis upon recruitment and complaints about emigration, surprisingly little is known about the engineer himself or his career.

In the rapidly changing industrial climate of today it is by no means certain that the best use is being made of the very limited time available for formal education. Nor is it certain that graduate engineers are being trained or used to best advantage by industry. The investigation reported in this article was undertaken in an attempt to redress the balance by providing some facts about mechanical engineers and their profession which may provide the basis for more authoritative discussions.

The education and training of mechanical engineers can be viewed correctly only in the context of national needs. For instance, we are just realizing that Britain's future need for skilled technicians may be even more urgent than that for technologists.

The general solution can only be tackled by some national body with the aid of reliable factual information about economics, manpower, resources and all other important factors. A start in this direction has been made by the National Economic Development Council. A first step would be to decide what proportions of unskilled, skilled, and professional people of all types will be required, and whether these can be provided without interfering with the plans for other sections of our economy. How many engineers are wanted compared with scientists, economists, accountants, sociologists and arts graduates? What proportions of engineering graduates, Dip. Tech., H.N.C. and part-time students will be required, and how many technicians per engineer?

Need for a survey

The inauguration in 1960 of a separate Mechanical Engineering Department at Cardiff provided a good opportunity for rethinking at least the aspect concerning professional engineers.

It was not clear that the traditional lines of education were necessarily the best. One of the first questions arising was: if the object is to educate future mechanical engineers, what is a mechanical engineer and what are the main types of work he will be expected to do at various stages of his career? There was a dearth of reliable information.

—"surprisingly little is known about the engineer himself . . ."

Therefore we decided to study the careers of professional mechanical engineers in order to see how well their education had fitted them for their work and how efficiently they were being utilized by industry. It was hoped that the present situation might give valuable clues about future trends.

With the limited facilities available, it was possible to study only one branch of engineering. Mechanical engineering was chosen, because a mechanical department had initiated the enquiry and because the Institution is the largest of the engineering institutions.

While there is no reason to assume identical patterns in other branches of the profession, the analysis of one relatively homogeneous occupation affords a clear frame of reference. This article is only a brief account of some of our findings which may throw light upon the relationhsips between the background of engineers and their subsequent careers.

The study comprised a detailed examination of career patterns and the retrospective evaluations by professional engineers of their academic work, technical training and

* This article is condensed from the authors' paper at the Engineering Education and Careers Conference.

Table 1—University graduates

Institution:	Percentage of total
Civil Engineers	61
Mechanical Engineers	27
Electrical Engineers	42
Chemical Engineers	65

These figures were taken from the 1959–60 Engineers' Guild Salary Survey, which provides only a rough guide.

Table 3—School attended, by age and graduate status

	Graduates				Non-graduates			
Age group	−35	35–44	45+	All	−35	35–44	45+	All
Total no. in group	95	179	113	387	180	263	147	590
School, % of no. in group:								
Secondary technical and modern	7	14	14	12	42	35	28	36
Grammar	63	53	46	54	43	44	39	43
Public, independent	28	30	37	31	9	9	16	11
Other	2	3	3	3	6	12	17	10

Table 2—Father's occupation and son's graduate status

Father's occupation	Graduates, per cent of total (387)	Non-graduates, per cent of total (590)
Professional, high executive	25	12
Managerial, executive	15	8
Inspector, supervisor etc (higher)	20	11
Inspector, supervisor etc (lower)	7	17
Routine non-manual	9	6
Skilled manual	19	35
Semi-skilled manual	3	8
Routine manual	0·5	2
Retired, unemployed, etc.	2	1

experience in industry. On the academic side, the survey was designed to reveal those subjects which proved to be most useful to engineers in practice, thus providing a guide for designing the optimum syllabus.

The descriptive account of career patterns was itself of great interest. In this context the results were analysed under three main headings.

1. Mobility and recruitment: family background, channels of entry into the profession, types, and evaluation of, training received.

2. Career: history, aspirations and expectations, professional commitment, authority structure, hours, income and relations with colleagues.

3. Leisure: hobbies, recreations, etc.

With this general outline as a basis, we designed a questionnaire which was checked initially by pilot interviews and then suitably modified. We decided to attempt a nation-wide survey by personal interview because it can yield so much more information. Not only is the response rate much higher but many open-ended questions can be followed up.

A specialist survey firm assisted with the design of the questionnaire and carried out the interviews, coding, card punching and computer tabulation. Interviews lasted between two and three hours. Respondents were on the whole most helpful and genuinely interested in the study.

The sample was drawn from the IMechE's Corporate Members and Graduates of more than two years' standing. Because of the low proportion of university graduates (shown in Table 1), compared with other institutions, and the special interest in them, it was decided to augment this group by 50 per cent. A random sample of 387 university graduates and 590 non-graduates of all ages, from all parts of Britain were interviewed in the spring of 1962. This represented a response rate of 77 per cent.

In the following pages, whenever a statement is unqualified it applies to both graduates and non-graduates, or has been weighted to allow for the correct proportions. The term 'graduate' will mean university graduate, and not necessarily a Graduate of the Institution.

Social origins and schools

The social origins of graduate and non-graduate engineers (as shown in Table 2) are quite different. Both streams of the profession are, however, made up by a majority who have advanced beyond their father's occupational status: 58 per cent of the graduates and 79 per cent of the non-graduates have moved up. Nevertheless, the largest single aggregate of graduates (40 per cent) originates from professional-executive strata, while for non-graduates manual working-class backgrounds are most common (45 per cent).

Although only 4 per cent of the total are sons of professional engineers, the degree of recruitment from the skilled manual category suggests indirect correlation: a father who was a skilled craftsman had provided a technical interest and some knowledge of the world of engineering.

The contrast between the older and younger engineers reveals considerable changes in social origins of both graduates and non-graduates, as shown in Fig. 1. Whereas the oldest non-graduates were mainly recruited from middle white-collar groups, the youngest are mainly working-class in origin. Recruitment from the professional-executive group has also declined among the non-graduates. The trend is for middle-class entrants to pass through a university.

There is as much upward mobility among the youngest graduates (72 per cent) as among the oldest non-graduates.

Since the proportion of graduates among mechanical engineers does not appear to be increasing, the trends for non-graduates (the great majority) are especially important.

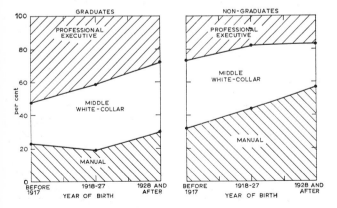

Fig. 1. The changing trend in social origins (based on the father's occupation): more from the manual working class, at the expense of the professional classes (graduates) and the white-collar occupations (non-graduates). It is possible that sons of professional families tend to go into pure science, but there are no figures to prove it

Corresponding to changes in social background are those of the types of schools attended. It is, of course, largely as a result of the relatively recent changes in educational opportunities that a wider basis of recruitment has been possible.

Professional mechanical engineers as a whole are more likely to have attended grammar schools than are comparable groups of managers examined in other studies. Second to grammar schools are public schools and secondary modern and technical schools.

Between the various age groups of graduates, the major change, shown in Table 3, has been the increasing proportion from grammar schools, making up the decrease from public and secondary technical schools. For the non-graduates the greatest change has been more attendance at secondary technical schools.

The increasing number of working-class entrants are those who have received their education in secondary technical schools. The decline in public school entrants, on the other hand, corresponds largely to the decline in recruitment from professional classes.

The type of school attended has a very strong effect upon the likelihood of attending university: less than a tenth of those from secondary technical or secondary modern schools got a degree, against a fourth of those from grammar schools and half of those from public schools. The relation with types of university is even stronger. Three-fourths of the 'Camford' (mostly Cambridge) graduates in the sample had been to a public school and the rest to grammar schools. On the other hand, over two-thirds of provincial university graduates had been to grammar schools, with less than a fourth to public schools, and 8 per cent to other types. London graduates also included a fourth from public schools, only half from grammar schools, and the largest proportion from secondary technical and secondary moderns (most of whom obtained external degrees).

Within the category 'public schools' different patterns exist. Again there is a graduation from the top nine 'Clarendon Schools', through the other major public schools, other Headmasters' Conference schools, to other independent schools. The position in this sequence largely determines the proportion attending university; the proportion at Camford; the likelihood of a position in management; and the likelihood of financial success. In most cases those who had attended 'other independent schools' were indistinguishable from those at grammar schools.

Choice of profession

Half of the respondents had decided to become engineers by the age of 15, and by 16 three-fourths had decided upon their future careers. While grown men cannot remember all the factors that influenced their choice of profession (any more than their choice of wives), the aspects of the decision that were mentioned in interviews reveal notable contrasts. A third of all respondents mentioned technical predispositions—being 'good with one's hands' or 'mechanically minded'; another third mentioned the influence of persons—most frequently members of the family connected with engineering, but also school-teachers and relatives not in engineering. Graduates made reference to their academic aptitude at technical things or a general interest in engineering much more than did the non-graduates. The latter, on the other hand, stressed particular opportunities, e.g., local industry offering good prospects.

Whereas the older engineers were more influenced by persons, the younger stressed their technical predispositions.

Table 4—Proportion of graduates by age groups

Age group	Under 35	35–44	45–54	55+	All
Graduates (per cent)	18	23	23	27	22

"*Half of the respondents had decided to become engineers by the age of 15 . . .*"

Although the image of the engineer in society may be somewhat unclear, one would expect that the schoolboy choosing this career would think in terms of creativity and making things. Yet, only 40 per cent of the sample thought specifically of work as an engineer when they first made their occupational choice. Over one-fourth claim to have anticipated a position n management.

The early predisposition to management is related to social origins of higher status and is more likely among the younger respondents. Interestingly, the non-graduates were more likely to think of management than the graduates. Among the latter, those with the best degrees were least likely to have thought of managerial careers. Similarly, those who attended grammar schools expressed less anticipation of management careers than those from secondary technical and modern schools. But those from public schools had the greatest managerial anticipations.

Early anticipations of managerial positions tend to be fulfilled. Forty-four per cent of those who expected to enter management did so, against a third of the others.

Higher education

The actual proportion of graduates among members of the Institution sampled is 22 per cent. The comparison of age groups in Table 4 shows that there has been a slight decrease in the proportion of graduates despite the increased number of university places.

Although the decline is hardly significant statistically, the fact that there has been no increase is important. Whether this is a healthy trend is questionable. There is undoubtedly a need for less highly qualified engineers in industry but whether the Institution should open its doors to all of them is worth discussing. The Dip. Tech. will, however, tend to redress the balance. There has been a large increase in the proportion of non-graduates taking H.N.C., which largely explains why there are not more graduates. Almost a third of non-graduates under 35 have H.N.C., compared with only 10 per cent of those over 45.

Among the graduates it is surprising that there has been no increase in the proportion taking higher degrees, 7 per cent having Masters and only 3 per cent Ph.D. degrees. The fewest higher degrees are found among Camford students, the most Masters among provincial graduates and the most Ph.D.'s among London graduates.

Table 5 shows that most graduate engineers came from provincial universities. Over the years their proportion has increased and that from Camford has declined.

Although 82 per cent of the graduates took their qualifications by full-time study, only one-fourth of the non-graduates did so, while 43 per cent pursued part-time day courses and 31 per cent evening classes.

There has been a marked decrease in the proportion of evening class students, together with a steady increase of part-time day students. The backgrounds of both are similar, but with a tendency for more of the evening scholars to come from working-class families and to have been to elementary and junior secondary schools. Thirty-seven per cent of both groups attain high success, but night students have more of a technical orientation and work in research, development and design, whereas day students are to be found more in managerial and non-industrial posts.

Eight per cent of the graduates had non-engineering degrees, mostly in natural sciences. Only 47 per cent of the non-graduates and 44 per cent of the graduates were educated solely in mechanical engineering, as can be seen in Table 6. The degree of specialization has, in general, increased with time. Non-graduates are more likely to have studied English, languages, economics, and industrial administration during their engineering course.

For both groups the most common non-engineering subject is industrial administration, studied particularly by the younger generation. Among the older people English was apparently studied less often but a foreign language more often during higher education. Graduates from provincial universities are more likely to take non-engineering subjects than are London and Camford students. More than half the sample had no desire to have studied additional subjects but among those who would have liked to, the most common, in order of popularity were: foreign languages, economics and accountancy, social sciences, English and technical report writing.

About half had no wish for further engineering subjects. The most popular subject, particularly among non-graduates, was electrotechnology. Next were automatic control and vibrations, mainly among those under 35.

Table 5—Universities attended

University	% of total
London	35
Oxford	2
Cambridge	14
Bristol	2
Birmingham	3
Durham	4
Glasgow	7
Other Scottish	3
Leeds	1
Liverpool	3
Manchester	10
Nottingham	3
Sheffield	3
Cardiff	2
Swansea	1
Other provincial	4
Foreign	1
Not ascertained	2

Table 6—Other branches of engineering studied

	Non-graduates (%)	Graduates (%)
Mechanical engineering only	47	44
Also trained in:		
Civil	7	23
Electrical	30	37
Marine	10	4
Chemical	4	3
Aeronautical	6	6
Automobile	7	2
Mining	1	—
Other speciality	5	2
General, combined	—	2

Table 7—Value and use of subjects studied

Subjects	Used once a week or more	Deemed essential
	(percentage)	
Mathematics	70	74
Engineering drawing	60	70
Technical report writing	48	50
Applied mechanics	43	60
Properties and strength of materials	41	55
Industrial administration	39	27
Production engineering	31	25
Principles of electricity	30	34
Fluid mechanics	28	37
Heat engines	25	37
Theory of machines	24	39
Metallurgy	23	21
Automatic control	22	16
Thermodynamics	21	32
Heat, light, sound	21	27
Electrotechnology	20	17
Statistics	19	10
Fuel	18	23
Work-study	16	14
Theory of structures	16	26
Vibration	10	13
Foreign languages	8	4

Value of education and training

One aim of the survey was to assess the relative importance of the various engineering subjects studied. This is very difficult but an attempt was made to compare respondents' subjective opinions about relative importance with the actual frequency with which these subjects were used in their work.

The general results are summarized in Table 7. First-class honours men tended to rate many technical subjects more highly than did second-class honours students who, in turn, tended to underrate these subjects by comparison with pass-degree and non-graduate men. The relative usage, however, did not show the same trends.

The correlation between the assessed importance and the actual usage was fairly close. Although those aged 45–54 tended to use all technical subjects most frequently, the oldest engineers rated most subjects more highly, perhaps because they tended towards administration and had encountered more types of problems.

Mathematics was quite clearly the most important subject to all classes and age groups, followed by Engineering Drawing and Applied Mechanics and Properties of Materials. It is noteworthy that Technical Report Writing also came within the first five.

A few subjects, such as Industrial Administration, Production Engineering, Automatic Control and Statistics ranked considerably higher on the use list than the subjective one. Others, such as Theory of Machines, Thermodynamics, Heat, Light and Sound, and Theory of Structures were, in contrast, overrated.

Despite the limitations of the method of questioning, such rankings might be used to decide on the allocation of time to some groups of subjects within existing curricula. It is clear, for instance, that more time must be devoted to technical writing and the broad field of communication, involving both writing and drawing. There was also general agreement that languages, administration, economics and social science generally receive insufficient attention.

About 70 per cent thought that the education provided by universities and technical colleges was fairly adequate, the merits of the former being a complete and broadly based education, and of the latter a practical course in close contact

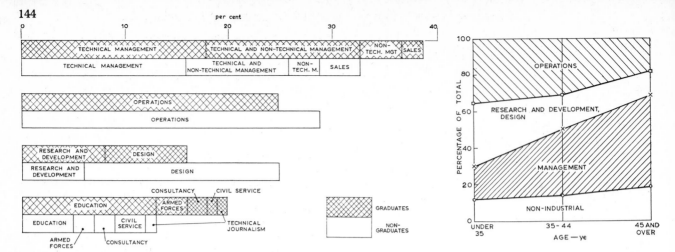

Fig. 2. Proportions of graduates and non-graduates in various engineering functions: the non-graduates preponderate only in operations and design

Fig. 3. Career type by age: there is a tendency for older men to move into management

with industry. Nevertheless, there was also dissatisfaction, the most common criticisms of technical colleges being that time was limited, and that courses were too vocational and not sufficiently fundamental; and of universities that courses were too academic and specialized and that there was insufficient contact with industry. About 10 per cent mentioned deficiencies in teaching.

There was more dissatisfaction with industrial training than there was with education, and 51 per cent felt that best use had not been made of the time available. The size of the firm made little difference. Training was too narrow, insufficiently supervised and unplanned.

Seventy-two per cent thought works training best obtained during academic education or, failing this, before. This opinion was held firmly by all ages and career types and 69 per cent had, in fact, been trained during the period in which they received their education.

Eighty-seven per cent had served an apprenticeship of some kind; 64 per cent of non-graduates and 34 per cent of graduates for five years. Another 34 per cent of the graduates had served a two-year graduate apprenticeship. It is remarkable that 36 per cent had never worked in a drawing office. The fact that 13 per cent of the whole sample had served no apprenticeship seemed to make little difference to their financial success or career.

Industry and type of work

Having considered the engineer's origins, how he enters his profession, and how he is trained, we next examine him actually at work.

Slightly less than two-thirds of the sample are employed in commerce and private industry. Another 14 per cent work for nationalized industry or local authorities (excluding teaching). Those not in industry include: 8 per cent with universities and technical colleges, 7 per cent in the Civil Service, 4 per cent in consultancy, 2 per cent in H.M. Services, and the remainder in such careers as technical journalism.

Mechanical engineers are widely spread among a large number of industries, with concentrations in mechanical equipment (23 per cent), electronic and electrical firms (13 per cent), chemicals and plastics (11 per cent). Other industries in descending order are: aircraft, metals extraction, power, automobile, construction, shipbuilding, materials, food and drink, and railways.

Table 8—Publication activity

	Graduates (per cent)	Non-graduates (per cent)
Publications, lectures and inventions	6	4
Publications and lectures . .	9	3
Publications and inventions . .	5	2
Lectures and inventions . .	6	3
Publications only	10	4
Lectures only	11	10
Inventions only	8	12
None	45	62

Some three-fourths work for large organizations which employ at least 500 people. They are fairly evenly divided among unit, small batch or mass production processes.

The types of work on which engineers spend most of their time vary little with age group, graduate status, success, or particular career group. The most common types of work (in descending order) are technical administration, design, routine office work and research and development. There are more people working on design than is apparent from Fig. 2, and in all jobs, administration and routine office work absorb a higher proportion of the time available.

How hard does the engineer work? Compared with his American counterpart, the British engineer puts in relatively short hours. The average working week, including work taken home, comes to 42 hours. A third work less than 40 hours, over a third between 40–45 hours, a quarter between 46–54 hours, and only 7 per cent work 55 hours or more. It is the

" — administration and routine office work absorb a higher proportion of the time available"

" — two-thirds are inactive members . . ."

older man who tends to put in the longer hours and he is likely to be employed in a non-industrial setting or, if in industry, he is a manager and is very well paid.

The distribution of functions shown in Fig. 2 raises some serious questions about the utilization of training. American figures show only half the proportion in operations and twice as many in research and development. The British concentration in operations suggests undue emphasis upon keeping things going at the expense of innovation.

Working in operations and in design is negatively correlated with degrees and with class of degree, while positions in management, research and development, and in non-industrial careers—above all in education—are positively correlated. In management, however, the graduate with a lower second- or third-class degree is more highly represented than is the man with a first or an upper second. The preponderance of unassuming Seconds over pretentious Firsts in the ranks of management may be due to the predisposition of a First to research, but it may also reflect the preference of the large organization for men without embarrassing brilliance.

Fig. 3 shows that in industry there is a tendency to move from operations, research, design, and development into management. Those who enter after the age of 45 tend to come from positions in operations. If a man has not reached managerial level by the age of 55, he is unlikely ever to do so.

Mobility

Much of the career movement that takes place occurs within organizations, but there is also considerable movement between industrial organizations and between industry and other types of employment. Eighty-three per cent of those in private industry began their careers there, but at the same time, half of those in nationalized industry, half of those in the civil service, and two-thirds of those in other non-industrial occupations also began their careers in industry.

Including changes of firm, the engineer has had, by the end of his career, an average of seven transfers or promotions. Already by the age of 35 there have been four moves on the average.

Mobility between organizations is, of course, less frequent. It is, nevertheless, surprisingly high in comparison with other studies of mobility between industrial firms. Only 16 per cent of the sample had remained with their original employers (18 per cent of the graduates and 14 per cent of the others). Most movement takes place before the age of 35 but it does not stop then. By the end of his career, the typical engineer will have worked for some four different organizations; two-fifths will have worked for five or more.

The disparity in rates of mobility between engineers and others can be explained by both positive and negative factors.

On the positive side, the engineer holding a professional qualification can use his skill in a variety of settings. His position in the industrial market allows him to explore several greener pastures. On the negative side, it may well be that many firms are not able fully to utilize the engineer's abilities, thus forcing him to move on. This phenomenon is more likely to occur in smaller firms, where the rate of mobility is, in fact, higher.

Satisfaction with work

The engineer is confronted with the problem of deciding where his real interest lies. Promotion up the managerial hierarchy is not entirely compatible with technological interest. Since senior posts on the technical side are limited, he will at some stage, if he is ambitious, be faced with the choice of ceasing to be an engineer and becoming an administrator. Half of our sample are to be found in the managerial ranks by the age of 55, but many engineers do not aspire to this type of work.

Like others, engineers appear to over-estimate the standing of their own profession. At the same time they are quite concerned about raising its prestige. Less than a third are satisfied with it. Among doctors, solicitors, university lecturers, research physicists, company directors, dentists, chartered accountants, primary school teachers and works managers, engineers rank themselves fifth while the general public ranks them eighth, below research physicists and chartered accountants.

Is satisfaction derived from the task performed or from associated activities, such as human relations? For the vast majority of engineers, it is the work that is its own reward. Satisfactions that derive from position in a hierarchy are by comparison very far down the list. Seventy-eight per cent would, if they had to do it over again, choose engineering, rather than a different career but a fourth would prefer another branch of engineering.

When asked whether, on equal pay, they would prefer an administrative position or a technical advisory position, half expressed their preference for the technical side and only a fourth for the administrative, the remainder desiring a combination.

There is considerable disparity between the type of work done and the desired ideal, even though preferences had influenced the former and experiences shaped the latter. Over a third of the managers expressed a technical preference. More than half of those in operations and almost two-thirds of those in R.D. and D. would, if the pay were the same, prefer a technical advisory post. But the pay is not the same and most promotions reduce the technical work content.

Professional activity

Whether or not an individual is a Corporate Member of the Institution is more than anything a function of his age. But for the younger man, Corporate Membership is related to graduate status, to the class of degree, and to having a position in management or in a non-industrial setting. The same factors identify the individual who is a corporate member of more than one professional body.

Combining the number of institutions belonged to with the degree of activity in each reveals the usual organizational phenomenon of a small number of people doing most of the work: some two-thirds of the sample are inactive members, 15 per cent are only moderately active. Activity tends to increase with class of degree and with age. The industrial manager collects a larger number of corporate institution memberships but it is the man in a non-technical setting who is most active in institution affairs.

"— reading three or more journals is associated with success"

The amount of publications, inventions, etc., is greater than the activity in professional institutions but less than half of the individuals do all of this work, as is shown in Table 8. Only among authors whose sole publications are patents does the non-graduate exceed the graduate, and the younger man the older.

Keeping up with technical progress requires some effort. Although 8 per cent of the sample claim to read more than 10 professional journals regularly, 20 per cent admit reading none at all. The remainder are almost equally divided between those who read one or two, three or four, and between five and nine. Strangely, academic standing makes no difference, one-fifth of even the first-class graduates reading no professional journals regularly. But perhaps the word 'regularly' does not cover reading in attempts to solve particular problems.

The range of journals read is quite extensive, but foreign-language journals are read by only 0.3 per cent, suggesting that even those people of foreign origin in the sample have adopted British insularity.

Success

The assessment of success is not an easy matter, for there can be little agreement upon the ultimate criteria. Since engineers, as we have indicated, are employed in a variety of settings, position in an organization cannot be the measure of achievement. Accordingly, we have utilized that crude index, income, as common to all work situations. While we do not make the assumption that virtue is always rewarded, money shows what factors are rewarded. Since incomes are closely related to age, our measure of 'success' allows for age, being based on the median income for an age group.

The effects of various background factors are very much intertwined. Type of school attended is very highly correlated with social class of origin. Both in turn affect the likelihood of attending a university, and its type. But the crucial factor of success, more important than family background and more important than the type of school attended, seems to be the possession of a good degree from a university. It can be seen from Table 9 that, whatever their respective social origins, whatever the type of school attended, a larger proportion of graduate, as opposed to non-graduate, engineers are successful.

The pass graduate from Camford does little better than the provincial pass graduate and worse than the London pass graduate. However, honours graduates from Camford do better than those from London who, in turn, do better than those from provincial universities. The proportion of the successful honours graduates is 70 per cent, as compared with 54 per cent of pass graduates. Of external degree holders 57 per cent are successful, against 65 per cent of other graduates.

Table 9—The effect of university and degree on the percentage of financially successful

University	1st or 2A	Other honours	Pass	Total
Camford	83	79	54	75
London	71	69	59	66
Provincial	65	67	48	58
All Graduates	70	71	54	64
All Non-graduates	—	—	—	39

In spite of income differentials between various work settings, our crude measure of success shows very similar opportunities of reaching the median age-income level, with chances being only slightly less than even in educational institutions, and considerably better only in consultancy.

Lower rates of success are associated with aircraft, railways and mechanical equipment and high success with chemicals, the electrical industry (but not electronics) and metal production. Success rates are quite similar in medium and large organizations and are slightly higher in smaller units.

Process-type industries have the highest success rates, largely due to employing a greater proportion of graduates.

In all types of careers, graduates have a better chance of success than non-graduates. For both groups the managerial ladder is the one most likely to lead to success. Graduates do least well in R.D. and D., where their chances of success are no more than even. Moving more than three times between organizations is negatively associated with success.

Being a Corporate Member, belonging to more than one institution, and high degree of activity in the institutions are all associated with success, as are publication activities.

Although those who admit to reading no professional journals are as likely as not to have a high degree of success, reading one or two journals is associated with low success, and reading three or more with high success.

Looked at from a variety of ways, there is no doubt that the achievement of graduate status among engineers is highly rewarded. Considering the other correlates of success, one might add that professional virtue also seems to be rewarded.

Conclusion

The results obtained are remarkable both for the similarities and the differences they reveal between graduates and others. Although their social origins tend to be somewhat different, there has been an increase in upward mobility for both. In this respect the engineering profession, in Britain, as in other countries, appears to be in the forefront.

But this does not necessarily imply that manpower is being used to the best advantage. Both graduates and non-graduates are professional engineers but are they the same sort of animal, and indeed should they be?

We cannot claim to have isolated all the essential differences. For instance, are the graduates simply the better men, whose talent would have been revealed in any case? Or are they better men because of their education? What are the implications for the engineering profession if the proportion of non-graduates continues to increase?

Similar considerations apply with respect to graduates having various classes of degree. The issue is one of priorities. Is the relatively small number of Firsts going into industrial posts desirable? What is it about the combination of a good degree from a good university that results in the high success rates? The proportion of Camford men is decreasing and the implications of this decline must be faced.

While present courses at universities and elsewhere seem broadly satisfactory in their technical content, they are not producing any breadth of education in either non-technical fields or even in technical communications. The ability to understand men should be taught much more than it is.

The similar representation of graduates and non-graduates in various types of work is striking. Yet why is it that in all of these the graduates are more successful? The large proportion of engineers working in operations does not seem to indicate the best use of professional training. The exodus from operations and R.D. and D. into management reveals how few opportunities of promotion there are in technical work.

The most striking indication of misuse of trained men is that half the respondents felt that parts of their work could be done by someone with less technical training, this being especially true in operations and R.D. and D.

There may well be a need to reconsider the criteria and classes of Institution membership. If there is a qualitative difference between university graduates and others, should they both be included in the same broad class of corporate membership? If there is no qualitative difference, why the disparate rates of success?

We have thrown out more questions than answers. If, however, we have succeeded in isolating some of the crucial questions, and supplying a few tentative answers, then our enquiry has been worth while.

Acknowledgements

Financial support for the research was provided by the DSIR and Social Surveys (Gallup Poll) Ltd collaborated.

Prof. S. P. Hutton *graduated with first class honours at Liverpool University in 1941 and after training with Tarslag Ltd worked on subsonic and supersonic wind tunnel problems at RAE, Farnborough. In 1946 he became a lecturer in fluid mechanics at Imperial College, London, and was awarded a PhD for research in hydraulic machinery. In 1949 he joined the then Mechanical Engineering Research Laboratory, later becoming Head of Fluid Mechanics Division. When he left the re-named NEL in 1960 he was Deputy Director. He became the first Professor of Mechanical Engineering at University College, Cardiff, and in 1969 moved to Southampton University where he is Professor and Head of the Mechanical Engineering Department. Prof. Hutton is the author of a number of papers published in the Institution 'Proceedings'.*

Prof. J. E. Gerstl *was born in Prague and educated in the U.S.A. He obtained an M.A. in 1955 from the University of Colombia and was awarded a Ph.D. by the University of Minnesota in 1959. After teaching at that University and also at the University of Michigan, he was appointed a lecturer in the Department of Industrial Relations at Cardiff in 1960. He spent two years in Wales followed by a year with the Industrial Management Group of the Department of Engineering, University of Cambridge. He is now Professor of Sociology at Temple University, Philadelphia. He specializes in the sociology of occupations.*

How to Get that Job

by W. I. Harrison, MIMechE

'Many are called but few are chosen' even in this age of comparatively plentiful jobs. Notwithstanding more scientific methods that have been put forward, the application form and the interview are still the main techniques of the personnel manager. The candidate who really wants the job must learn to master these.

We all read of company mergers, companies being taken over and, in many instances, redundancies through various other reasons. Inevitably, during his career, some or all of these may happen to an engineer; or he may find that it is time to spread his wings, to take on extra responsibilities, look elsewhere for increment in salary; or he may have to move to another town due to circumstances beyond his control.

Whether or not it be the first time he has to look for a new position, the question arises: what shall be my next move? shall it be a similar job with another company? shall it be something different but within my scope? and what salary shall I now ask for?

If you are in a job at the time of applying, your bargaining power is maintained; however, if you are seeking a new position while currently unemployed, it may be difficult to obtain a higher salary than the last one. This is unfortunate but true; it could well be that you may even have to accept less.

"—he may have to move to another town due to circumstances beyond his control"

Looking at ads

Almost any newspaper or journal carries advertisements for new positions, the *Daily Telegraph, Sunday Times* and *Observer* being amongst the foremost. For mechanical engineers there are few better places to look than the classified pages of the CME.

Many advertisements are over box numbers but advertisements under a company's own name are the most common practice. They let the prospective applicant know where he stands to begin with; without any waste of time for all concerned, he can decide if he would like to work for this company. Other questions which can be immediately answered from a good named advert are the apparent prospects, job description, salary and address.

Management consultants advertise on behalf of their clients to relieve them of the initial selection. Some consultants now make known the company they are representing. A brief note stating one's interest in such a position and brief details of career—or even a telephone call—will, in such cases, bring a job specification and some company background, plus the usual application form.

A word on box numbers. These usually involve applicants in lots of unnecessary writing and, in many instances, nothing further will be heard from them, whereas consultants and named advertisers will at least acknowledge an application. There are numerous and obvious reasons why companies do not wish to advertise their name; staff may be wanted from competitors; staff may be required for development work which the company does not want advertised; a company may not wish its own employees to know that outsiders are being considered. It is not unknown to find oneself applying for one's own job, so at least read the advert carefully. But, by all means, if there is an attractive job with a box number, apply for it. Apart from the other reasons, it could well be the company's policy to withhold its name.

Having seen a position advertised which seems attractive, read the advert a few times, be sure your experience and qualifications fit the requirements and ask yourself the following questions.

'What about leaving my present job—do I really want to? Is it because I am upset about something which will all be ironed out, over and forgotten in a few days' time?'

Remember, a wrong step is often difficult to put right and not only jobwise, domestically too. 'Do I want to move house?'—you will of course have to consider the cost of removal, legal fees, selling and buying houses. If removal expenses will be met by the new employer, this particular worry is relieved but what about schools for the children? How will they and your wife react to a new environment? No doubt you have discussed it all with your wife!

"—you have discussed it all with your wife"

If, after considering all the aspects of a change, you are still of the same mind, watch the run of the advertisements in the press. The pattern could be three consecutive days, or alternate days for one week. Management consultants could place one advert during the week and one in the week-end press. One way is to wait until the run is finished and then write. In this way one's application may be the only one to arrive on that particular day and may stand more chance of being read fully, instead of being one of the initial influx of applications which may only be glanced at in the first instance. Obviously all will be read at some convenient time but the odd late one might stand a better chance since the selector's expectations tend to drop with time.

The letter

A letter of application should get down to the facts without a lot of padding, stating age, qualifications, marital status and probably present salary. Even if this is not requested in the ad and terms like 'an appropriate salary will be paid' are used, you may be well advised to put forward your present or expected salary. This may, in the long run, save both yours and the advertiser's time; your present salary may be more than is to be offered. On the other hand an increment in salary may not be essential to you but factors such as security, responsibility, job satisfaction, may decide the issue; in which case, to quote present salary might jeopardise the chance of a job which you really want. Employers like to beat you down a little but not too much, lest you remain disgruntled.

However, stating the salary expected is always tricky. You could quote too high or too low, and miss out on the chance of an interview for both reasons. A way around is to mention the present salary, and suggest 'a reasonable increment'. At an interview this point would no doubt be pressed; but, after being interviewed and having learned more about the job, perhaps having seen the environment, it may be easier to put forward a figure.

To continue now with the way in which experience should be presented. This could be in the form of a short synopsis, followed by a chronological list of previous positions. One point to remember is that, when 'brief details' are asked for, it is most likely that an application form will be sent; also statements like 'applicants must have had previous experience' usually mean what they say. For a medium-salaried post advertised nationally, some 60 to 80 applications may be received and obviously only experienced applicants will get on the short list.

Never lie about your experience; it may gain interviews but good interviewers soon detect any lack of subject knowledge. If lucky enough to get the job, will you be able to hold it under the pressures of business, if your knowledge of the subject is inadequate?

A trap it is easy to fall into is presenting a standard, type-written, duplicated *curriculum vitae*, which might be all right now and again but cannot relieve you of letter writing. It can become the easy way out and eventually, because it is so easy, the applicant just writes a brief note with his standard sheet to dozens of companies, in the hope something will turn up. If such a method is adopted, some interviews will be offered, but will they be the right ones? After all, this is your future and it is not just a matter of getting any job. The job you want may need a special approach.

Advertisers like to receive tailor-made applications; a letter offers several variations in the way experience is presented; it may gain an interview where a standard form

would not. Furthermore, it conveys the impression of a man tempted by the particular job, rather than of a rolling stone.

It is widely said that engineers cannot present themselves in letter form when applying for new positions. This is not a topic for discussion here but, should the presentation be a difficulty, time spent on planning a letter of application is time well spent.

Not many advertisers specify that the letter of application should be typed; a number ask that it should be in the applicant's handwriting. To type a letter of this nature is not always possible, although it is obviously much clearer and quicker to read; a written letter, as a general rule, is quite acceptable and from the way it is written, its presentation, neatness, spelling, etc, an advertiser can assess a considerable amount without actually seeing the writer.

Application forms

Consider now the application form. Having applied for a position, it is very likely that the applicant will be asked to make such a return if his initial application was sufficiently interesting.

The receipt of an application form and its ultimate return is the second stage along the path to a new position. Up to two hours and more can be spent on some of the larger forms; if a completed form still does not result in an interview, you may be tempted not to fill in the next one and hope that the initial letter will suffice. But the completed forms are not passports to certain interviews; they may eventually become a nuisance; but they are a necessity.

The initial letter of application will not take the place of forms if this is not the company's policy; the non-return of a form tends to indicate that there is no longer any interest. From the employer's point of view, forms have the advantage of describing all applicants in a standard way.

A single-sheet form is normally quite straightforward: personal details, education, qualifications, details of previous companies, present salary, in some cases expected salary, a space for hobbies and a space for any other information you wish to add.

A list in chronological order of positions previously held, with dates of start and finish, name of company, salary at time of leaving and reason for leaving, are usually required. It is advisable to keep a list, adding to it, as it becomes necessary, of the dates of joining and leaving companies, the positions held, the salaries at start and finish and reasons for leaving. If, during one's career, there are numerous changes, it can become a major job to back-track to recall all the required information. But leaving gaps makes a very poor impression.

Hobbies and spare-time activities are also usually covered.

On some, but not all forms, references are asked for, with a

"Hobbies and spare-time activities are also usually covered"

note that such people will not be contacted without your prior permission. It is as well to contact your referees in the first instance to see if they will in fact give you a reference, if ultimately called upon to do so.

Many companies and, in particular, management consultants are now sending out quite large application forms, running into a number of pages. The information previously mentioned is requested; plus details of duties, responsibilities, number of personnel in your charge and to whom you were responsible for each position held. Answers to other questions asked amount to writing essays on what your present job is, its function, responsibilities, etc. A write-up on what you expect to be doing or to have achieved in ten years' time; an essay on past experience, good and bad, and how you expect to apply such experience to the current position, if successful, are also often asked for.

A practice most helpful to applicants is becoming increasingly popular: along with an application form comes a 'job specification', completely setting out the duties involved in the new post, and a description of the company, its products, its future commitments, its expansion policy, size, history, location and also the general housing situation in the area.

Sometimes duplicate application forms are sent, one for the applicant to complete and one to keep for reference. In any case, it is advisable to keep a record; at least it will help you not to contradict yourself during an interview!

The interview

The offer of an interview has arrived and you may now be confused, especially if applications were written to more than one box number at the same time. The letter may only refer to your letter of a certain date; if you have not kept the original adverts to which you replied, it may be difficult to decide which job you are being interviewed for.

If the date and time are not convenient to you, acknowledge the letter with a request to change these; most companies agree to reasonable alterations. But in the case of a 'group interview' alterations cannot be acceded to. All short-listed applicants are requested to attend at the same time and subjected to tests and interviews by various senior executives.

The letter offering an interview may state that you are 'short-listed'.

This will mean three to four applicants to be considered. Or it may say a 'preliminary interview', and you would probably be one of a much larger number from whom those short-listed will be picked and asked to return for a second time.

Before attending the interview it is as well to do a little research into the company, its products, size, directors: consult the buyers' guides, directories of firms, adverts from papers and magazines. A favourite question is 'What do you know about our company?' It would show interest and intelligence if the applicant were in a position to answer this question positively.

When going for an interview, look as though you wanted the job. With respect to dress, look smart and 'crisp'. Many interviewers judge at a glance, as soon as you enter the door and before a word is spoken, whether or not you are the person who is required for the position—all other things being acceptable, of course.

Make sure, before going, that you know your own background and also remember what you put on the application form, as you will obviously be expected to expand on the answers given there. You may also be asked to repeat them,

Mr W. I. Harrison was educated at the Peter Webster School, Chesterfield, and then served an indentured apprenticeship at the Chesterfield Tube Co., while continuing his education at the Chesterfield College of Technology. After two years in the RAF he gained his HNC and graduate endorsements in 1957. He has held engineering positions with Humphreys & Glasgow, Davy United, Rolls Royce, and W. C. Holmes. He was Chief Engineer with Thomas Green and Son, Leeds, and is now Managing Director of Hardave Enterprises of Ellesmere Port, Cheshire, who are Engineering and Management Consultants. Mr Harrison became a member of the Institution in 1967. His article is based on actual experiences.

not only to verify their accuracy but because the interviewer may not remember all the answers, even when he has the form in front of him. So don't get irked by repetition.

The interviewer will try to put you at ease but do not take him too literally and relax to an extent which might appear insolent. Furthermore, do not give the impression that you think the interviewer is a fool and adopt a superior attitude—that you know more than he does. Do not give the impression that, if successful, you would reorganise the company in the next few months.

It is important to get on the same 'wavelength' as the interviewer very early on. This means trying to understand what he wants to know about you and answering his questions in the way he wants you to. Do not waffle on and pad your answers with superfluous information in which he will not be interested. If, for instance, you were dismissed from your last position, say so, along with brief reasons; but do not go a long way round the facts without actually saying why—remember, if he is not satisfied with your answer, it is an easy matter to check with your former employer.

Upsets before the interview can put the applicant on edge so that he fails to present himself in the way he would like to. Therefore, try to forget anything which might have caused you to become ruffled, for instance, being kept waiting after you arrived on time; rudeness, perhaps by a secretary, when you arrived; the fact that, through some unfortunate reason, you arrived late for the interview, etc. Make such apologies as might be necessary but do not ramble on about it. You have come to be interviewed, not to recite a list of reasons why things were not quite as they should be.

When attending for an interview, it is just as well to take along your documents of qualifications, letters of reference, etc; it is not the usual practice to ask for them on the spot but on the odd occasion you may be asked to prove statements made.

If it is a position where extras are involved, such as cars, removal expenses, etc, and, if the interviewer has not already broached these subjects, you might well tackle them when fundamentals about the job itself have been dealt with. But if you bring up these subjects earlier during the interview, with the attitude of 'what's in it for me', this obviously gives the impression that you are more interested in the perquisites than the job; in consequence you will be unsuccessful. The object is to sell the idea that you are the man for the job.

A favourite question asked of candidates is, 'Why do you wish to leave your present employment?'. It is advisable to have this firmly in your mind in order to give a straightforward and reasonable answer, instead of having to think about it, with long, embarrassing pauses; which latter obviously indicates that you have not thought deeply enough about the move involved and, consequently, about the new job. Adequate reasons should be to hand, eg, dissatisfaction with present employment, lack of progress, broken promises, salary increment wanted, and many other reasons known only to an applicant. Express confidence that this is the position you have been looking for and that you feel you could fulfil the requirements of the job, as explained. The more factual the reasons, the better.

Many other questions may be asked of you, such as 'do you consider yourself to be a good mixer?' 'How do you get on with people at different levels?' 'Do you have a sense of humour?' 'Could you discipline subordinates?' 'Are you prepared to stay and settle down with this company?' 'What position are you ultimately looking for?' The questions may seem naive but it is the manner, rather than the substance, of the answers the interviewer is interested in.

You may be shown round the factory or various departments with which you are likely to be involved. Afterwards there may be another short discussion on what has been seen, further questions being asked, such as 'Is there any question you would like answering on what you have seen?' 'Could you suggest any improvements?' This latter question may seem a bit unfair after only half an hour's looking round but, quite often, suggestions, however simple, can be offered by a newcomer with an open mind. They should be offered in a suitably modest manner.

There will obviously be questions you would like answers to—status of position and duties; in charge of how many personnel; normal working hours; etc, etc. These questions should be left to the end of the interview when they will probably be invited. It is important to be quite clear about the essentials but anxiety about too many details does not convey the impression that your heart is in the prospective job.

The salary may depend on age, experience and qualifications and an offer might be made only in the letter of appointment, after a successful interview. It is then up to you to accept or refuse. A rough guide for a medium-salaried

"—such as . . . 'Could you discipline subordinates'?"

position is about 10 to 20 per cent above the present salary, assuming the jobs to carry similar responsibilities.

Conclusion

The changing of a job can be a long drawn out procedure, sometimes running into many months; the earlier letters to which there were no replies, the forms filled in which came to nothing, the interviews and the rest. All this could be the cause of even more unsettlement in the job you now hold; the constant waiting and expectancy of replies could affect your present work.

To change one's job is hard work and a mental strain: one could lose one's current job as a result, ending up without either.

If a job is offered verbally, do nothing about it until you have the offer in writing, also make sure you write a letter accepting the position. Before accepting reflect again; is this the job you wanted? are all the conditions what you really require?

Many positions have been accepted where the job is right but the conditions are not; if both do not run together to the satisfaction of the applicant, the full job-hunting procedure may very well have to start again in the not too far distant future, with the additional handicap of appearing to be a 'rolling stone'.

Conversely, it pays to devote much thought and effort to getting the right job with the right prospects.

White-collar Unions in Industry

by R. L. Clarke, MA, CEng, FIMechE

Unionism today is a vital factor. To understand it one must rid one's mind of Trotsky and the Tolpuddle martyrs; not to understand it is not to understand modern industry at all. But who can find his way through a maze of initials, such as DATA, ASTMS, UKAPE and ASEE?

There are 155 unions affiliated to the TUC and 376 registered with the Registrar of Friendly Societies. The unaffiliated unions include the Institution of Professional Civil Servants, over 50 manufacturers' associations like the British Motor Trades Association, and many professional associations like the Society of Town Clerks.

Unionism and striking are too readily associated. Even among the affiliated unions, the majority hardly ever strike. The ones that are continually in the news number little more than one dozen but they are large and powerful.

A Trade Union is defined as a combination, the principal objectives of which are the regulation of the relations between workmen and masters, workmen and workmen, or masters and masters, or the imposing of restrictive conditions on the conduct of any trade or business and also the provision of benefits to members. It is formed under the Trade Union Acts rather than the Company Acts because it thereby gains immunity from being sued for damages when dealing in trade disputes and agreements. This power is just as necessary to its functioning as the power of limited liability is to a company. It no more means that a trade union is going to behave dishonourably than that a company is going to go bankrupt.

On the other hand, there are always people who exploit a privilege to the full and the unions which treat the strike weapon lightly take the most understanding.

Throughout history there has had to be a means of voicing discontent. Opposition to the ruling hierarchy can bring about beneficial changes if sensibly organised. If not, it can break out in violence. The natural organisation for opposition is the anti-hierarchy where commands flow from the bottom upwards. Like any other feedback system, this enables messages to be passed back to the point of decision, the ruler of the hierarchy, provided the two organisations work together as a matching pair.

In the golden age of republican Rome tribunes were set up to represent the people but, proving troublesome, the office was later conveniently incorporated with that of Emperor. To be effective, the Ombudsman must owe his position to the people and in no way to the hierarchy he has to criticise. At the point of contact between hierarchy and anti-hierarchy there will be friction and wear; but this means no more than that the machine is functioning.

History shows repeated efforts to prevent hierarchies from misusing their power. The success of such efforts (the Reformation, the Commonwealth and the Reform Bills) heralded prosperity while failures led to disasters like the French or Russian Revolutions. The most confirmed Cavalier must see the correlation between prosperity and the wider distribution of power. Little though the communists may be liked, few people would wish to see the Czar and his boyars back in power.

A hierarchy is brittle by nature. Without independently organised feedback, it lacks the knowledge of what changes are necessary and when.

It might be thought that our democratic constitution was sufficient guarantee of power distribution. But five years to the next election is a long time. Modern problems cannot wait and the vote does little for people who want to negotiate with their employers. The only effective way of doing this is to join a union. So unions we must have; they are our safety valve. The measure of discontent can be assessed by the number that go militant and the extent of their militancy.

The effect of inflation

Paradoxically industrial unions are at their most militant today when the standard of living is high and a Labour government at the peak of its power. Before the war, when people really had something to complain about, the unions were in no position to fight. Men with jobs were too concerned to keep them and men without had no money to pay union subs. Resentment against the then governments' deflationary policy was carried forward into the next generation. So, after the war, both political parties adopted the alternative policy of inflation. Briefly this means that, instead of saving up the necessary money to spend for various purposes, the government decides what it wants to spend to expand the economy and 'prints' money accordingly.

For ensuring full employment, inflation works. For strengthening unions it works twice over, once by giving union members ample funds and a second time by giving them a continual source of complaint. They see their wages steadily shrinking in real value. The government expects that they will make and justify claims to restore their purchasing power. There is no practical way for 20m. people to do this except through unions.

Confirmation that unions have this constitutional duty was provided in the White Paper *In Place of Strife* and in the Donovan Report on which it is based. The TUC has become a pillar of the establishment and its member unions have acquired stakes in private industry as equity shareholders. This picture of bourgeois respectability is marred for some by the suspicion that these funds are stored up for future use as strike pay.

Mr R. L. Clarke*, *educated at Cheltenham College, the Royal Military Academy and Cambridge, served in the Royal Engineers and retired with the rank of Lt-Colonel. He took a two-year course in armament technology after his war service and then worked at the War Office in Technical Intelligence and at RARDE, Fort Halstead. After a period as military assistant to the Controller of Munitions, he retired from the Army to become R and D Manager with the Hoffman Manufacturing Co. In 1965 he became a consultant and in 1968 Director of R and D Management Advisory Ltd. He was elected Vice-President of UKAPE in 1970 and President in 1972.*

It is quite true that a least one union recruits by boasting that its financial position allows any of its members the luxury of coming out on strike whenever they feel the inclination. It finds that arguments with managements are so much more productive when both sides share this knowledge. Strong-arm methods, as in other days, allow small sections of the community to do well for themselves—for a time. But there is a governor in the system that brings power groups back into balance. The scales are by no means weighted against the hierarchy.

The hierarchy can count on the support of the establishment and the uncommitted spectators. Even a left-wing government soon loses patience with the unpredictability of union spokesmen. It can deploy the greater wealth and the better brains. It holds the power of initiative. Above all, it has no disciplinary difficulties.

Militancy

The unions are bedevilled with the discipline problem. Instead of the one hierarchic boss they have to satisfy millions who are not at their coolest when they need help. Implementing the will of the majority is a commitment that unions take very seriously indeed. Small wonder it is that union officials seem to turn somersaults in their efforts to meet their agreements as well as please their members.

The greatest difficulty seems to afflict those that have only one weapon in their armoury and that a crude one—the withdrawal of labour. Its effectiveness depends upon how quickly it hurts; the star performers are the production operators who can make a strike bite at the first whistle. Unions who have their strength in the back room where a stoppage may pass unnoticed for weeks are anxious to woo the support of the shop floor. This brings results but they must not be too damaging to the firm. Because every union knows that its investments, its existence and the future of its staff depend upon the continued well-being of industry. No symbiosis is satisfactory which damages one partner.

The appeal of the militant union is directly to the pocket. Investment in a union subscription has paid off better than the stock exchange, and it will continue to do so until the party under compression will yield no more. Some other appeal will then have to be found. As with other investments, there is a risk to the investor. He may be called upon to strike. Even with strike pay this makes a heavy demand on discipline, especially when called in support

* *This article expresses his personal views*

of some other interest to which his union may have undertaken a commitment.

Union officials have no legal power to enforce discipline over recalcitrants; pressure can only be brought by the majority of the members. This results in the rather distasteful actions of the many against the one which can be dismissed too lightly as bullying or blackmail. It is a necessary part of the system which we as a nation have accepted.

Neither can the forcible stopping of production be dismissed as necessarily an evil thing. The government does the same thing when it raises the bank rate. The best drivers do not drive on their brakes but the fact remains that you get along very much faster when you do.

Stoppages waste a lot of time on the shop floor and negotiation wastes even more of the time of management. But too much time spent on considering the point of view of employees is probably better than too little.

Both points of view command respect—the hierarchy presses for higher production and higher investment, while the union presses for wider distribution of spending power. Without strong unions, an industrialist would give less thought than necessary to the effect of his actions on the community: one man's employee is another man's customer.

Allowing for the human failings of idleness, greed and stupidity which are always with us, the two sides of industry have been in reasonable balance so far.

White collar unions

For many years the staff stood aside from union disputes. Identifying itself with management, it enjoyed a satisfactory wage differential and other privileges to reinforce its respectability and held aloof from picket lines and kangaroo courts. But after repeated wage awards to craft unions it became evident that the differential was shrinking to vanishing point.

Employers played into the hands of the militants by economising at the expense of those who were too proud or too loyal to bring pressure to bear upon them. By the middle 'sixties foremen found themselves earning less than their semi-skilled piece-rate machine operators. From this the staff learned two lessons: that wage claims had little to do with productivity or equity or anything else except the power to force the issue; and that the wages of sin were likely to be substantially higher than the wages of virtue.

The draughtsmen, already working in groups and feeling themselves the least identified with management, were the first to act.` Their union, the Draughtsmen and Allied Technicians Association* had been active for many years and was able to give them a flying start. They brought to the negotiating table a high degree of intelligence, a pressure of pent-up bitterness, and a rate of subscription which exceeded that of most engineering institutions.

Militancy is not so much the attribute of a union as a reflection of its members' state of mind. The vanishing differential over 20 years in itself provoked no more than a moderate reaction. But the industrial reorganisations of the last two years have changed staff attitudes to an extent which has yet fully to show itself.

Consider the case of the engineer who has been wrapped up in his work for many years. He has had little interest in unionism whose methods he finds distasteful. He has a high, if unexpressed, regard for his firm and for his colleagues. He devotes much extra time to keeping it one jump ahead of its rival, and for this he asks no reward. But he feels subconsciously that his efforts are noted with approval, that he is reaping the benefits of goodwill and security, and that one day when his firm can afford it he will be given material recognition for his loyalty. Instead of this he wakes up with a shock to realise that a take-over has amalgamated his firm with its rival, that the project on which he has been working is to be wound up, that his value is assessed at the salary he is still earning, and that the people to whom he looked for leadership and recognition have been pensioned off. He feels desperately in need of help.

His workmates are in no better shape than himself, his firm has proved a broken reed and his institution has no interest in his personal problems. But a militant union will give him a sympathetic ear for his past and comforting promises for his future. Joe Muggins, he reckons, is not going to be taken for a ride twice.

But the way of the militants is not all roses. Their success lasts as long as the distress and no longer. Firms which are big enough to ride out financial storms have generally less trouble internally. The white collar worker's interest in unionism wanes when he gets a fair wage. He wants to respect his boss; he enjoys his work and the contacts that it brings; he wants to be able to place his trust in his firm. But once he has been let down he will place it elsewhere for a whiff of insecurity is remembered for a lifetime. He cannot feel himself to be part of an organisation which is readily prepared to dispense with his services. Loyalty must work both ways, and when only the union offers loyalty in return there is nowhere else he will turn.

DATA demonstrated where loyalty was due by achieving a 50 per cent increase in wages in five years. Employers who maintained that they would have given it anyway were not believed. In fact they met wage claims with a steady resistance, granting them grudgingly from positions of weakness which aroused contempt rather than gratitude. It is difficult to see how they could have played their hand in any way more likely to encourage militancy.

DATA rapid'y extended its membership from among technical staff who anxiously watched the draughtsmen overhauling them. It obtained national agreement to extend its representative rights over allied technicians such as planners, estimators and testroom staff. Today, taking finance into account, it is probably the strongest white-collar union in industry. It has 16 divisional offices and a national youth committee. The basic unit is the office group which is empowered to negotiate directly with employers.

The local organiser trains new groups in the art of negotiation, starting with some minor issue like the opening times of canteens in order to give them confidence. Naive negotiating methods are frowned on—like asking for just what you are entitled to, or appearing satisfied when you get it. Excuses are left for reopening negotiations, such as unclear points and verbal agreements.

Members are forbidden to take a job below the recommended union rate. The funds from the political levy are applied in such a way as to favour the union rather than the Labour party generally.

The success of DATA has brought its own difficulties. With the basic draughtsman's wage raised to over £1500 per year, there is little scope for promotion unless the next higher range too can be raised. DATA has therefore put its shoulder to the wheel and made considerable inroads into the £1800 to £2000 per year grade. At a recent meeting of company technicians no fewer than one third of those present opted to join. A 40 per cent membership in any organisation is generally considered adequate for claiming negotiating rights.

There is no sign that employers have learned from their failure with DATA. Few seem to realise how much their technical staff feels underpaid though, when taking on new staff for similar work, they usually have to pay much more. This too is greatly resented by existing staff who may feel too old to move elsewhere.

Meanwhile DATA is thinking ahead. It is obvious that professional engineers have lost so much salary differential that they almost block the career progress of draughtsmen and technicians at about £2000 per year. In several engineering firms DATA has claimed to represent all engineering staff, irrespective of qualifications.

This is a challenge to the Association of Scientific Technical and Managerial Staffs which has signed the Bridlington Agreement against poaching on other TUC-affiliated unions. It is also an affront to the recently-formed United Kingdom Association of Professional Engineers. Its recruiting would not be affected, since it is not affiliated to the TUC but its recognition by the employers would be in jeopardy.

DATA denies that a professional qualification is relevant to a union and in this is much assisted by those employers who take the same view as regards remuneration.

DATA is confident of the continued supineness of the employers and it knows that UKAPE is a non-militant professional union which would be reluctant to use force or the threat of force.

The professional engineer

The professional engineer in industry is vulnerable because he is a specialist, because he is one of few in most firms, and because he is often isolated. When accountants recommend economies they naturally seek them in long-term, speculative and to them less comprehensible activities, all of which point to the highly technical research and development work. This is the first to suffer when a firm with a reputation for engineering excellence becomes junior partner to a firm with a short-term profits policy.

But it is time to put in a word of sympathy for the business man. He is employed by shareholders to make a profit and, in some cases, to save the firm by whatever means he can. He is not employed to look after the personal interests of his employees and it is unfair to load him with this extra responsibility. If they want a job done for themselves they should employ somebody to do it; that is why they should pay subscriptions to protective societies. This, stated baldly, may come as a shock to people who expect to be protected for nothing as of right. A business flourishes by buying what it needs as cheaply as possible; it is unpleasant but true that this principle also covers the supply of the services of chartered engineers.

In negotiating for any services businessmen assume that they are dealing with people who know their job. They want to pay the lowest salaries compatible with an assured long-term supply of satisfactory quality. But there will be no such assurance if they are dealing with individuals who have as much idea of looking after their financial interests as children. Responsible union backing is therefore very much in the interests of industry if the supply of good engineers is not to dry up. Why then has so much been done to discourage it? The reasons are prejudice, confusion and inability to see the broad picture.

Unionism is a jungle holding a wide variety of beasts, with

motives ranging from the subversive to the altruistic. Negotiators are like wrestlers striving for holds. Every advantage is used to the full; none is surrendered without recompense. As they become more practised, bouts become longer and more inconclusive. Wage claims are flung from all directions, some genuine, some tactical, some frankly speculative, most confusing and all of them adding to the load of overwork. Mountains are made of petty differences, emotions are worked up artificially and the ever-lengthening history of tangled histrionics recounted.

Every factory has its special conditions which can be made to prop up rickety evidence. Officials make case law as they go, spinning complications of custom and practice which it takes a full-time expert even to comprehend.

There is scope for argument on even the simplest agreement. For instance, a recent national wage award gave DATA members an 8 per cent cost-of-living rise in exchange for a wage standstill for a certain period. It would be illogical for employers not to give this to non-DATA members too but, not having agreed to a standstill, these would be at an advantage over DATA members whose funds had been used to get the award. But suppose they did not get the rise, what would happen if they were to join DATA subsequently? What about those who would have got a rise anyway but without a standstill?

To DATA the obvious solution is a closed shop in the DATA areas, and this is what its growing power will be used to achieve. A federation is being planned with the Amalgamated Union of Engineering and Foundry Workers to provide the necessary support. The different parts of this federation, to be called the Amalgamated Union of Engineering Workers, will make their own decisions but support one another. It is uncertain how far this plan will aid the recruiting of higher technical staff, but it may be that DATA is more concerned about raising the salary 'ceiling' than about who does the raising.

Up to £5000 a year

The Association of Scientific, Technical and Management Staffs claims to be the appropriate union for higher management generally, including professional engineers. No agreements have been reached with employers on the latter and ASTMS have actively opposed DATA's ambitions in at least one instance; but it would probably prefer DATA to UKAPE which it may regard as an unknown factor. ASTMS has had its roots in the universities since the days when it was the Association of Scientific Workers. By combining with ASSET it became the union for militant foremen and launched an 'abrasive' policy in October 1968 on behalf of managers earning up to £5000 per year.

It commands excellent publicity through an active Joint General Secretary but lacks DATA's roots in industry and is probably less powerful financially. It has recently been handling claims for computer programmers, higher technical officers and research officers at London University. If doctors wearied of the BMA they would not have far to look for effective support. DATA proposed amalgamation with ASTMS a few years ago but was turned down. This still rankles a little but on the whole officials work in friendly rivalry.

ASTMS is now trying to get involved in Government commissions and training boards. It is steering a delicate course between respectability and the economic viability that militancy has brought up to now. It does not see why professional engineers should have preferential treatment.

This is not the view of UKAPE which sees 200 000 professional engineers as a unit well worthy of separate representation; any other course would mean surrendering a professional status as yet barely won. It sees its role as strengthening the engineering profession rather than competing with existing unions. UKAPE's interests are complementary to what is available at present, ranging from career structure and university recruitment to registration, no-poaching agreements, transfer of pension rights, patents and inventions and mid-career training.

A scale of salaries coupled with responsibilities is shortly to be issued. A role similar to that exercised for doctors by the BMA/British Medical Guild is sought without the duplication of any non-union type services CEI itself decides to undertake.

Lest is should be feared that the formation of UKAPE might prejudice the position of the engineering technician or technician engineer, the Association of Supervisory and Executive Engineers has agreed to provide a similar service for non-members of the CEI institutions, and the two organisations intend to work in harmony. Although UKAPE has powers to strike, it does not offer militancy. There is no reason why it should, because there is nothing in its constitution to stop its members from belonging to any other union as well.

Conclusion

The position today is confusing and the possibility of conflict through misunderstanding cannot be ruled out. But at least there are several organisations anxious to protect the professional engineer which is more than could have been said ten years ago. The options must be kept open so that he can choose freely between them.

One can guess how things will settle down when the fire of militancy has faded for lack of fuel. UKAPE will find its essential role as the professional engineers' BMA in a friendly but outspoken relationship with CEI. ASTMS will become respectable as the quasi-professional association outside the engineering field and DATA will become the driving force of a vast engineering combination from the sub-supervisory grades downwards. Negotiation will become a profession in itself and its exponents will move from union to union and back to management, practising their art with as little real emotion as barristers.

Given prosperity and a general feeling of industrial security, there is no doubt that a peaceful compromise will be reached.

No Escape for Johnnie

by John Winton

Johnnie Daniels read the news of the take-over offer in his morning paper. It had been rumour for weeks, and now it was fact. The board had recommended shareholders to accept. So much for the Chairman's defiant statements on television a month before. So much for the other side's bland disclaimers in the financial press.

Johnnie himself was safe enough. Whatever purge followed the take-over, only a madman would dismiss the Projects Manager of the Engineering Division, a qualified engineer with twenty years' experience in the firm. But Johnnie knew there were bound to be changes and even now he could sense a new atmosphere in the building. The footsteps outside his office door were urgent, hurrying. The mail was late, the switch-board more ham-fisted than ever. 'Take-over fever', it was called.

At lunch in the managers' dining room the conversation lurched from one heavy-handed 'eve of execution' joke to another. Brewer, from Personnel, raised his glass. Departmental managers were allowed one free bottle of light ale at lunch.

'Let's have a toast, fellows. We who are about to die salute you.'

'The condemned man ate a hearty breakfast, eh Tony?'

Somebody laughed uneasily. Johnnie looked around the table. He had known most of these men for years. Now, one or two of them would not meet his eye, afraid they might expose their private fears. Johnnie knew that like himself they were all family men, with bills to pay, commitments, mortgages, a way of life to keep up. But their worst fear was not financial. To be dismissed after a take-over struck at a man's pride. It implied that maybe he had never been necessary.

'When shall we twenty-three meet again? In thunder, lightning or in rain? At the Labour Exchange, more likely.'

'Oh drop it for God's sake, Tony.' They always ate lunch together, and they knew him too well. Their meals here, like those in the canteen on the sixteenth floor below, were heavily subsidised; a paternal firm did not want employees fanning out into the Westminster pubs for beer and sandwiches.

'I'm sorry.' Tony Brewer looked hurt. 'Just because we've been taken over doesn't mean we have to huddle round like a funeral breakfast.'

'What do you think about all this, Johnnie? You've been here longer than most.'

'It's much too early to say yet.' Johnnie picked his words carefully. 'I don't think we need assume... assume the worst. It might mean expansion, more opportunities than ever.'

The faces round the table brightened. Johnnie had said what they wanted to hear.

* * *

After lunch Johnnie stood for a time at his office window watching the traffic, minute from this height, travelling along by the river. In spite of his cautious words, he expected the telephone to ring, a summons to arrive, some repercussion at least from the take-over. But the afternoon passed uneventfully and Johnnie at last went down to the employees' underground car park with a sense almost of anti-climax.

He always drove to work. True, the traffic was terrible, as everyone said, but there was that secure feeling in a car. It was good to sit in privacy, warm and dry, whatever the weather, sealed off from the world.

He had the luck of the lights and arrived home in twenty-nine minutes. He noted the good time, six minutes less than normal. Doris was in the front garden, weeding the rose beds. Watching her from the path, Johnnie had a sudden disturbed sense of precognition. But, in fact, how many times had he stood there, just home from the office, with his wife in the garden, straightening her back when she saw him, brushing away the lock of hair just as she was doing now, just about to ask the question she always asked?

'Did you have a good day?' She never read the financial news.

'So so. Children not home?'

'No. Carol's rehearsing for her end-of-term play and young Johnnie is at the chess club. It's his evening. He said he'll be home late. I'll just finish this, and then I'll get supper.'

Johnnie did not mention his news until they had finished eating. Doris cared nothing for 'economies of scale' or 'elimination of areas of wasteful competition'. Only one aspect interested her.

'Is it going to make any difference to us, Johnnie?'

'It might.'

'Will it mean you might have to get another job?'

'Good gracious me, no.' Johnnie dismissed the idea. But had there been a wistful note in Doris' voice? Why?'

'It's a very natural question, Johnnie. One reads about these take-overs. I just wondered whether you might be thinking of a change yourself.'

'Not me. What, after twenty years with Turret Turvills? I've done too much for that firm, and given them too much of my life. We should know whether it affects us very soon.'

* * *

They knew a fortnight later. The letter was in Johnnie's morning mail, in the characteristic envelope used for the firm's internal correspondence. It was not even marked 'Confidential'.

Johnnie read it once, without taking in its meaning. He read it again. 'Considerable reorganisation in all departments . . .' 'Major changes in personnel structure . . .'

Johnnie became aware that Tony Brewer was standing in his doorway.

'I said, have you got the chop, Johnnie?'

'I don't know.' Still shocked, Johnnie read the letter again. 'I'm not sure.'

'Judging by that envelope it looks as if you have. They're chucking them about like confetti this morning. Cor, talk about the night of the long knives.'

Johnnie shook his head disbelievingly. 'But I'm Projects Manager. I've got a . . . I've got a department of two hundred people.'

'Empire-building is out of fashion, these days, Johnnie.'

'Empire building . . . How dare you . . .'

Brewer held up a hand. 'Don't get on at me, Johnnie. I'm on your side.'

'But I've worked for this firm for twenty years, twenty years of my life.' Now that the full enormity of it was upon him, Johnnie was talking wildly, tears not far from his eyes. 'More than twenty years. They can't just . . . I can't just get my cards like a . . . a casual hop-picker. With no warning. It's not fair. It's not right.'

'Don't upset yourself, Johnnie. It's only a job after all. You can get another one.'

'But I've always worked for this firm.'

'All good things have to come to an end, Johnnie.'

Johnnie's hurt and bewilderment turned to rage. 'Get out of my office!'

Brewer shrugged. 'Keep your hair on, Johnnie. If you're so upset about it, why don't you get on to El Supremo himself?'

When Brewer had gone, Johnnie picked up his telephone. Clearly, there had been some mistake. Mr Sanderson, known as El Supremo, was Head of the Division and a member of the board. He could put it right.

Mr Sanderson's secretary was audibly flustered. It had obviously been a bad morning. But Mr Sanderson's voice was cool and remote.

'What can I do for you, Johnnie?'

'What can I do for you? Johnnie could hardly believe his ears.

'It's this letter . . .'

'Oh that.' Johnnie could almost hear Mr Sanderson composing his face in suitably solemn lines. 'I'm sorry, Johnnie, believe me. It was a Board decision. Naturally, I fought to . . .'

Johnnie dropped the receiver on the voice, extinguishing it. Suddenly, he remembered Godfrey. They had been sappers together in Italy, and Godfrey had won the Military Medal for dismantling a booby-trap in Florence. They had been fellow students at Imperial College, graduated together, and joined Turret Turvills on the same day. Godfrey had had a future with the firm, but he had resigned, to start his own consultancy. Johnnie had strongly urged him against it. But Godfrey had gambled and now apparently, he was prospering.

What a coup, if he could walk straight out of here into a new job with Godfrey!

Godfrey's voice was warm and welcoming. Something in its very buoyancy depressed Johnnie.

'Johnnie Daniels, well I'm damned, it's been a long time. How are you, and how's me old flame Doris? And the children? They must be grown up by now. How are you all?'

'Fine, fine, Godfrey, we're all fine. Godfrey, do you remember some time ago asking me to let you know if ever I had anybody I could recommend to you?'

'Indeed I do. And I meant it. I'm always on the look-out for talent. Who do you have in mind?'

'Me.'

'Ah. Well now, Johnnie.' The short silence, and the subtle, shaming change in the tone of Godfrey's voice told Johnnie all he wanted to know. 'It's that take-over, is that it?'

'Yes.'

'Well. To be quite frank, Johnnie, you weren't quite the sort of chap I had in mind. It's difficult to explain just like that, but . . .' There was another silence, which made Johnnie cringe. 'But look. Tell you what. Would you like to come and see me when you've finally got shot of that old saltmine and we'll see what we can do?'

'No, don't bother, Godfrey. I'm sorry I telephoned you.'

'But Johnnie . . .'

Johnnie could not bear any more talk. He got up from his desk, emptied the papers from his brief-case, took his pipe and tobacco pouch from the top drawer and put on his hat. The telephone was ringing again when he reached the door, but he walked on without stopping.

* * *

It was years since Johnnie had driven through their part of South London in the middle of a working day. With a stranger's eyes, he noticed now what Doris had been hinting at for a long time.

Their district was going downhill. The roads were always being widened, and now they were building a huge fly-over to take traffic to the South Coast. The quality of life was changing, the small services disappearing. There were four Italian restaurants now, 'trattorias' they called them, within half a mile, but nobody had taken on the cobbler's shop after the old man died two years ago. There were boutiques but nowhere to have a watch repaired. Many of the larger, older houses had already been converted into flats.

Johnnie had locked the car and reached the front door before he remembered that Doris was in town for shopping. There would be nobody at home.

In all the years they had lived in the street, Johnnie had never been inside the public house only three hundred yards away: *The Three Jolly Ploughboys*, an incongruously rural name.

There was one other man in the saloon bar. Johnnie recognised him as Arthur Price, a neighbour. He too, Johnnie recalled, had once lost his job, in a brewery merger.

'Good morning, Arthur. Slack day?'

The man turned, and Johnnie saw that he was slightly drunk.

'My name is Arthur and it is a slack day. So slack I've just given myself the sack.' Arthur snapped his fingers, with a surprisingly resonant sound. 'Want to buy a car-load of women's underwear samples?'

'No thanks.'

'Please yourself.' Arthur looked more closely at Johnnie. 'You seem to know my name. Have we met before?'

'I'm a neighbour of yours. John Daniels.'

'Ah yes. I remember.' Arthur's eyes showed no sign of recognition. 'So you are. Good chap.' He drained his glass, and held it up in front of him. Johnnie took the hint. 'Have one with me.'

'Thanks, I will. A large Scotch, Daphne, if you'd be so kind.'

With a meaning glance at Johnnie, the barmaid took a fresh glass from the shelf.

'And what brings you here, Johnnie Daniels, on such a fine working day?'

Johnnie steadied himself, surprised by his own nervousness. He gripped the handle of his pint tankard. This would be the first time he had told anybody.

'I've just lost my job. Firm's been taken over.'

Arthur spluttered with laughter. 'Join the club, boy. Prepare to have your soul destroyed.'

'Did you have much . . .'

'. . . Difficulty in finding alternative employment, as they say? It only took me three bloody years, that's all.'

'But you did get another job?'

'Oh yes, I got a job. By dropping my standards. I've had six since and now I need another. How long did you work for your firm?'

'Twenty years. More than twenty years.'

'Just like me.' Arthur shook his head sadly. 'It's too long, mate. You get so that you want to go on in the same old way, for the same firm. You get insti . . . institutionalised. Let me give you some advice, for free. From now on, use your friends all you can. The best jobs aren't advertised, you have my word for it. If a chum offers you a job, for Pete's sake take it and take it quickly.'

Johnnie remembered Godfrey. 'Provided it gives you some scope for your previous experience, surely?'

Arthur blew out his cheeks derisively. 'Never mind your previous experience, my friend. If someone offers you a job that's anything like a job, you take it. Otherwise, you know what it means?'

'What?'

'Answering advertisements. A refined form of torture, my friend. You won't believe me until you've tried it yourself. There you are . . . ' Arthur gestured at Johnnie. ' . . . Johnnie Daniels. In the prime of life. Years of business or administrative experience, or what have you. Should have no trouble at all getting another job. So you think. But half the firms don't even reply to your letter. You wonder why the hell they bothered to advertise in the first place. If you do get as far as an interview, you'll find a lot of firms don't even know what sort of bloke they're looking for. Some of them don't have a real vacancy anyway. They're just casting a bit of bait over the waters to see what comes up. I expect they'd be amazed if they got anybody.'

Arthur smiled but, looking into his eyes, Johnnie began to have an inkling of the measure of the tragedy.

'After you've been at it for a long time you get to the stage where everything you do is wrong. Write your qualifications in longhand on the back of your letter and they say, 'we don't want anybody as casual or offhand as that. Type them out on a separate sheet and they'll say, 'Oh yes, professional jobhunter.' Wear an old suit and they say, 'We don't employ scruffs.' Buy a new suit and they'll say, 'God he's richer than we are, he doesn't need the job.' You can't win. It's a vicious circle, once you get on it. The longer you're out of a job the worse it looks and the harder it is to get a job. How old are you, Johnnie?'

'Forty-seven.'

'Same as me.' Arthur shook his head again. 'I'm fifty now. There's a sort of a . . . ' He had difficulty with the word. ' . . . A what do you call it, a watershed, at the age of forty. Before, easy. After, bloody difficult. At my time of life, you know, little things begin to bother you. One day I was going up for an interview and I suddenly noticed I had odd

John Winton is the usual pen-name of Lt-Cdr John Pratt, RN, CEng, MIMechE. He was educated at St Paul's, the Royal Naval College, Dartmouth, and the RN Engineering College, Manadon. He joined the Royal Navy in 1949 and served as an engineer officer in the Korean war, at Suez, and in submarines, before retiring in 1963 to become a full-time writer. He has published six novels, including 'We Joined The Navy' and 'HMS Leviathan', and has written many technical handbooks and instruction manuals. He is 40, married with two children, and lives in Cheshire.

socks on. That threw me completely. Needless to say, I didn't get the job. Life terminates at forty.'

Arthur drained his glass, but this time he pushed himself away from the bar. 'The worst of it is, that it finally gets to you. You begin to believe that you really are no good, unemployable.'

Paradoxically, Johnnie was encouraged rather than depressed by the encounter. Arthur Price plainly had a paranoid sense of persecution. The fault was in him, not in his stars. Johnnie could see no comparison with himself.

* * *

But he was startled by his family's reaction. Carol clapped her hands delightedly when she heard the news.

'Daddy does this mean you're on the dole?'

'I suppose it does.'

The dole. Johnnie felt the full impact of the word. It was curious how the young generation had no shame about it. What would Old John Daniels, the stout independent grocer of Finchley, have had to say about a son of his on the dole? In Old John's eyes, to be unemployed was worse than going to gaol. And his mother, now living in testy retirement in Worthing, what delicate euphemisms would she use at her weekly whist drives for a son on the dole?

Johnnie was disconcerted, too, by his son's attitude.

'How splendidly non-middle-class to be on the dole, Dad. That's what social equality is all about.'

Johnnie frowned. 'And just what exactly do you mean by that remark?'

'When the bourgeoisie become the State-subsidised layabouts, that's what I mean.'

'I didn't realise we were bourgeoisie.'

'Dad we're the typical bourgeoisie.'

Doris alone seemed to share Johnnie's baffled feeling that a whole era of normal existence had suddenly and inexplicably come to an end. Johnnie lay awake that night, tormented by memories. Doris leaned over and switched on the light.

'I knew you were awake, Johnnie. I wonder why they did it? We've always gone where they wanted us to go. We went to Newcastle and Reading when they wanted. We were even ready to go to Trinidad, until they changed their minds.'

Trinidad.

That had been the big chance. The West Indies venture had failed eventually, because of political difficulties, but it had been a golden opportunity. Sanderson, who had been placed in charge, was now on the board.

'There's always Richard. I'm sure he'd offer you a job, Johnnie.'

'Your brother? What a thought.'

Doris' brother was a retired Army officer who had used his gratuity to start a small firm making agricultural machinery in the West Country. Johnnie could remember scoffing at the idea—'He'll find things a bit different now. No polo or pukka sahibs here, what, what?' To Johnnie's surprise and somewhat to his chagrin, Richard's business had not only survived but thrived.

Johnnie heard an echo of Arthur Price's advice. 'All right, I suppose it won't do any harm to go down and see his works at least.'

Richard's 'works' was quite outside anything in Johnnie's previous experience. It was a row of huts along the edge of a disused wartime airfield on a Wiltshire down. There were rough patches of concrete where more huts had once stood, a short line of parked cars, and a three-ton lorry with a load of small castings standing by the nearest hut. There was no sound, except the wind blowing through the acres of grass on the airfield.

Richard was walking from the door of what had been the airfield control tower, the only two-storeyed building in sight. Shaking hands, Johnnie felt his old hostility rising up again when he saw Richard's pepper-and-salt tweed suit, his pink carnation buttonhole, and his regimental tie.

'Glad you could come, Johnnie.'

'Glad to be here.'

Somewhere overhead, there was rippling, soaring bird-song. Richard followed Johnnie's gaze.

'Skylark. Sounds as if he's got a full order book, doesn't it? Come on, I'll show you the set-up.'

Richard led the way, between clumps of blazing red poppies. 'At the moment we specialise in disc harrows, drillers, insecticide sprayers, anything cheap and cheerful you can hitch on the back of a tractor.'

'You make money from that sort of thing?'

'God yes. Blokes are leaving the land in droves these days and every farmer's got to invest in more and more machinery. Mind you, we've got some other ideas, too. Duffy, he's our chief designer, he's a genius. We've even had one or two larger firms sniffing round us. 'Take-over' if you don't mind my using what must be a dirty word.'

'Is there any chance of that?'

'No fear. Perhaps we are a bit under-capitalised. That's inevitable, I suppose. But I want to work for myself. I handle all the sales side, I'm good at that, but what I badly need here is somebody with some technical expertise. Someone who can talk to Duffy in his own language. He's always baffling me with technical gobbledygook. I want somebody who can chase up sub-contractors, someone who's used to shifting paper-work about. We're snowed under with bumph these days. In fact, I want somebody who can handle everything, answer the telephone and talk sense to whoever it is on the other end while I'm away. Here we are.'

To Johnnie's eyes, this was industry on a primitive, almost feudal, scale. There were not more than forty employees. Many of them were women, who sang to the pop music coming from the radio, giggled, and called Richard 'Major'. There were four welders, and two men and a young boy in the paint-spraying bay. Duffy was a dishevelled-looking youth of about twenty-one, in spectacles and polo-necked sweater. He spoke in monosyllables, and hardly looked up from his drawing board when Johnnie was introduced.

'Is there a canteen?'

'Canteen?' Richard stared. 'For Christ's sake Johnnie, what do you think this is, ICI?'

The last hut was full of completed items, standing in neat sharp rows like chessmen, gleamingly painted, labelled and ready for dispatch.

'I'm still not sure where I would fit in.'

Richard stopped in his stride. 'Look Johnnie, I can't give you a neat label with a title to hang on your office door. The job will be what you make of it. I'm not asking you to put any capital in. I want your technical know-how, especially on the admin side. How much were they paying you, if you don't mind me asking?'

Johnnie felt himself going red. 'Two seven five.'

'Is that all?' Richard grimaced. 'Thought you'd be getting more. Well, I'll offer you the same, for the first year, and then we'll see.'

'I'll let you know.'

'Don't call us, we'll call you, you mean.' Richard's face was rueful. 'All right, you must do what you think's best. I'm offering you a job if you want it.'

Doris listened in silence until Johnnie had finished describing his trip.

'But I don't see why you should think it all so comic, Johnnie. It's Richard's own firm, and he is making money. In a way, he's offering you an escape.'

'Tucked away down in the West Country, miles from anywhere, working for some tinpot little firm? I was Projects Manager for Turret Turvills, remember?'

'Is it so very important to you to have the same sort of job you had before, Johnnie?'

'Whether it is or it isn't, I'm not going to work for your brother.'

'He's not asking you to work for him. He's asking you to work with him.'

'What about the house, and the children's schools?'

'That can all be fixed, Johnnie, you know that. None of us would mind moving.'

'The answer's still no.'

'I think you're making a mistake.'

Johnnie rounded on her, in exasperation. 'I know you'd love me to go and be a country gentleman like your splendid brother. Have all the tenant-workers touching their forelocks and calling me 'Mr John'.'

'Johnnie, you know that's not true.'

'You've never really got used to the idea of me being an engineer, have you? There's something infra dig about it. You want me to stride around with a carnation in my buttonhole. Your family have always thought of me as some kind of artisan . . .'

'I shan't say any more. Are you going to tell Richard?'

'You tell him. He's your brother.'

'Johnnie!'

* * *

The omens for the interview seemed to be very promising. He was reassured by the familiarity of his surroundings. The building was only a few streets away from his old office. The decor was almost identical in the entrance hall. The commissionaires could have been brothers.

'Mr Watkins is engaged at the moment, sir. Would you like to take a seat over there?'

Johnnie sat down, trying to compose himself for the interview. It was not until he reached forward to take a magazine from the low glass-topped table that he noticed, with a sharp pang of dismay, that he was wearing odd socks.

Too Old at Fifty?

by I. D. Campbell, BSc, MIMechE

Being out of a job at 50 can be a harrowing experience even in the midst of a so-called shortage of engineers; and even for people of proven ability. Can we afford to waste the professional skill and experience of older men?

> The man appointed to this top position will be about thirty-two years of age and will have had experience in the design and construction of chemical plants, followed by some years in general management.
>
> *advertisement*

To some people, war in retrospect seems more tolerable than it must have been at the time. When it is all over, you only hear from those who more or less survive. Of these, the most battered have the least to say. Similarly, competent senior engineers who have survived the fantastic experience of unemployment in this age of supposed manpower shortage put the past behind them and can hardly bring themselves to say what ought to be said about it.

And then, of course, there is the thought that nobody, even if they are prepared to listen, will really believe what you say. The circumstances have a sort of lunatic improbability about them. Is it not better to forget about it?

Selling yourself

This is a matter that should have a great deal of thought, especially from those who are, or think they are, least likely to suffer similar experiences. If qualified men over 50 cannot find work there is something seriously wrong, and it will take more than a few brief anonymous letters in the press to put it right.

The use of trained men is too often a story of indifference and waste.

Of course, the problem of the older man who is actually unemployed has its special difficulties. The unemployed of any age, few though they may be, are liable to include the unemployable, the unstable, the incompetent and the generally impossible.

Rather than choose someone who is actually out of a job, the employer generally thinks it safer to choose from those already in employment. There is, too, a sort of sales resistance to the unemployed man. The interviewer experiences a kind of embarrassment, so does the man, and this tends to diminish and demoralise the applicant. Only a conscious effort on the part of the interviewer will clear the air.

Being out of employment is not necessarily to a man's discredit. Sometimes it may be quite the reverse, but to prove it may call for a kind of salesmanship that the applicant has not got. However, to do the job for which he is applying he may not need salesmanship. The ability to secure a position in his unfavourable circumstances is therefore very likely to be quite unrelated to ability to fill the position satisfactorily. The responsibility for appreciating this must rest with the employer. Society demands that this be treated not

Mr I. D. Campbell *was educated at Darlington Grammar School and King's College, University of London, where he graduated with 1st class honours in 1926. He then followed a College Apprenticeship with the Metropolitan-Vickers Electrical Co. He was for six years Assistant Switchgear Engineer with the English Electric Co and entered the electricity supply industry in 1934. During the war he was at Hull, first as Technical Assistant, then as Generation Engineer, and became Deputy General Manager of the Sheffield undertaking in 1945. At nationalisation this position ended. He remained for a short time in the Yorkshire Division of the new Authority, and then went to the National Coal Board as Chief Power Generation Engineer. He returned to the manufacturing industry in 1952, and in 1954 joined the B.T.H. Co. His position there ended with divisionalisation in 1960. He took up consulting work shortly afterwards. In 1966 and 67, at the age of 62, he was in East Pakistan, now Bangladesh, as industrial adviser to the Government.*

merely as a matter of finding a man for the job, but also of finding a job for the man.

To sell your own virtues is always a difficult business. You may praise anything you have to sell except your own services. If a man has skill and experience, if his deviation from mediocrity lies in the direction of engineering sense and insight, he is unlikely to be unaware of it. But if he says so, although there is at least a chance that he may be right, it will not be counted to his credit. The dice are loaded against him. If he is good, how can he be out of employment? When Disraeli was striving to achieve recognition, one of his bitterest opponents said "Disraeli is a man without a position, and a man without a position has to do things that are not necessary to a man for whom a position is made." Employers would do well to remember this today.

The alternative to stating your particular capabilities is to leave the employer to find them out in due course, always assuming you are given the job. This may not be much good either. To many managements, staffing is a matter of pushing supposedly standard types into undoubtedly standard pigeon-holes. Too big or too small, it is tough on those who do not conform; but those who are too big come off worst. Seniority and the qualities as well as the defects that go with it, can be equally disqualifying. The older man will come to be seen as having an inflexible purpose, or as being set in his ways, according to whether his ideas coincide with those of his directors or not. The younger man would perhaps prove more amenable.

How people react

I have never set much store by security. How could one not despise the established system which usually leads any man over about 35 to accept the most unreasonable and humiliating treatment from his employer rather than sacrifice his pension and security? You can get security in the grave.

Having seen a great deal of this problem from both sides, I should like to set down some of my experiences in trying to re-establish myself after my apparently secure position had disappeared.

I tried to maintain my freedom of choice. It has been said that you can either have victory or you can have success: I certainly did not have success. At 50, I was for ten months out of a job. At first, I was not greatly concerned. I was well known and had a reasonably successful record. I was soon to learn better.

The first thing I discovered almost at once was that many of my supposed friends had been devoted to me for my previous position. Once I had nothing to give, they crossed the road when they saw me coming. I had many friends in influential positions who had seemed helpfully disposed. Their reactions to my approaches are interesting.

No. 1 wrote: "It is a sure sign of the times that a man of your qualities should be actually out of a job." He seemed genuinely concerned, but had no suggestions to make.

No. 2 sent a helpful reply, tentatively suggested a position in his firm, and even suggested a reasonable salary. He passed me down the line to another, who suggested a lower salary. This process was repeated until the salary had dropped by about £600 *pa*. At this point I suggested that it was time to get down to a definite offer. I then received a letter saying that the firm felt that, if they offered me such a low salary, I should be unlikely to remain with them. They therefore preferred to withdraw.

No. 3 looked grave and said I was in a very dangerous position. "However," he added cheerfully, "I'm sure you'll be all right," and hurried off.

No. 4, who was a very prominent man in his way, explained that all my troubles were due to my having so foolishly given up my job. Gently I explained that I hadn't given it up. It didn't exist any more. I had, however, rejected the alternative that I had been commanded to accept. He was sympathetic. "Tell you what. Would you like me to write to Sir Blankety Blank and Sir Thingumy Bob?" I said I should be grateful if he would.

As I was leaving, he asked after my family. I showed him a photograph. He asked if he could borrow it to show to his wife. "I will send it back. I shall be writing to you." I never saw him nor my photograph again.

No. 5, who was noted for his vociferous profession of high ideals of comradeship, teamwork, and the like, was very sympathetic but assured me that he had absolutely nothing to offer. Shortly afterwards, he remarked to a mutual acquaintance that, of course, he could have offered me a job, but was afraid of upsetting his happy team. Clearly a man of the highest integrity.

No 6, head of a Government Department, thought there was an appointment coming up for which I should be suitable. He named a salary and thought the post might carry a house. At the interview (the salary was lower, and there was no house) I was asked what guarantee I could give that I would stay in a job at nearly £1000 *pa* less than I had been getting. I said I would comply with the terms of the appointment and there was no justification for asking more from me than that.

I did not get the job. I heard later, 'on the grapevine', that I had been turned down because they could not understand why I was willing to accept such a low salary. You can't win, can you?

Nos 7 and 8, after some negotiations, both offered me specific positions. Salaries and conditions were agreed. I waited in vain for the confirmatory letter. On enquiry (yes, I had to enquire), I was told that in each case the Board had insisted on making the appointment from within the firm.

These are only examples. While all this was going on, I applied for innumerable advertised posts, without the least result. I saw innumerable people, many of whom promised to do various things which, for the most part, they never did. You see, I had no gun in my hand. They felt no obligation, no responsibility for me or my troubles. One of them was frank enough to say so. He said that it was not reasonable that I should ask him to solve my problem for me. I suppose he thought I should appoint myself to a new position.

Finally, an unexpected opportunity was offered, and I accepted it at a salary £1000 less than I had had before. This was a successful proposition. Within a year or so my status and salary were returning to their proper levels, I was building up my organisation and doing good business. A few years later, just as I achieved solid success, due to a take-over, my job disappeared. I went into consulting work and have so far managed to survive. I am now past my 60th birthday.

Facts of life

This is a brief summary of a true experience. Like Alice, by running very fast, I have succeeded in staying where I am. Has it all been a futile waste of time and effort? Is there something useful to be learned from it, or is it something we have known all along and refuse to learn any more about because we think it can't happen to us? Does it matter to the safe majority or is it just too bad for an insignificant few?

I think it matters very much. It is high time to take a new look at many of our accepted habits of thought. For many, perhaps for a majority, there is no problem. Not because they can easily change their jobs, but because they would never dream of trying. Long before 50, probably at about 40, most men assume that they are now fixed for life. Indeed, to think so is in itself something of a mark of success. Employers are happy to have them think that way. Bargaining power is reduced, and if familiarity with the ways of the firm counts for more than professional distinction, that makes the man easier to live with, even if he is not likely to be bubbling over with new ideas. It also makes him practically unfit for employment elsewhere. That is accepted as a fact of life and who wants to battle against the facts of life?

The belief that at 40 a man can be too old for any change in employment has implications that are even more serious for the profession than the actual difficulty in changing jobs.

An engineer returning to work in this country after some years abroad once remarked to me that nearly all the men he met here seemed to be absolutely fed up with their jobs. There was more than a grain of truth in this. Men soon come to believe that they are too old for any but the same old jobs, and become conditioned for failure. Professional knowledge, which ought to be the means of earning professional respect anywhere, means too little to too many.

The way we play it, engineering skill is something to grow out of, not into. People who are enthusiastic (or perhaps rather naive) like to show or simply to exercise their skill. They will accept, or even seek out, professional responsibility.

Meanwhile the worldly-wise will look after their own position so that they do not find themselves out of a job at 50.

For most British engineers security means security of employment in a particular job. In this respect the more restless American can give us points. If he did not think he could get another job, he would not feel secure. For him, security means the ability to move around on his own feet within his profession. That means freedom. Security, the British way, too often means goodbye to professional freedom. In a good safe job, professional excellence is less important than to know about the customs of the firm and the whims and fancies of the boss. Don't worry too much about engineering, you won't have to compete with any but the same old crowd. You hear them say, almost with pride, 'it's years since I did any engineering'. Is it any wonder that, after 20 years of this, a man's middle-aged services are not very negotiable?

Anno Domini

But does not a man's capacity decline with advancing age? This resolves into two separate questions. Firstly, at what age do the established customs and prejudices of industry and the effects of past misuse of the man begin to make re-employment difficult? Secondly, at what age does the natural deterioration in a man's capacity reduce his professional value and make it unprofitable to employ him?

We have seen what industry does for a man. What about Anno Domini?

We are considering whether a man can be too old at 50. What happens at a later age when his faculties really begin to diminish is quite another question. Let us consider briefly the significance of age for the engineer.

It has been widely maintained by responsible authorities that there is no basic deterioration in the mental faculties until quite late in life. The acuteness of many well exercised older minds is to me a constant source of admiration. In the Twelfth Fawley Lecture, Sir Denning Pearson said:

> Engineering is constantly developing and changing. It offers a potentially interesting and intellectually stimulating life right up to the age of retirement. There need be no falling off in either usefulness or interest with increasing age. . . .

There is no issue in an engineering life that cannot benefit from mature experience. We are right up against the forces of nature, and there are problems all the way. It is no use thinking that you can make a parcel of engineering problems and leave it to the boys. This work and its supervision call constantly for the application of mature judgment.

Not all the older men have enquiring minds but is the proportion among the younger men any better? At 50 a man may show remarkable inflexibility, made more evident by the possibly aggressive display of his experience. But if, at 50, a man thinks he knows all the answers, the chances are that he never had much capacity to learn. The late Henry Ford once said that when a man became an expert it was time to give him the sack. I know what he meant.

So let us try to keep the older man in useful employment. After all, he will always be part of the industry. Fob him off with silly answers if you like, but in the end someone will have to do something. What shall we do? Shoot him? If he were a dog, most Englishmen would think it their duty at least to answer that question. Instead, as he is a man, he receives the stock answer 'No suitable vacancy', only to see the same firm advertising exactly the right vacancy next day.

Pension rights and provision for old age may present problems with the older employee; but state provision is beginning, though inadequately, to alleviate them. The employer who rejects a man simply for this reason is shirking an issue which has to be faced in the end; someone has to do something about it. Oddly enough, firms will often more readily employ a man of 65 than one of 50 because they feel that the pension question does not then arise.

Some people are needlessly concerned about the imagined difficulties of putting older men to work under younger superiors. Barring the obvious objections to putting a senior man to work under his own previous subordinates (and even that has been known to work quite happily in rare cases) a man who has reached the age when he would like to see his responsibilities being trimmed down will usually take quite kindly to this condition. When he reaches the age at which he wants to reduce his work load it is not necessary, nor humane, to push him over the edge. It is quite feasible, natural, and of benefit to the nation to give him some less exacting duty in which he can, perhaps, exercise a skill that the pressure of other work had prevented him from using properly for many years.

Above all, let us remember that we are talking of human beings, in fact of colleagues who often taught us what we know. There is no lack of employers who pay lip service to the principles of social responsibility and proper conditions of employment but the number of those who practice them in the case of older or more difficult men is rather smaller.

Even if we take the view that human considerations should not interfere with business decisions, some questions remain to be answered: if we are short of professional engineers, can we afford to get rid of some at 50? And if this becomes common practice in the profession, can we expect the most promising young people to enter it?

Problems of the Young Graduate

by A. M. Salek, BE, and W. P. Julius

If the average young engineer is dissatisfied with his job, both recruitment and status of the profession will suffer. The London Graduates' Committee of the IMechE carried out a survey which, among other things, brought to light the principal professional problems and discontents of their members; on its basis, the Section's Chairman and a member of its Committee discuss the difficulties of young engineers and suggest some remedies.

The Committee of the London Graduates and Students Section has recently become increasingly disturbed about the low professional morale and general dissatisfaction apparent among many members of the Section. In order to seek a wider consensus of opinion we decided, therefore, to send out a comprehensive questionnaire, part of which is reproduced in Table 1. All our members were asked to return it anonymously. We also took advantage of this opportunity to include questions which would enable us to improve our service to our members.

A gratifyingly large number of members returned completed questionnaires—over eight hundred out of 2950—showing that the junior members are genuinely interested in their Institution. This article is based on the initial analysis of the returns.

It is clear that there is a very large degree of dissatisfaction

Table 1—The Relevant Parts of the Questionnaire and Results (percentages of those who replied)

PERSONAL DETAILS
1. *Class of membership:* Graduate 77   Student 23
2. *How old are you?* 21–25: 34·2, 26–30: 43·9, 31–38: 22·5
3. *What educational qualifications have you, or are you pursuing?* (See Table 2)
4. *Have you completed all the educational and practical requirements for Associate Membership?* Yes 67   No 33

RELATIONSHIP WITH THE INSTITUTION
6. *Do you belong to the Institution—*

	Yes	No
To be associated with a professional body?	83	5
To take advantage of the exchange of technical information?	75	7
To benefit from established educational standards?	45	22
So that membership can help you to obtain good employment?	74	11

7. *What effect has your membership of the Institution had upon advance in your career?*
Considerable 5   Some 25   A little 30   None 40
8. *Have you made any engineering contacts through your Institution membership?* Yes 15   No —

EMPLOYMENT AND CAREER
19. *Indicate the category, or categories, of your present occupation:—* (See Table 4)
20. *How many staff does your organisation employ?*
Under 100 :11   100–499:17   500 or over: 72
21. *In what category does your employment fall:—* (See Table 3)
23. *Bearing in mind your graduate status, indicate how satisfied you are with your present employment:—* (See Table 5)
24. *Do the following have adequate (in your opinion) appreciation of technical matters?*

Your immediate superiors	Yes	77	No	23
Your top management	Yes	61	No	39

25. *Do you consider your present employment to be:—*

A permanent career?	33	Only for experience?	16
A stepping stone to a better position elsewhere?	54	Part of a training scheme?	10

26. *If you have not changed your job within the last two years, is it because you:—*

Are you quite satisfied?	16
Not sufficiently dissatisfied?	40
Have thought about it but done nothing?	12
Have not seen a suitable position advertised?	79
Have applied for one/several positions without success?	28

27. *If you have changed your job within the last two years—*
Did you have difficulty in obtaining another one?

     Yes 26   No 74

Were there many applicants?
Over 30:16   Over 10:15   Under 10:12   Unknown 57

Did you get the job you really wanted?	Yes	78	No	22
Was the change advantageous overall?	Yes	93	No	7

28. *If you have applied for jobs and have been unsuccessful do you feel it was because:—*

Your engineering qualifications were not good enough?	6
Your engineering qualifications were too good?	4
Your experience in the new field was inadequate?	30
Your management experience was inadequate?	12
Of possible personality reasons?	6
Of monetary reasons?	12
The advertisement was misleading?	12
You turned down the job?	18

29. *Leaving aside the specification of the job as advertised and explained, do you feel that the methods of interview and personnel selection adequately assessed your ability, especially your technical attributes?*
Yes 40   No 60
30. *Do you feel generally—*

	Yes	No
Suitable posts for graduate engineers are hard to find?	60	40
Employers seek engineering staff for work which could be done by less qualified people?	76	24
You would have been better off in another occupation?	37	36

EDUCATION
31. *Do you consider that in the widest sense, the educational requirements for Graduate membership are adequate?* Yes 84   No 16
34. *How useful is your technical education in your job?*
Invaluable 23   Considerable 34   Occasional 36   Incidental 7
35. *Do you feel that you have been technically over-educated?*
Yes 15   No 85
Had too limited a general education?
Yes 47   No 53
36. *Is the Graduate education period*
Long enough? 83   Too long? 11   Not long enough? 6
37. *Do you consider that your practical work was:—*
Very useful? 71   Too long? 14   Too short? 8
A waste of time? 5   Satisfactory in content? 49
Unsatisfactory in content? 23
38. *Looking back, do you feel that the educational establishment where you were taught was sufficiently in touch with the outside engineering world;* Yes 64   No 36
39. *Do you consider the facilities for postgraduate courses adequate?*
Yes 59   No 41
Have you ever taken advantage of them?
Yes 49   No 46   Intend to 23

163

with the employment situation, the training and the status of young members.

The age range of those who replied was from twenty-one to thirty-eight and showed a Gaussian distribution, the majority being between twenty-four and thirty. The majority were using the HNC route to Corporate Membership as shown in Table 2; as many as 17 per cent were not following any of the usual courses.

It appeared that 72 per cent of our members were working for organisations employing over 500 so that the results of this analysis basically reflect the feelings of those employed in large or medium-sized concerns. The type of concern worked in followed the pattern listed in Table 3.

Apparently the large majority of members are ambitious. In answer to the question "What position do you hope to occupy by the age of 45–50", 19 per cent in large, and 36 per cent in small, firms wanted to be directors. In all sizes of establishments 40–49 per cent wanted to be senior executives. Obviously many young engineers' ambition is high-level management.

We thus have a background to the report of a very large sample of young members who are mostly under thirty, who mainly work in large organisations and who hope, in 15 to 20 years' time, to become the leaders and senior managers of industry.

Wrongly employed

The most serious aspect of the employment situation is the fact that 78 per cent feel that they are doing work which could be done by less qualified people. If this waste is occurring throughout the country, it represents an enormous misuse of the nation's brain power and it is not surprising that so much is heard of the 'brain drain'.

Why do 38 per cent feel they would have been better off in another occupation? This fact certainly amplifies the impression of the Section Committee, that many members feel generally dissatisfied and their being underemployed must account for much of this discontent. Table 4 shows that, in view of their technical underemployment and high ambitions, they are taking to administration as a way of changing their occupations. Therefore their technical training has been wasted to some extent and they have not been properly prepared for the employment to which their ambitions may eventually lead them.

Various facts brought to light that young engineers are dissatisfied with their present jobs, or with the prospects which these afford but that they are determined to get on in spite of their difficulties. Eighty per cent regard their present jobs as mere stepping stones, means of gaining experience and training.

Here are two representative case histories in the words of the persons concerned.

On leaving University I was accepted by one of the large motor firms for a Graduate Apprenticeship. With a dozen or so other graduates we embarked on 2 years' practical training around the works. However, as the 2 years passed I began to suspect what a bunch of misfits we were going to be. Comments from various departmental managers we came into contact with varied from "I don't know what we could do with you in this department" to "I'm a practical man I don't need all this academic nonsense".

As forecast, placing us in a job proved extremely difficult and subsequently we ended as a dissatisfied and disillusioned group. Promotion and prospects looked bleak and generally we were paid less than unqualified people at the same age.

After 3 years of this drudgery I managed to escape and received a post as chief engineer of a plastics firm. Here at last I have the responsibilities and status that my professional training and education have been directed to; and hence a new satisfaction and attitude to my job.

Table 2—Educational Routes	
Route	*per cent*
ONC	14.7
HNC	29.4
HND	9.8
I Mech E	10.9
Dip Tech	4.8
BSc (Eng)	13.7
Other	16.7

Table 3—Type of Employer	
Route	*per cent*
Commerce or Industry	71
Nationalised Industry	7
Teaching Establishment	7
Teaching	7
Consultancy	6
Civil Service	5
Local Authority	4

And again:—

After an expensive and far too long apprenticeship in the aircraft industry, I worked alongside many other graduate engineers on a drawing board for two years without a chance of a change. I left to join another industry, still on the drawing board, but responsible for my own projects as one of a small team.

Still there was no chance of promotion with a non-technical DO manager. So I joined a nuclear group for three years—very good experience, but still no possibility of promotion. I struggled off the drawing board after 18 months to work on stress and heat transfer.

This was interesting work but I had to leave to improve both my financial position and prospects. Eventually I have managed to become a manager of a design development office in a medium sized company. This is a quite different type of work to my previous jobs. However, the non-technical upper management still do not give me the full responsibility that the position calls for; the policy is to wear belt, braces and elastic waistband as well.

Fifty per cent of them have changed their jobs in the past two years, and 93 per cent of these have found the change advantageous. This certainly suggests that many employers fail to plan satisfactory careers for the young engineers whom they are so eager to take on.

Table 5 amplifies the picture of discontent or disillusionment and sums up the employment situation in a nutshell.

What can be done?

Many Graduates and Students feel very strongly that the Institution should obtain a revision of its Charter to enable it to take an active and aggressive part in improving the social and financial status of its members. It could then wage a publicity campaign directed at employers, with the object of reducing the underemployment of a vital part of the nation's technically trained manpower.

Personnel managers should be capable of assessing the technical ability of applicants and matching it to job requirements. It appears from the survey that 60 per cent of those who replied felt they had not been adequately assessed in this respect when they attended interviews.

On the other hand, young engineers should do far more themselves to remedy the situation. All too often Graduates and Students give the impression that they do not deserve higher status or faster promotion: they dress, speak, write and behave in a slovenly way and are too willing to work in very bad conditions. It is partly up to them to make their employers aware of their potential value to them. With this in mind, the Institution could help its young members to attain a more professional attitude by encouraging informal meetings between them and successful senior engineers at the peak of their careers.

Since, for one reason or another, so many engineers aspire to management, it would save a lot of expensive retraining of new recruits to allow ambitious young men of high potential to undergo some form of management training after a few years' practical experience. Following this, they should, if they wish, be allowed some practical managerial experience. In this way employers could retain men who would otherwise leave them and ensure a competent second string management.

Employers should recognise that a man's outlook and behaviour are very much dependent on his environment. If he is treated as a responsible person, with the status called for by his education, his slovenly attitude towards work and his general disinterest—which are so often criticised—would largely evaporate: a good man always grows into the job he is given but he doesn't easily shrink into it!

No management training

We would have liked to split up all our findings to a much greater extent, to investigate whether the different educational backgrounds of our members produced widely different outlooks on the various aspects of employment, education and status. However, there has been insufficient time to do this in more than a few cases.

The most frequent criticism of the responder's course of studies was insufficient management training. Lack of this is a very serious gap in an engineer's education, as even those who do not eventually turn to management would benefit a great deal from a broader understanding of its problems and points of view.

There is much to be said for giving engineers a wider education. Forty-three per cent of those who returned the questionnaire found their technical education of only occasional use, if any; and 47 per cent felt that it had been too narrow, particularly those with an HNC.

It was generally considered that there should be a closer relationship between college syllabuses and practical engineering. This view was summed up by Mr Bosworth in *The Engineer* of July 10th, 1964:

> The aim of both students and teachers, therefore, tends to be away from the real world of engineering and directed towards a stylised science which is neither scholarship nor useful art.

A very large number of our members felt that, for various reasons, the present workshop training was so much time wasted. They recognised it as a potentially important feature of an engineer's education but felt that there was too little supervision of such courses by the Institution. They were

A. M. Salek *was born and educated in New Zealand and graduated from the National School of Engineering, Christchurch as a BE (Mechanical). He joined the local agents for Martonair, and visited England in 1951 for training. He has since been employed in the application of compressed air equipment in industry. In 1956 he returned to England and was subsequently appointed General Sales Manager of Martonair Ltd. He has just completed a technical liaison tour of Canada and New Zealand which included a full lecture programme. In 1961 he joined the Committee of the London Graduates Section and has held the offices of assistant Hon. Secretary and Hon. Secretary, and is now Chairman. During his term of office he has taken a particular interest in the problems of Graduate employment and the presentation of Graduate views within the Institution.*

spending too much time watching, rather than operating, machines. Shorter and more intensive courses would save time and frustration.

A large proportion also felt that laboratory work at college took an excessive proportion of their time. They want fewer tests, done more thoroughly and with more attention to report writing as an exercise in the art of self-expression.

Forty one per cent did not consider the facilities for postgraduate courses adequate. This was a well informed reply, as 49 per cent of the whole sample had attended such courses and a further 25 per cent intended to do so.

In view of the very great concern with management training, we feel that this subject should take a more important part in the examinations. Also, it should be better taught, and more relevant to real life. Companies should be encouraged to release their engineers after they have had industrial experience to attend a full-time course in management.

Colleges and their staff should keep in closer contact with industry, and staff should be encouraged to take advisory positions in industry.

Time could be saved on laboratory work to enlarge the young engineer's knowledge of the Arts, and the Humanities in general. Such laboratory work as is done must be up to date and be relevant to the theory known to the student. How to make good use of the library and the art of report writing should also be taught.

The narrowness of the HNC suggests that the engineers who qualify in this way find themselves at a disadvantage, both as regards performance and status. The Institution should require a higher standard of literacy and general knowledge from candidates for election.

Status and the Institution

Corporate members may not fully understand the status problem of younger members as many of them have already achieved their goal by their own efforts; but among younger people the criticism is often heard that 'the Institution does nothing for me'. Asked what they expected, 27 per cent of respondents volunteered the opinion that the Institution should fight for the status of its members. It is surprising how often a parallel was drawn with the Law Society.

In addition to this came many pleas for an appointments board which would deal only with approved employers, and

Table 4—Breakdown by Occupation

Occupation	% Sample	Occupation	% Sample
Basic research	3	Consulting engineer	7
Applied research	13	Design engineer	20
Technical administration	25	Development engineer	12
Non-technical administration	8	Test engineer	8
Project leader	9	Instrument engineer	5
Sales	8	Production engineer	11
Design draftsman	10	Works engineer	3
Detail draftsman	2	Maintenance engineer	5
Teaching	8	Entirely non-technical	1
Technical writer	2	Full time student	4
Postgraduate student	3	Day release trainee	0·6
Graduate apprentice	3		

Some members gave more than one description of their job in this section

Table 5—Degree of Satisfaction with Work

Aspect of Work	Very	Fairly	Not
Technical content	39	47	14
Professional responsibility	31	47	22
Staff responsibility	22	42	36
Promotion prospects	28	43	29
Status	28	49	23
Working conditions	40	42	18
Remuneration	22	56	22

a careers advice bureau. Many members also thought that there should be a bar or catering establishment at Institution headquarters to make it easier to meet older members in an informal atmosphere.

There was a strong feeling that the Institution regarded Graduates and Students as 'second class members' rather than as the future Corporate Membership.

It is interesting to note that 83 per cent of the respondents joined the Institution to be associated with a professional body and that 74 per cent thought, at the time of joining, that it would advance them in their employment. Only 5 per cent, however, felt that it had actually been of considerable help; 25 per cent felt it was of some help to them.

Many felt that the Institution should. rid itself of any restrictions in its Charter so that it could take an active interest in the prosperity and status of its members. They agree, however, that they must also help themselves in this matter. The public's image of a student engineer is an illiterate, shaggy, beer drinking rugger player, quite oblivious to any civilization around him. This naturally makes it difficult for the same public to regard an engineer just qualified with the respect he feels is his due! Particularly, since most engineers find it difficult to communicate with one another, let alone with the general public. If young engineers behaved as though they had status, this would help the Institution in trying to make the employer realise that in his graduate engineer he had someone worth having.

It seems, then, that the average young member, when he joins the Institution, is looking for something more than that which is written into the Charter at present; and that he feels strongly about deficiencies in his training.

Older and more influential members may find it difficult to appreciate the discontent of young engineers in their firms—or even to believe that it exists. For it is difficult to remember in full one's own early struggles and in many ways the path of the young engineer today is a great deal smoother than it was in the past. Nevertheless, the recent survey bears out our long-standing impression that many young engineers are dissatisfied with their lot.

If this article helps to bring this fact—together with some suggested remedies—to the attention of those who are in positions to help it will have served its purpose.

Incentives to Innovation and Invention

by J. C. Duckworth, MA, MIEE*; sketches by A. E. Beard

The creation of a society which welcomes innovation—a prerequisite to a steadily rising standard of living—depends on a reorientation of our sense of values. The Government must support projects which—while of overall national benefit—would not pay for individual firms. But in order to do so efficiently we must develop a quantitative technique for assessing the national benefit.

"—the pure scientist appears to be held in higher esteem than the engineer ..."

There has in recent years been increasing public discussion of the part which science and technology have to play in the modern world and of the importance to the economic welfare of the nation of science and technology and its application. I have had the good fortune to work with a public corporation, which has for some time been concerned with this subject, and particularly with encouraging the application of invention and innovation to civil, as opposed to military, uses.

I am convinced that the future national welfare of our country depends largely on the speed with which industry can turn to new, commercially viable, processes and products. The rate at which this is done must depend on the incentives offered both to individuals and firms.

Since one's approach to a problem is necessarily subjective and dependent on past experience, I should like to explain my particular interest in this subject by reference to my own history as shown in my biographical note. Throughout my working life I have had the good fortune to be concerned with a number of technologies at their most exciting stage of development, but I have pursued that type of career best described as knowing less and less about more and more, rather than tending to specialise, as is done in the academic world. I consider myself reasonably representative of the scientist, technologist or engineer who becomes increasingly concerned with problems of management and finance more than with the practice of his own profession and I hope that my remarks will be of interest to those who follow this path as well as to those whose work is of a more professional character.

It is now often said that it is important for the welfare of the country that more people should turn to technology and engineering rather than pure science, and that more trained scientists, engineers and technologists should enter management. It is also a well known fact that at the same time this policy is largely frustrated for, in this country, the pure scientist appears to be held in higher esteem than the engineer and technologist. Much effort is now being devoted to a change in this attitude of mind, and one can only hope that the efforts of the engineering institutions in this respect will be successful. In many European countries, notably perhaps in Sweden, the Academy of Engineering has long held at least as high a social rating as the Academy of Science, and has been able to select as students the very best young brains. There are still too many relics here of the era when it was thought to be ungentlemanly to do anything which was profitable as well as useful, and in our post-imperial role this is a luxury we cannot afford.

Status

It is interesting to reflect to what extent the impreciseness in the English language of the word 'engineer'—covering, as it does, a vast number ·of non-professional activities—contributes to the apparent lack of status of the professional engineer as compared with the scientist. Perhaps one of the most helpful contributions we could ask from the classicists is that they should coin a new and socially acceptable single word to replace the clumsy expression 'Chartered Engineer'!

For my own part, however, I have no regrets whatever at having deserted the more academic scientific pursuits and I would advise any young scientist or engineer who has other than purely academic abilities to move unhesitatingly towards application and management. In my view, it is wrong to say—as is so often done—that it is a waste of a scientist when he enters management. Indeed, if he has the qualities necessary to a management executive, it is usually a financial loss to the nation if he remains a scientist, and he is likely to find a more exciting and spiritually rewarding life in the business and industrial world with its wider interest and richer human relationships.

Before returning to my subject of incentives, I must make it clear that I do not believe material benefits alone are man's primary consideration. In an acceptable developed society individuals must have the greatest possible freedom to

Then Managing Director of the Nat. Research, Development Corporation. This is a slightly condensed version of the author's 11th Graham Clark Lecture. For a career and portrait of Mr Duckworth see page 156, March issue.

select the career of their choice. Young people are greatly influenced in choice of a career by the social status attached to it and of all the highly developed countries we have the highest regard for the pure scientist and the lowest for the business executive or salesman. Material incentives alone will not reverse this attitude.

However, I have not found amongst individuals in the industrial and business world with which I have been concerned any greater devotion to material ends than in more academic and professional spheres and there is certainly no less high a sense of duty towards the national weal. We have to recognise that there has been a vast and fundamental change in our position as a nation, from one in which we could benefit from our long early lead in industrialisation and from the trade advantages accruing to the pre-eminent imperial power, to that in which we have to compete among equals for our share of the world's wealth.

In these circumstances, we must recognise that cultural objectives can only be pursued to the extent that the nation can afford them and, this being so, the creation of national wealth must surely be given a more honoured place in our society! I need make no excuse, therefore, for concerning myself in this lecture primarily with material incentives and rewards, because we now have to think first about how to make money before we argue about the best way to spend it!

Incentives to individuals

Adequate incentives to individuals might be less important than in the past. Significant technological advances now tend to be made by teams rather than by individuals. These teams are employed by industrial and other organisations for the sole purpose of making such technological advances, and many individuals can therefore have a satisfactory and rewarding career in science without actually participating directly in any financial benefits arising from their work. In the same way the innovation, i.e. the application of the invention or new technique, is more likely to be made by an industrial concern rather than by individuals to whom any direct rewards will flow.

Even for the employed technologist there are wide variations in different countries in approach to the question of personal reward for inventive work. The original purpose of patent law was to create rights for the individual inventor which would safeguard his position when his ideas were commercially exploited. It is now common practice in all countries for the technologist to assign to his employer the rights in any invention which he may make in the future. The original concept of personal responsibility is maintained in that individuals and not, for instance, corporate bodies, are recognised as first inventors for legal purposes.

In this country, however, when the inventor has assigned his rights to his employer, the latter has complete freedom in deciding the magnitude of any reward, no matter what the benefit may be to the employer himself and any possible direct incentive arising from the hope of reward is largely removed.

In some countries, notably, for instance, Germany, the position of the employed inventor is a good deal more favourable. He is guaranteed a certain minimum proportion of the benefit arising from any invention, even if it can be shown that he was employed to make such an invention and has undertaken the necessary work entirely in his employer's time. If he can demonstrate that his invention showed, for instance, greater originality than might have been expected from him in his normal employment or that some of the work was done in his own time, then he has a legal

claim to a greater proportion of the benefit. Special courts have been set up in some countries to assess the degree of individual responsibility of the inventor and to decide the appropriate portion of the reward accruing to him as a result of the benefits of his invention. As a consequence many employed inventors in Germany have amassed private fortunes and

"*—there are wide variations in different countries in . . . personal rewards . . .*".

there is little doubt that knowledge of this has its effect on workers in the more applied areas of technology.

One objection often raised to the latter system is that it is likely to lead to a greater secretiveness amongst workers in a field of possibly important application. However, Germany's record in the application of scientific technique to industry is an excellent one, and though this factor is probably a very minor one in the achievement of this record, it does at least seem not to have hindered their prowess.

Patents

There is an old story, the accuracy of which it is now difficult to check, of a meeting in Cambridge between Lord Rutherford and a leading German worker in the field of atomic physics. The German professor proposed to Rutherford that they should jointly file patents deriving from some of their recent work and he was astonished and dismayed when Lord Rutherford flew into a rage and dismissed him from his office. Lord Rutherford, in keeping with the attitude of his time, despised such commercialism in the approach to science. But Germany has had far more experience than this country of the need to live on the efficiency of its industrial processes and the high moral attitude towards the purity of science is one which, as I have said before, we could afford only in our era of imperial grandeur.

At a later date and, it is true, for a somewhat different reason, there was a further illustration of the prevailing British attitude to the exploitation of scientific work when, after the discovery of the properties of penicillin, it was decided that basic patents should not be filed on the initial inventions. Following this, essential production processes were duly patented in America—where they were developed—and as a result, there has been a constant outflow of royalties from the country of initial invention to that of commercial exploitation. It was this case which provided a major argument in favour of setting up the National Research Development Corporation. Recently the same laboratories at Oxford have carried out the original work on cephalosporin derivatives but this time the initial invention has been covered by patents and, as a result, a situation commercially favourable to this country has been established. One cannot believe that the public, either in this country or abroad, has in any way suffered from such a policy of energetic patent filing and exploitation, even for medical products. Indeed, the fact that some degree of exclusiveness can be offered may well lead licensees to devote more money and resources to the development of products than they would do if the scientific results were available to all; so that the availability of useful products may well be expedited by patenting.

Although the research team may be relatively more important now and in the future than it has been in the past, we must by no means discount the importance of the individual working on his own. In addition to the true private inventor, a number of universities leave the patenting and exploitation of results of their work to the individual scientists and technologists in their laboratories. Qualified patent agents and such organisations as the National Research Development Corporation are available to ensure that inventors obtain the patent protection which is available to them.

In general, patent protection, so long as it is properly used, provides adequate reward for significant invention, although one can, of course, always point to exceptional cases. The attitude of workers in British universities towards the patenting of their inventions has certainly changed radically, even in the short period since the last war. Even if the university worker is troubled by ethical considerations as to the propriety of receiving financial rewards for his work, it is most important that he should protect his invention properly by the filing of appropriate patents, so that the national interest may be safeguarded. There is no particular merit in repeating the commercial history of the development of penicillin!

Industrial incentives

I will now turn to the question which is at present the subject of much discussion and investigation—that is, the extent to which Government or its agencies can—and should—encourage innovation in industry to achieve increases in productivity. A whole army of intellectual brainpower has been directed at this subject. At the time of writing, little in the way of results had been forthcoming but it is quite possible that, before this is published, such profound conclusions may have been reached that any observations of mine will have been rendered trivial.

It is true that the activities of the NRDC have been limited to a fairly small scale by the finance provided by the Government. However, the scale on which operations have been conducted presumably reflected the importance which Government and the public accorded at the time to its direct concern for the activities of civil industry. Indeed, this concern is relatively new.

Throughout the ages Governments have, of course, offered inducements to merchants and traders to contribute to the local wealth but, in this country for many decades the policy has been one of *laissez faire*—industry and trade will best thrive with minimum interference from Government.

"—an output per worker of between 2½ and 3 times that in British industry . . ."

The direct interest of Government in industry first arose for military reasons, when it became apparent that, because of the increasing difference between military and civil product requirements, industry was not able on its own to meet military needs from its normal production, and that specialist Government ordnance factories were insufficient on their own to meet military requirements in time of war. Hence arose the system of Government peacetime protection for industries important in war, of development contracts and of orders for prototypes.

The interest of Government in encouraging scientific research for application in civil industry was recognised after the first World War in the creation of the National Physical Laboratory, followed progressively between the wars by the Research Councils and all their research establishments. However, at that stage the Government took no practical steps to ensure that advances in technology were in fact applied by industry. Indeed, to some extent, the research establishments followed the universities' lead in 'pursuing knowledge' rather than working on problems of interest to industry. After the last war the beginning of Government interest in the application of scientific developments in industry was illustrated by the encouragement of the industrial research association movement and by the creation of the National Research Development Corporation. In recent months the full spotlight of political and public interest has been turned on this problem. It is at last realised that our material welfare depends on the efficiency of British industry and services and the Government is anxiously investigating the means by which it can encourage increased productivity.

The American lead

We must observe, however, that inventions and innovations are not necessarily meritorious in themselves but only in so far as they contribute to higher efficiency and enable us to compete more effectively in world markets. The fact that knowledge of modern technology is only one factor is borne out by the observation that United States industry in general has an output per worker of between 2½ times and 3 times that in British industry, whilst at the same time it is clear that there is no significant technological information available in America which is not also available here. This is true even in industries where, on the face of it at least, there does not appear to be any marked difference in the degree of application of modern technology.

It is clear, therefore, that our immediate problems are concerned mostly with management and with the attitude of workers towards their jobs. It is surprising that very little seems to be known about the reasons for this disparity between US and UK productivity figures, because it would appear to be amenable to quantitative investigation. Such reasons as the high capital investment per worker in the United States, or the relative freedom there from restrictive practices, are often quoted, but there must certainly be a number of other contributory factors. Since we could more than double our productivity without the application of new technology, if we could emulate the United States in this respect, there would seem to be a very good case for a thorough study of the reasons underlying this difference, and it would be difficult to conceive a more potentially rewarding research programme!

In the long run, however, the achievement of a steady increase in productivity, upon which a rising standard of living must depend, can be achieved only by the introduction of new processes and products. In a traditional free society such increases in productivity take place as a result of normal competitive processes in industry. The new product or process will be introduced by a particular firm only when it pays that firm to do so. Ideally, one would like to see a situation where this maximisation of profit for the individual firm also produced the maximum return to the national economy. If this were so then the only case for Government intervention would be that Government knew better than industrial organisations what was good for them—a dangerous and doubtful supposition! Unfortunately, it appears that this identity of interest between the nation and the individual firm frequently does not exist, and I will give some examples later.

I am in danger at this stage of straying into the field of economics, in which I can claim no expertise. It does appear to me, however, that one of the main purposes of taxation should be to ensure that, as far as possible, individuals and firms, whilst looking after their own interests, act in the national interest also. If the Government is convinced that the rate of technological advance achieved by ordinary competitive processes is not the optimum for the national interest, then it is clear that the taxation procedure is not achieving this aim and it would seem that the main remedy lies in fiscal changes.

However, quite surprising difficulties appear in this approach; for instance, a recent survey has shown that only 16 per cent of firms make prior allowance for all available tax incentives in their calculations when deciding whether or not to install a new product or process. Our system of investment allowances and of fiscal encouragement of research and development is as good as any in the world: but if 84 per cent of our industrial firms do not take full advantage of such taxation incentives it is somewhat difficult to see what the next step should be. Inability by firms to take full advantage of available tax reliefs is a matter of individual attitudes and abilities and the general climate of opinion towards innovation and change. One can only hope that processes of education, better publicity about taxation procedures and the beneficial effect of true competition will in due course have its effect.

Direct Government action

In the present state of the national economy there appear to be many cases where Government action is, or would be, advisable in the national interest and I now propose to examine some of the reasons for this, based on the experience of the Corporation in its work. I will not consider special interests such as, for instance, defence, public health or safety, where there are obvious Government responsibilities but only those areas where normal commercial criteria apply.

The first class of projects in which industry requires Government assistance is small in number but of great significance and of public interest, namely, those where the research and development expenditure, or the capital cost involved in new plant, are too great to be shouldered by any individual firm or even by a group of firms. In this category the most obvious examples are the development of civil aircraft and of civil nuclear power stations.

Even in the United States where industrial concerns are generally considerably greater than in any other part of the world, these projects are now beyond their scope and Government action has been decisive. It seems likely that Government will always have to make the decisions in such major technological developments and the surprising thing

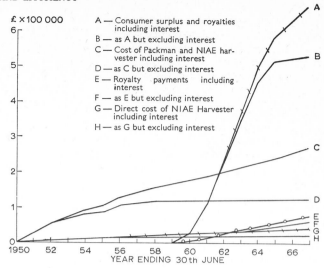

A — Consumer surplus and royalties including interest
B — as A but excluding interest
C — Cost of Packman and NIAE harvester including interest
D — as C but excluding interest
E — Royalty payments including interest
F — as E but excluding interest
G — Direct cost of NIAE Harvester including interest
H — as G but excluding interest

Fig. 1. Cumulative cost of NIAE and NRDC investments in a potato harvester project, compared with total benefits accruing from it

is that Government in this country has equipped itself so inadequately to do so. I am not concerned here with the inadequacies of cost control and project management which have recently been the subjects for public discussion but with the prior assessment of the merits of a development.

Economists have done little so far to develop methods of deciding whether or not a particular major development is in the national interest. It is clear that we need far more information, in particular, better methods of assessment of national benefits, before Government can arrive at rational decisions on any particular major project. It is a fact well enough known to scientists and technologists that the provision of adequate information must be the forerunner of all significant advances. One great contribution which scientifically trained executives could make to Government is to see that scientific method is applied in Government procedures, in that the maximum of information is made available before decisions are taken. A proper assessment could then be made of the incentives to proceed with any particular project and decisions arrived at on a far more rational basis.

A somewhat analogous class of cases arises when the project is of medium size but there is no particular industry already in existence with the potential or expertise to develop it fully. For example, the hovercraft requires not only technological development but also a quite new approach to the whole problem of operation which is

"—there will always be a strong tendency for investors to look for immediate profitability . . ."

outside the normal experience of either aircraft or shipping companies. Moreover, new regulations governing the operation and safety of hovercraft have to be developed and it was beyond the means of any one firm to make a proper appreciation of all these matters and so decide whether or not there were the necessary incentives to proceed with the development.

The third area in which there appears to be a difference between the interests of the individual firm and of the nation is in the 'time of pay-off'. The directors of a company acting in the interests of their shareholders must, quite rightly, take a relatively short-term view in this matter. Shareholders can with relative ease switch their investments from one company to another, and there will always be a strong tendency for investors to look for immediate profitability. As citizens, however, we cannot so easily switch our nationality to follow the highest standard of living and we must always look to returns over a longer period, both in our own interests and in those of our descendants.

Determining national benefit

The fourth main reason for Government action, and perhaps the one which is of greatest interest in the present context, is that many technological innovations bring benefits which are spread widely amongst firms and individuals, so that, though the benefit to the national economy as a whole may be considerable, the estimated benefit to the innovating industry may provide insufficient incentive to proceed. Considering hovercraft again, these vehicles provide a completely new means of transport. Assuming that a viable operating and construction industry is created, benefits accrue both to the manufacturer, to the operating company, and to the public who use the service. In addition there is the direct national benefit due to any foreign currency earnings from royalties or the export of craft and services.

But not one of those concerned may feel that the benefit accruing provides sufficient incentive to undertake the whole development. Under these circumstances it is obviously highly desirable to have means of calculating the overall national benefit so that the relative merits of projects can be compared. Unfortunately, as I indicated previously, the methods of doing this have been surprisingly little developed. Even after development and exploitation has been completed, little appears to have been done to determine whether or not a particular project was in the national interest. One can see, therefore, how much more difficult it is, before a project is started, to make a reliable quantitative assessment of its possible future benefit, bearing in mind the additional uncertainties that have to be taken into account, such as the size of the market, development cost, and development period.

We have recently made a start on the problem of attempting to assess the national benefit of a number of our projects, beginning with some simple cases where cost/benefit analysis should not prove too difficult. As an illustration I will discuss briefly the development, sale and use of two potato harvesters. I do not wish to go through the arguments and calculations in detail, and will merely present the results, for which I am indebted to the two economists who have undertaken this

"The second was more complex . . ."

work, Mr Heath of the Board of Trade and Mr K. Grossfield of NRDC. In Fig. 1 financial outlay or financial benefit, as the case may be, is plotted against time. To make a true assessment of benefits (or otherwise) from national expenditure, we have to take into account two developments on which public money was spent. The first harvester, initiated by the National Institute of Agricultural Engineering, was further developed by industry with assistance from the NRDC. The second, which was more complex and expensive, was developed by industry, with considerable support from the Corporation since it appeared, at the time, that the additional cost would be outweighed by the advantages, particularly for larger acreages.

In the event, the first harvester found a considerable market whilst the hopes for the second were not realised. If one considers the development costs of the cheaper harvester alone, then these have been adequately covered by later royalty returns. The development costs of the two potato harvesters, however, including the one which was not a success, have not been met, and the national expenditure has therefore not been covered by royalties. The other benefit to the nation—apart from exports, which were considerable—is in labour savings to farmers, releasing labour for other jobs.

Making reasonable assumptions about the benefits to farmers who bought the harvester, one can arrive at an overall national benefit resulting from national expenditure. This is strongly positive, even making no allowance for export earnings. It may be noted that the cumulative benefit—excluding interest—on the graph levels off. This is because it is assumed that economic pressures, and particularly labour shortages, would in due course have forced the development of a harvester by industry either here or abroad, so that the national expenditure has only bought time.

However, the timing of new developments, as regards market requirements and foreign competition, is a matter of the greatest importance. It could well be in the buying of time in appropriate cases that Government assistance is of the greatest benefit to industry. If this is the case, the calculations clearly illustrate the advantage of the national expenditure on the potato harvester, even allowing for the failure of the larger project.

This brief illustration provides evidence for the view that the national interest may require developments which are not yet attractive to an individual firm or indeed may never be so.

Qualitatively, this view is now generally accepted. Recent history appears to demonstrate that wealth accrues to the innovating society but in too many cases in this country the accountants of an individual firm can demonstrate that a particular innovation is not in its own interests. On the other hand, Government appears to have no quantitative methods of deciding where the national interests lie, and consequently adopts two extreme criteria to decide its actions. On the one hand, it adopts the strict accountancy view of a project, which tends to lead it to the same conclusion as the individual firm; on the other, when political pressures build up to force action on some major project, the defensive walls of the Treasury are finally breached and the money pours out for a particular project with no commercial criteria being applied at all! This situation is obviously nonsensical and some reasonable quantitative criteria must be developed to act as a guide to Government action.

Returning to the four reasons previously given for Government action in individual cases, there is rarely a clear-cut situation in which only one of these is decisive. Indeed, there may well be other reasons, particularly an element of more direct Government interest, such as defence in the case of the aircraft or nuclear power industries. In the case of major industries such as these there is also the much debated value of technical 'fallout' or 'spinoff'. The value of this, as a proportion of the amount spent, seems to have been much exaggerated—and United States experience indicates that it becomes less as the technological gap widens between civil requirements and those of military or quasi-military projects. This is not to say that it is insignificant in real value when vast sums are being spent, but it should be relatively easy to obtain the same results with much smaller expenditure on better chosen goals.

Conclusion

The engineer or scientist who turns to management has a vital role to play in modern society, where a steady improvement in the standard of living, based on technological advance, is now an accepted requirement. Increased spending on academic research alone does not achieve this advance—it is the application of scientific method and results to management and to new products and processes which yield the increases in productivity on which a higher material welfare must depend. It is often said that, as a nation, we are good at fundamental research and invention but not so good at putting results to profitable use. It may well be that one follows from the other—for many years there has been too great a tendency for the better men to turn to pure science, and so to reduce the availability of talent in engineering and technology where it would produce more material benefit to the nation. A reversal of this trend depends on factors other than material ones—mainly a change of outlook throughout society as a whole and particularly in educational establishments.

The creation of a society which welcomes innovation—an essential prerequisite to a steadily rising standard of living— also depends largely on a reorientation of our sense of values. The driving force towards innovation, however, must largely be a material one and the rate of progress will depend on the incentives offered. The scientist entering the field of management associated with applications of technology can only be struck by the inadequacy of the methods of assessment available or in use, particularly at the national level. With the aid of modern computers it should be possible to develop far better quantitative methods for the assessment of projects and so to make decisions on a more rational basis.

My final plea is that more effort should be devoted to this problem—it would be difficult to conceive a way in which relatively small amount of money could be better used to the national advantage.

The Proper Use of Professional Men

by B. T. Turner, MSc, AMICE, MIMechE

While much can be done to improve the quality and quantity of future professional engineers, the only way of overcoming the current shortage is by much better use of those we have already. In the process of furthering this we shall find ourselves rationalizing professional work and applying method study to it; developing existing abilities to the fullest and promoting greater job satisfaction; and obtaining many other benefits, both for those directly concerned and for society.

Mr B. T. Turner was educated at King's College School, Wimbledon, and London University where he took his B.Sc. and M.Sc. degrees. Previously he had been articled to a firm of consulting civil engineers. During the 1939–45 war he worked on radar and gun-laying devices with Nash and Thompson, and other firms. After a period with Ascot Gas Water Heaters, he joined I.C.I., Billingham, in 1947. In 1950 he worked with Saunders-Roe on aircraft equipment systems. He became a project engineer with English Electric Aviation (B.A.C.) on guided weapons until 1960, when he became Assistant Director of Engineering with the English Electric Co. His work was largely concerned with ensuring better utilization of technical manpower. He became Principal of the Dunchurch training centre in 1964 and is now a director of Industrial and Commercial Techniques Ltd.

In the complex affluent society in which we live today the professional engineers and scientists have become key figures, largely because they are responsible for controlling and changing man's environment. One of the depressing features about this society is the fact that very few engineers reach the top ranks of the controlling hierarchy. This applies both to industry and to Britain as a whole.

The dichotomy between the teaching of the humanities and life sciences on the one hand and the physical sciences on the other has been recognized for some time. However, the training given to young professional engineers which should enable them ultimately to control industry still does not adequately cover many of the disciplines now required. For too long engineers and scientists have been enquiring into the 'how' rather than the 'why' of their work. They have become preoccupied with structures and processes, but have not considered sufficiently the purposes of their projects. The practice of engineering involves men and money as well as materials, machines and energy, all of which have to be applied for the benefit of mankind. But men are not as predictable as machines and consequently engineers should not remain mere experts in controlling hardware, who are themselves controlled by non-technical managers. Engineers should become managers able to control both people and projects, for the problems of organizing industry to the best advantage today demand more than a nodding acquaintance with engineering science and technology.

The supply of technical manpower is a vital national problem and many reports have been written to stress it. By contrast, very little has been published on how to use the existing technical manpower effectively. Although engineers and scientists are being turned out in increasing numbers, as shown in Fig. 1, the education of many is wasted because they spend far too much time on trivialities and routine. It will not suffice in the future to train more and more specialists. There is also a need for 'generalists'.

Industry should now be giving considerable thought to various tools and techniques which can be used to improve the utilization of professional engineers—such as computers, calculating machines, mechanization and automation of experimental work, and the use of more non-professional specialists.

Efficiency in the intellectual plane can ultimately be improved, as it has been in the physical plane, by increasing mechanization, standardization and improved organization. Under the growing influence of economic pressure and fierce competition, Britain, with its few natural resources, large population and high standard of living, will have to utilize and develop its technical manpower to the full.

Continually increasing productivity is now a necessity for Britain, if full employment is to be maintained with stable prices and a rising standard of living. This national problem will only be solved if each employee is utilized and developed to the full.

The effectiveness of all manpower and the willingness with which men and women work depends ultimately on their management; therefore, this subject is very timely in our country's present situation.

Education must change

It is pertinent to ask if the present method of educating engineers, with its emphasis on analysis and specialization, is really conducive to their best utilization, for it tends to restrict their vision and confine their abilities to a narrow field.

Criticism has rightly been levelled at our present curricula. In particular, a different approach is necessary to design. For too long designers have been poorly recognized by industry and not afforded a sufficiently high status. A designer needs to be thoroughly conversant with manipulative processes and materials of all kinds as well as with the results of relevant research. He must be able to use modern design tools such as computers and to appreciate cost and time factors.

Designers are creative people and work in the realm of

shape and size where proportions and movements are determined by both engineering and aesthetics. This is a type of work which requires the high intellectual capability expected from professional engineers.

On the other hand, there appears to be a need for a broader concept in engineering education and training to produce engineers who can manage a whole project, as well as specialists who deal with one aspect of it. It seems a pity, therefore, that the CAT's vie with the universities. Would it not be better to concentrate the teaching of design and manufacturing techniques, which depend largely upon empiricism and synthesis, in the CAT's; and to concentrate on research and development work, requiring an analytical approach, in the universities? Indeed, is this not what Sir David Eccles implied when he said that the CAT's were 'the nurseries of the men who will develop the inventions of the scientists'?

The disparity between the humanities and the physical sciences has been recognized. It has been argued that there is insufficient time to teach additional subjects, such as the elements of economics, patent law, history of engineering, human relations, communications, etc., and that these should be regulated to postgraduate courses. Might it not be better to reduce the penetration in depth of some subjects in favour of greater breadth? The early practical training and education of professional engineers which would fit them to manage product divisions of industrial concerns does not appear to be adequate, nor does it cover many of the disciplines now required.

Are long periods on the shop floor, learning a few crafts, desirable for future professional engineers? Some acquaintance with the shop floor or site is necessary, if only to learn something about the reactions of human nature in the industrial setting. A portion of this training could usefully be given before entering college. But project work in a team which designs, manufactures, and tests a complete machine* would seem a better approach.

In the past the education of professional engineers has given them confidence in manipulating and understanding mechanical principles; therefore, they tend automatically to apply such principles to people, which results in frequent failure to achieve successful relationships. Their preoccupation with physical laws leads them to expect the same kind of predictability in people. Engineers often shun problems which are ill-defined and abstract, where there are no clear-cut answers from traditional practice. Thus, they back away from leadership, from dealing with people and ideas, and neglect management principles. Durrant's epigram needs to be remembered:

'Under every system of economy men who can manage men manage men who can only manage things.'

Organization for innovation

Having educated your engineer, you have to utilize his capabilities to the full and this requires an adequate organization. Whatever form this takes, it is the individual who, clearly, makes or mars it. Because of the complexity of modern technology, team effort is required for most projects but ideas generally originate with individuals. The problem is how to develop, within a given framework of groups, the originality and maximum productivity of the individual. In modern engineering the best way of achieving this appears to be by breaking down projects into a number of vertical specialist activities and co-ordinating the work of these

---

* An article on project work appeared on page 430 in the Sept. 1961 issue of the Chartered Mechanical Engineer.

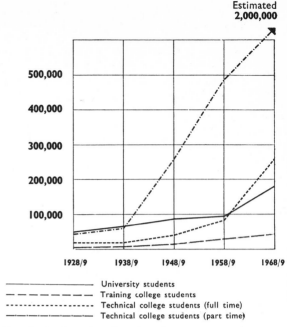

Fig. I. How the total number of students has grown over the years

University students

Training college students

Technical college students (full time)

Technical college students (part time)

groups in horizontal teams, led by engineering or project managers.

While certain 'mechanistic' types of organization may well be suitable for stable conditions where there is little or no innovation, an 'organic' system becomes necessary in rapidly changing conditions. Jobs tend to be less clearly defined and lines of authority less clear cut. In any case, an organization does not really operate according to the line chart in the managing director's office but in accord with a flow of information such as is sketched in Fig. 2. Good communications between people at all levels are therefore essential for efficiency.

In respect of innovations, one of the major problems facing British industry as a whole is the excessive time lag in getting from a formal technical solution or 'software' to a detailed practical design which may be turned out as hardware at a reasonable price. Modern techniques of cost and time studies, critical path scheduling, etc., might well pay handsome dividends here.

A number of factors are essential in reducing this time-lag. The professional engineers concerned should be really project-minded and, in effect, managers. They must be all-rounders who are time and cost conscious, who have a sense of urgency and are sound in human relationships. Designers, adequately trained, must have an appreciation of R & D work as well as of design and manufacture. An agreed programme should have been scrupulously prepared, taking into account money, manpower, materials, machinery and all the relevant factors of research, design, development, manufacture and/or construction and commissioning. Finally, there must be adequate control of modifications: optimization of resources can be as important as optimization of design.

The control of modifications deserves more attention than is generally given to it. Engineers are adept at thinking up new and seemingly better ways of doing things. In projects of high innovation content it is essential carefully to restrict modifications if programme key dates are to be held and engineers used effectively. Of course, where failures occur in development testing, a modification may well have to be put in, but not until it has been closely scrutinized by a competent committee representing design, development, manufacture and construction interests. Failure to control modifications causes untold delay throughout the translation process from an idea to hardware.

Developing managerial talent

All professional engineers have a duty towards the profession—and towards the whole society—to develop themselves and those under them. Top management ought to ensure that this is being done at all levels. Speaking generally, an engineer below 35 years of age needs moulding; above 35 he needs stimulating.

It is vital that the professional engineer should cultivate a wide-angle view of his problems; that is, he should know what is happening in other spheres of activity and how his work fits into the social picture. Only this enables him to correlate the work of others.

The knowledge of how to manage people is developing only slowly; the distinguishing feature of management is the use of forces in people, as opposed to forces of nature. It is at this point that most engineers seem to be weakest, principally through lack of training. People who have been students of the Arts can sense an 'atmosphere', or judge by general impressions. But scientific and engineering training frequently consist only of the study of detail. Small wonder that scientists and engineers who work for years on isolated pockets of knowledge tend to miss the wood for the trees. Engineering methods can be useful where a decision is based on the knowledge derived from reproducible experiments. But this is not the only way of discovering facts.

While pure scientists can live in detachment from personal feelings and wishes, the professional engineer often has to decide on matters which, by their very nature, involve emotions as well as intellect and here detachment itself becomes a distorting factor. Cold intellectual appraisal of a human problem, however necessary for certain limited purposes such as the administration of the law, can never lead to a true understanding.

It was F. W. Taylor and other practising engineers in America, who first showed that certain management principles could be identified and taught. Although several universities are attempting to do this at the postgraduate stage, it would seem that there is room for more effort in this direction. A serious limitation is the lack of a real laboratory. Industry can provide the only adequate laboratory for the study of industrial management. Industry itself, therefore, must help by running special courses for potential engineering managers. Such courses can deal with the writing of reports, the presentation of facts, efficient employment of engineers, control of engineering departments, value analysis, and the structure of an industrial organization. The main trouble here is that because the best men often cannot be spared, they are deprived of a useful development aid.

Another means of character development is to encourage engineers to publish. Bacon's familiar statement 'reading maketh a full man, conference a ready man, writing an exact man' needs to be remembered. Besides specialized technical articles there is a need for more general technical writing for the lay public. If engineers do not communicate

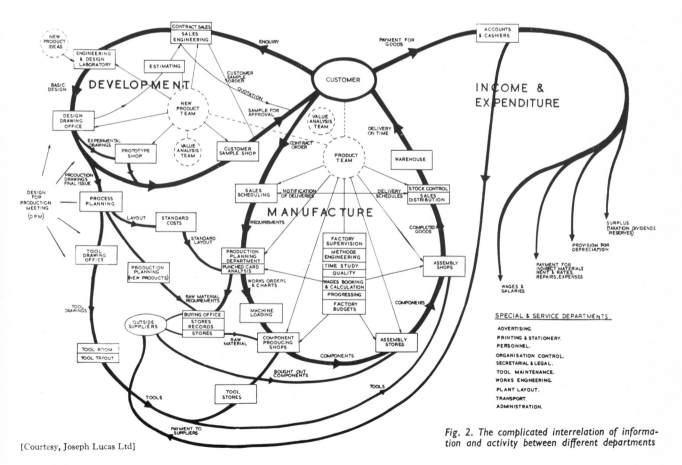

[Courtesy, Joseph Lucas Ltd]

Fig. 2. The complicated interrelation of information and activity between different departments

their technical accomplishments in a language that all can understand, then they will, as a profession, become decadent. They must propagate or perish, publicize or fossilize. *Technology* some time ago said:

'Never has so much been known by so many, nor so ill-expressed'.

We should encourage engineers to travel abroad to widen their horizons. In particular, they should visit the scenes of engineering failures. One tends to learn a great deal from a structure or mechanism that fails.* Much can also be done by selected and planned reading on technical management. Good bibliographies are needed for this, however, if fruitless reading is to be avoided.

A 'generalist' engineer will need to obtain experience in a number of allied fields, some of which are not necessarily technical, such as budgetary control, industrial law, accountancy, and presentation of engineering evidence. The only satisfactory way of learning these disciplines is in real situations, by job rotation. Young engineers should recognize that horizontal moves may be necessary before vertical moves become possible.

Perhaps too little emphasis has been placed in the past on the professional engineer's duty to society. More firms should encourage engineers to serve the community so that they may broaden their interests and indirectly advertize their employer's work. It is little use for engineers to complain that the status of their professions is not as high as that of the legal or medical ones, unless they contribute indirectly to the needs of society, by serving on local authorities and other public bodies.

In the final analysis, adequate development depends upon the individual engineer himself. Epictetus was right when he said at the beginning of the Christian era 'no man can do anything for us, we have to do it for ourselves'. But it is possible to create the right atmosphere in which others can develop.

Inadequate utilization

Training good engineers and developing their minds will not in itself ensure that they are employed to advantage. Several surveys have ascertained how professional engineers spend their time at work, usually, by the random sampling technique of activity analysis. The results are challenging; on the average, it appears, between 20 and 25 per cent of their total working time is devoted to really effective technical effort, the remainder does not utilize their specialized knowledge and skills. True, it may often be necessary for professional engineers to engage in peripheral activities but care needs to be taken to ensure that these do not absorb too much of their time; for his analytical training makes the professional man unsuited to routine work.

The question arises: what constitutes efficient utilization? If efficiency is input over output, what input and output are implied? There seem to be two factors here; one is the ability available compared with that required, the other is the time factor. The latter can be measured but an objective estimate of ability is not easy to obtain.

A further aspect should be considered here; efficiency should not just be concerned with profit made by the firm but needs to take cognizance of certain human elements such as personal satisfaction, happiness and health.

Perhaps the only general factor favouring good utilization of engineers today is the present shortage of them which inevitably imposes a greater intensity of effort in some quarters.

* *See paper by Sir Claude Gibb in Proc. Instn mech. Engrs, Lond., vol. 169, p. 511.*

Fig. 3. An advanced digital computer system used for solving technical problems by English Electric

In itself this does not mean better utilization, but a full load of essential and challenging technical work often provides a mental stimulus. Of course, it is necessary to guard against fatigue.

Delegate or die

Very often an engineer will become preoccupied in proportion to his ability because more people turn to him for advice and decisions. This preoccupation leaves him less and less time for study and experiments and increases with age and seniority. His actual knowledge in his particular field may well reach a maximum and then diminish. If he is to maintain his position he must delegate his work and exercise control by asking the right questions. Quite literally, in some cases, he must delegate or die. Refusal to delegate is one of the main reasons why some engineers are poorly utilized. They like to dabble in details which lesser men could deal with.

Another reason for ineffectiveness is the hoarding of engineers in certain department. The difficulty of balancing technical manpower against research and development programes can lead to this. Companies very naturally do not like to employ on the 'hire and fire' basis. Nevertheless, in large companies it may well be better to 'hire and move' rather than hoard engineers, where little important technical work remains for them.

Duplication of effort is expensive and may only be justified for life-and-death projects such as defence work. To eliminate it, frequent surveys should be made which calls for a wide sharing and dissemination of knowledge.

Inadequate communication is yet another reason why technical manpower is misused. In practice it is often difficult to define the problems. If engineers are not to waste time on solving the wrong problems there must be 'noise-free' communication between them.

While jobs are often defined with some precision it is rare to find performance guides laid down. What is expected from this engineer on this or that job in a specific time? Many technical man-hours are wasted on routine or repeat calculations and drawings. To avoid this, charts, manuals, tables, graphs and other formalized aids should be prepared. An excellent example is to be found in the data sheets of the Royal Aeronautical Society.

It is becoming increasingly important to match men to the jobs required to be done. In a large company there are very many different types of jobs but there are also very many different types of people, all of whom have their own particular bents. But men must not only be matched to their jobs but kept matched, as they both grow and change. Correct matching does not occur accidentally, it requires careful and continual tuning, for science and technology are changing rapidly. It may be easy to spot the extremes, such as the lone laboratory worker or the born leader of men; but other categories are not so easily identified. So often the standards and requirements of a job are fixed by the person who last held the post and it is necessary to review work continually to see what is actually required.

What can be done

Having mentioned some of the causes of ineffective utilization, it is worth considering some of the techniques which might be used to improve it. One of the most potent is rationalization of mental effort.

Much engineering work on conventional products consists in re-shuffling existing basic 'bricks'. It is possible, by breaking down the design process, to formulate calculation procedures so that the minimum number of technical decisions have to be taken. Such design sheets will require careful planning so as to avoid errors and ambiguities; but, once prepared, junior technicians, girls with O level G.C.E., can sometimes be used to complete them. All this frees the engineer for more original work.

While work study techniques have been applied to the production side of the industry, there has been little application to research and development. The apparent simplicity and common sense of method study deceives many but it is only when a written answer is demanded at every stage of an operation that errors and misconceptions are avoided and alternatives thrown up. Based on the Rudyard Kipling formula:

I have six honest faithful friends
They taught me all I knew
Their names are What, Why, When,
How, Where and Who

the critical examination sheet forces engineers to write down logically and methodically thought out answers to the What, Why, When, etc. This provides a useful chart for future work.

In comparison with the medical profession, engineers have been slow to employ technicians in responsible work. The demarcation is not always clear cut and there is a large area of work which may be done either by professional engineers or technicians. Often the technician possesses skills which the professional does not have; he is not necessarily a grade lower. The essential feature of the professional man is that he performs the diagnostic or prognostic analysis of the problem. Naturally, engineers, technicians and other assistants, work together as a team.

But it is no good giving a professional engineer technicians and other supporting personnel unless he can manage them and cultivate them. All engineers must learn to investigate, eliminate, delegate and cultivate. To cultivate an associate it is necessary to realize the importance of frequently presenting an accurate picture of the results obtained, as this has been shown to give a marked increase in personal efficiency.

Mechanical aids

After human assistants, mechanical aids rank highest among the methods of reducing the load on the professional engineer. They may take various forms of which the most lavish at present are the electronic computers. A typical modern digital one is shown in Fig. 3. These can save time and cost in research, development and design, and improve accuracy, with a consequent saving in technical manpower. However, when such machines are installed it is necessary to guard against a modified Parkinson's Law; 'Work increases in order to occupy the machines available'. One good feature of computers is that writing the programme itself necessitates rational and systematic thinking about the work.

The digital computer is extremely useful for optimization since it can perform iterative calculations very quickly. Often it will pay to have lengthy, repetitive calculations on relatively simple problems done by digital computers to save an engineer's time.

Sometimes large digital machines are unnecessary; quite simple computers may be built for solving specialized problems such as analyses of torsion stresses and thermal conductivity.

Desk calculators and analogue machines, varying from the slide rule to mechanical differential analysers, are also versatile aids. Special purpose analogue models, such as the electrolytic tank for studying heat transfer (Fig. 4), can give the experimenter a 'feel' of the problem.

Fig. 4. Using the analogy between electric current and heat flow in homogeneous conductors, a piston is being tested in an electrolytic tank. This permits automatic plotting of isotherms where direct measurement is impracticable

More use should be made of models for obtaining solutions to design problems. This technique is familiar in aircraft, ship-building and missile work. Models may also be valuable tools for proving the efficiency of any projected scheme or showing alternative solutions.

In development work where a large number of variables occur, necessitating a number of observations, much man-power may be saved by the use of statistical techniques. It is important that engineers should call in statisticians to help them plan their experiments well in advance so that the maximum information may be obtained from the minimum number of observations.

Where lengthy experimental work is involved, manpower can be saved by using more automatic recording equipment, similar to that employed in the process industries. By punching results into cards or tape and letting a computer analyse the data overnight, the results can be ready to be reviewed the next morning. In this case careful planning of the experiment is necessary. A typical example of this was the structural testing of a locomotive frame (Fig. 5). A complete stress analysis under varying load conditions was completed in three weeks. Normally this would take several months.

Dissemination of information

A great deal of time is spent by some designers in reconnoitring, searching and trying to find parts which have been used before so as to avoid the needless creation of a new part or tool, or the procurement of a new material. One of the main reasons for excessive variety is the absence of an accurate means of identification and locating of past designs. The first essential in information retrieval of this kind are well-defined codes, grouping similar items according to their common features and subdividing them by their differences. In all engineering work we need to identify, classify, and code past efforts to produce a memory bank for future use.

Again, the messages of science and engineering are recorded in articles, reports, and books and a variety of lesser sources. In the 1956 survey carried out on behalf of the D.S.I.R. to discover what use engineers made of technical literature in the electrical and electronic industries, a very low level of interest was revealed, which seems surprising.

It may be that because of poor communication techniques, the reader is unable to find useful information in a reasonable time.

Unfortunately, there are a number of barriers to communication which make it difficult for the unaided engineer to make effective use of technical literature.

The principal barrier is caused by the sheer volume of documents. Literature is doubling every ten to fifteen years, as shown in Fig. 7. A second barrier is the jargon used for scientific and engineering communication.

This situation is obviously deteriorating.

To overcome these barriers it is essential for professional engineers to have an efficient library service which must be able to identify and locate pertinent information on specific subjects by literature surveys, bibliographies, reading lists, etc.

Check list

In order to obtain a regular idea of performance, check lists should be given to engineers which would help to pin-point weaknesses. Such lists might contain the following questions:

1. Are there any engineering operations which could be changed in order to reduce or eliminate wasted time and energy?

2. Do you consider your communications with management and with your supervisors satisfactory? Or how could they be improved?

3. How can shop personnel be made more effective in assisting you?

4. Do your working conditions enable you to operate at peak efficiency? Or how could they be improved?

5. Is there any routine work which could be performed or made easier by mechanical aids?

Undoubtedly, the performance of engineers is affected by their surroundings.

Much attention has certainly been paid to factory layouts and the provision of certain amenities for the production side of engineering but further consideration is necessary on the research and development side; where there is concentrated thinking it is essential to have quiet surroundings for the work.

Provision for muting telephones should be made and consideration

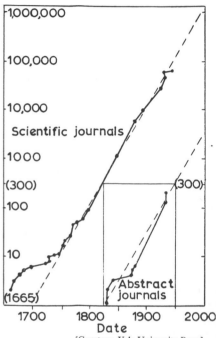

[Courtesy, Yale University Press]
Fig. 6. How journals have multiplied

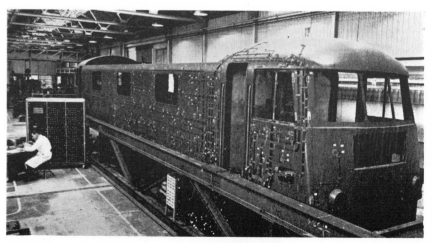

Fig. 5. Structural testing of a loco frame: a digital computer processes results of automatically scanned strain gauges

needs to be given to proper air-conditioning with control of effective temperature. The introduction of negative ions to induce exhilaration and reduce fatigue may well be beneficial.

More attention should be paid to the colour of surroundings and the possibility of obtaining a vista from the work room. It has been claimed that natural scenery helps to free the mind from trivial worries and permits it to concentrate. Good lighting and furniture is also a necessity; as are proper laboratories and working areas for design and development engineers.

Conclusion

The ability of man to control the rest of nature has increased enormously over the last century but in spite of this the present methods of utilizing the world's resources of materials and power are still inefficient. In contrast, man's control over human nature has been slow to grow and man's ability to work efficiently has also been poor. Over the past two thousand years the world has tried the systems of slavery, feudalism and private enterprise and each phase has increased efficiency and improved individual development, but there is still room for improvement. As a small country with few natural advantages, it behoves us to ensure that all our engineers are developed and utilized to the full.

Industrial management ought to give much thought to this subject and ensure that the various tools and techniques which can be used are put into operation as widely as possible.

Higher efficiency in the intellectual plane will ultimately be achieved, as it has in the physical plane, by increasing mechanization and standardization, and by improving organization. In taking this step, man is only following nature. The human brain includes certain automatic devices such as those controlling the heart and lungs; these are designed to use the minimum amount of effort and energy, which allows the mind to devote the greater portion of its energies to more complex activities. The brains of industry should be used in the same way.

The Communication of Technical Thought

by A. J. Kirkman, MA

There is much concern among employers about the poor command of English shown by engineering and science graduates; this is a serious matter when it results in failure to communicate technical information. The Department of English and Liberal Studies in the Welsh College of Advanced Technology has launched an investigation into the problems of scientific communication. Preliminary results show that the faults are by no means all on one side and that schoolteachers must take much of the blame; but barriers to communication are often raised artificially by the scientists themselves.

Protests about obscure, illogical and incoherent scientific writing come as much from scientists themselves as from non-scientists. For example, Professor Andrade protested[1] as long ago as 1948 about scientists who write

> . . . matter that is unintelligible not on account of its difficulty, but on account of the confusion of thought, the lack of sequence and the tangled, sometimes ungrammatical jargon in which it is expressed.

It is paradoxical that lack of orderly thinking and exact statement are the outstanding features of the writing of men whose lives are spent in a search for order and accuracy. We launched our investigation at the Welsh College of Advanced Technology in an effort to gain a clearer understanding of this paradox.

Our first aims were to find out whether current criticisms of scientific writing are justified, and to see what the strengths and weaknesses of this writing are. We set about this in two ways: by having wide-ranging discussions in schools, colleges and universities, in industry and in research centres; and by analysing the writing of our own students, scientific journals, research reports and industrial literature of all types.

So far, more than 100 companies, research centres, professional institutions and university departments have co-operated by providing detailed comments on the project, contributing specimens of the types of writing required from professional scientists and engineers, and by criticising present writing standards. Many individual scientists and engineers have made welcome contributions in response to invitations in the technical press and professional journals.

We have experimented with methods for testing writing skills, and we have begun a consideration of the philosophical problems involved in communicating scientific information.[2]

We hope to find out why particular weaknesses occur. Is it because there are special philosophical difficulties inherent in science and technology? Are there difficulties caused by the structure of the English language? Are these weaknesses a result of the way in which scientists are educated, of the vocabulary they are taught to use, or of the way they are taught to think? Or are we faced chiefly by a sociological problem, arising from the attitudes to the task of communication fostered in scientists and engineers during their training?

Our first year's work has been largely exploratory. We are by no means satisfied that we have collected all available evidence and we should welcome further co-operation from any source; but we now know the nature and range of the problems we have to investigate. This article is a short account of some of them. We do not necessarily accept all the suggestions put forward, but they have all been stated with conviction in more than one of our discussions, and we feel they are worthy of careful consideration.

—they have not been given an appropriate training in self-expression

English teaching in schools

In our preliminary discussions we have drawn attention to all the deficiencies commonly cited in scientific writing: deficiencies in logical thought, in discriminating selection of information, in careful and effective arrangement of information, in clear indication of relationships, and in ability to make orderly statements in lucid constructions and plain vocabulary. Hardly anyone has denied that the deficiencies exist; some efforts have been made to excuse them; and there have been a few attempts at explanation.

Foremost among the excuses offered has been one that distinguishes carefully between thoughts and their expression. It has been claimed that scientists are usually capable of thinking clearly and logically, of selecting information and drawing conclusions without confusion; but that they fail to translate this clear thinking into controlled, coherent writing because they have not been given an appropriate training in self-expression.

The blame for this has been directed at school English teachers. They have been accused of concentrating exclusively on decorative and emotive uses of language, and encouraging clear-thinking young scientists and engineers to take these as models of expression. 'Literary' styles are totally unsuitable for the succinct expression of scientific ideas, but English teachers have been unaware of this, and have not taught other styles which would be more useful in scientific contexts.

Undoubtedly, some English teachers in the past have misled some of their students into thinking that good English means the use of erudite vocabulary and elegant style. It is probably this that has led to much of the verbosity and over-loading of sentence structures found so frequently in scientific writing. But 'O' level language requirements are nowadays designed to discourage flowery essay writing, and most modern English textbooks hold a nice balance between the needs of evocative and factual writing. Indeed, a common complaint against English teaching is that it now spends too much time on the dull mechanics of grammar, punctuation and spelling.

In any case, to protest that English teachers have demanded expansive, imaginative writing in essays is no excuse for the grammatical weaknesses, loose use of words, ambiguous relationships, mystifying punctuation, and inaccurate use of connectives (such as *thus, hence, yet,* and *similarly*) that recur frequently in scientific writing. Though the vocabulary used is very different, the same basic patterns of word order, clause arrangement, indication of relationships, and punctuation are used to express ideas in meaningful statements, whether we are writing about Coleridge, cosmology or catalytic cracking.

The differences between imaginative and practical writing are in style and content, not in the degree of technical correctness.

Nevertheless, the consensus of opinion in our discussions has been that English teachers have failed to provide enough practice in writing on practical subjects.

This opinion has been paralleled by an equally strong feeling that scientists are not made to do enough general reading during their education. By not reading widely, even in scientific literature, scientists deprive themselves of the useful experience of seeing how other people communicate their ideas in cogent order, good style and effective phrasing. Our suggestion that scientists perhaps don't have time for general reading has often been met with scornful disbelief.

It is difficult to sift fact from prejudice (built on a wide variety of school experiences). However, we are studying closely the theory which underlies the suggestion that there is a distinction between ways of thinking and writing about literary and scientific subjects. If this theory can be proved, it will have important consequences for teaching methods in the sciences as well as English.

Two types of mind?

The theory postulates two types of thinking, each with its own appropriate mode of expression. One, which has been called sequential, is used in mathematics and science; the other, called associative, is used in history, literature, and the arts. It is suggested that these are incompatible, and that the weaknesses of much scientific writing arise when the form of expression appropriate to the arts is forced upon scientific material.

—scientists do not read enough . . .

Sequential thinking proceeds from one simple thought to another either by deductive logic or, because the facts themselves are related by cause and effect. The connection is expressed by words indicating logical sequence, for example, because or therefore.

In contrast, associative thinking proceeds from one thought to another by way of a variety of connections, perhaps based solely on the sound of one of the words used in the first statement, or on an emotion stimulated by those words, or on a simple chronological or spatial relationship of the facts concerned. The connection need have nothing to do with cause and effect; and the words chosen to make it need not indicate the nature of the connection, because the writer is not necessarily trying to convey an ordered sequence of information; he is only suggesting certain relationships by making one statement after another. He can use words like 'then', implying chronological rather than logical sequence, or 'rather', implying emotional preference and not objective judgment.

The important distinction is that sequential contexts call for comparatively inflexible lines of thought and rigid, impersonal forms of expression, whereas associative contexts permit random and diverse patterns of thought which can be variously expressed.

This is, no doubt, an over-simplification of the thought processes appropriate to the arts and sciences, but its implications deserve close attention. If we accept the sequential/associative distinction, we can expect scientists and engineers, who are trained almost exclusively in disciplines that require sequential thinking, to show up badly when they are called upon to manipulate ideas associatively. Also, as scientists have been trained by their English teachers to use English only in associative contexts, their writing in sequential contexts will inevitably be poor.

At present we are not sure if a distinction between associative and sequential skills is valid. Obviously no person possesses either skill to the complete exclusion of the other, and there cannot be a clear division between the vocabulary and styles of writing appropriate to the two modes of thought. But in many of our discussions scientists have strongly supported the general idea of such a division; and there is further supporting evidence in the work of Dr Liam Hudson,[3] of the University of Cambridge Psychological Laboratory, on intelligence testing.

Two types of test

Dr Hudson has been using so-called 'open-ended' intelligence tests* which indicate that students have widely varying abilities in different styles of thinking, and that it may even be possible to discriminate between aptitudes for arts or science courses by comparing performance in these new tests with performance in the usual types of intelligence tests.

The usual deductive (sequential) intelligence-test question calls for the selection of an obvious logical answer. For example: "glove is to hand as hat is to . . . . (top, hair, trick, head, box)".

In an open-ended test there is virtually no limit to the possible answers. For example: "How many uses can you think of for a hat?" Some of the answers to this question will be more sensible and more ingenious than others; but there is no single right or logically obvious answer. Clearly this question calls for a less logical, more associative kind of thinking than the deductive question.

* *cf. also p. 350 of the July 1963, Chartered Mechanical Engineer.*

Dr Hudson's results have shown that students of arts subjects are strong on open-ended tests and relatively weak on the usual I.Q. tests, and those specializing in the physical sciences are the reverse. They suggest that young physical scientists tend to be less flexible intellectually than young arts specialists, and more restricted emotionally. They are less likely than arts specialists to question accepted attitudes, to challenge authoritarian views about politics and school life, to show emotional involvement in their opinions, and to have a broad, rather than a narrow, range of interests and hobbies. Altogether, the young scientist shows a marked movement away from material that is imprecise, human and controversial.

—*the distaste most of them show for the task of communicating*

This fits closely the image of the young scientist or engineer which recurs in our investigation. It is very relevant to our study of communication among engineers; for, as I shall describe later, one of their outstanding weaknesses is the distaste most of them show for the task of communicating. No doubt this is because the communication of scientific ideas (especially in industrial situations) takes them away from precise, impersonal experiment and observation and usually involves assessment of personalities and entry into controversy.

Clearly it is of paramount importance for us to find out whether characteristic styles of thinking and distinctive emotional patterns are innate and *cause* students to prefer one discipline to another, or whether they are the *effects* of specialized training in one or other of these disciplines. If they are innate it will be impossible to train science students to think and write in a way for which they are not naturally equipped. But if bias turns out to be the result of exclusive training in certain disciplines—not innate bias as much as stultification of unused talents—we shall surely have to consider modifying our present training systems; for they will stand indicted of suppressing latent imagination and creativity in scientists and of discouraging logical thinking and a sense of order in our arts students.

Purpose and rate of thinking

The vagueness of purpose behind much scientific writing, (especially in industrial reports) has frequently been cited as a cause of obscurity and incompleteness. The writer who is not sure why he is writing or who are to be his readers cannot write effectively, and often produces the sort of report that is a series of personal reminders of steps taken and observations made. Almost inevitably his readers find steps missing or justifications omitted.

These weaknesses probably would not occur if the writer were face-to-face with his audience. We have been told many times that scientists are much better at communicating orally (but not necessarily in lectures) than in writing. No doubt this is because they then receive feedback in the form of visual elements of communication (the explanatory gesture, the glazed eye, the uncomprehending shake of the head).

The remedies for these difficulties are: more careful briefing by those who require reports, and conscious attention to the elements of communication during the professional training of all scientists.

Another reason that has been suggested for poor scientific writing is that scientists are not practised in slowing down their thinking to a rate appropriate to the writing process. Writing is essentially a slow business and demands a deliberate slowing down of the writer's thoughts; but just as an aircraft that works supremely at high speeds becomes unstable when required to move slowly, many people become uncertain and disorganized when they are required to think deliberately. This could account for the frequent discrepancies between what they obviously mean and what they actually write.

It would be extremely difficult to support or refute this theory by devising tests to distinguish between the speed of the scientist's thinking and the speed with which he can make some response to indicate the nature of his thinking. In any case, the validity of this argument is doubtful: for one of the central disciplines of science, the design and construction of experiments, requires just such a slowing down of the thought processes as is required for writing; and we can hardly say that scientists and engineers show significant weaknesses in this part of their work.

Two languages?

A fourth excuse made for inarticulate scientific writing maintains that the grammatical structure of English is so rigid that when an attempt is made to force scientific ideas into it, either the ideas or the grammar become bent and dislocated. A close examination of scientific writing indicates the opposite: that when scientists move out of the rigidly formulated patterns of mathematical and chemical symbolism into everyday language, it is the very flexibility of structure and the looseness of vocabulary which create difficulties for them.

It has also been suggested that ordinary English just does not have sufficiently varied structures and vocabulary to carry scientific ideas. In its extreme form, this theory suggests that the symbolic patterns of mathematics (and to some extent of chemistry) form a separate language of science, which is quite distinct from English. It is the impossibility of translating accurately from one language to another that is said to cause the breakdown in communication between scientists and nonscientists, and even between specialists in various branches of science and engineering.

This would-be explanation could perhaps be justified if we could show that mathematics is something we learn in contexts entirely separated from our everyday existence. But we learn counting, adding and subtracting in our everyday lives, and the words that we use to describe them form an important part of our everyday vocabulary.

Even when we come to more advanced mathematics at school, and have therefore to add a whole range of new expressions to our growing vocabulary, we do not keep them in separate compartments and label them 'the language of science'. On the contrary, they are carefully explained to us in words with which we are already familiar. It cannot be denied that the chain of explanations that links the sophisticated scientific uses of words like 'force' or 'space' with the first situations in which we came to understand them is a very long one indeed but it is unbroken.

Mathematical notation, then, is not a language separate from English. Perhaps we can describe it most accurately as a special part of our vocabulary, a shorthand system for experts, used customarily in esoteric contexts far removed from our everyday experiences. Our argument must centre on the possibility of transcribing mathematical vocabulary into everyday usage *without loss of meaning*. Must our attempts to translate the mathematical shorthand result in the use of esoteric words which will form a jargon as incomprehensible to the non-scientist as the mathematical symbolism itself?

Only a person who has intimate knowledge of both mathematics and the English language will be able to judge whether a transcription has been successful. He will do it by seeing whether he understands both the scientific original and the 'translated' version as meaning the same thing.

If he is conscientious about it, he will supplement his subjective judgment by examining the actions of groups of people who are using the separate vocabularies. He will know what a statement means to scientists when he sees them using it in the design or interpretation of other work; and he will know that non-scientists understand the transcribed statement when they make the correct responses or (better perhaps) ask some pertinent questions.

In all our discussions with scientists during our investigation so far, we have made a point of asking specifically whether it is possible to make transcriptions from scientific jargon and mathematical formulae into everyday English. The overwhelming response has been that it can be done. Granted that the intellectual and educational level of the non-scientist audience is comparable with that of the scientist writer (for the communication of scientific ideas to audiences of lower abilities presents another range of problems), and granted that it would be quicker to convey the same information in scientific shorthand, it is quite possible to communicate the ideas of science in good everyday English. References have been made time and again to the success of Huxley, Bondi, Hoyle, Pyke, Bronowski and most of the contributors to the *Scientific American* and *New Scientist* in doing just this.

When jargon is justified

It must be stressed that no-one considers the use of jargon reprehensible in itself. We must be clear what we mean by jargon and when its use is open to attack.

Jargon has two distinct elements—special vocabulary and special turns of phrase. These are created by experts in any discipline to convey among themselves clearly defined meanings. Usually these meanings could be put otherwise only at greater length. The use of jargon is quicker and more efficient.

In these circumstances, the use of jargon by specialist scientists is perfectly reasonable. Similar uses of specialist jargon in non-scientific subjects are widespread and valuable. But as soon as jargon terms

Writing is essentially a slow business . . .

and idioms are used outside their special contexts, and as soon as they are used within those special contexts in place of perfectly adequate everyday terms, simply because they sound more scientific, they become indefensible.

Unfortunately the conscious or unconscious use of unnecessary jargon appears to be widespread. The use of mechanical, cliché-ridden, stylised ways of expression, which gradually lose their exact meanings as they are cheapened by dull and thoughtless repetition, has a pernicious effect on scientists' ways of thinking. They try to use the pre-packed phrases and structures to express all their thoughts. Missiles no longer hit the target; they "impact with the pre-determined target area". The word "about" disappears because scientists are so used to saying mechanically "of the order of" (consider the meaning of: "Switzerland, Norway and other countries of this order . . . ."). A pile of pre-modifying adjectives is preferred to a relative clause, even though a clause is easier to assimilate. "Big" gives way to "of considerable magnitude", "high" becomes "elevated", and "utilized" replaces "used".

Perhaps this is because of a mistaken impression that unfamiliar, resounding words are more scholarly (or at least, more scientific). In fact, the use of ostentatious words where exactly equivalent simple words are available, gives the impression of pedantry rather than scholarship and makes the non-scientist feel that he is being unnecessarily blinded with science.

Another reason that has been suggested for the heaviness and obscurity of much scientific writing is the desire to make every argument formidable, to show massive competence, and yet to qualify and hedge round assertions so that there can be no suspicion of personal fallibility.

The cause of this is probably the desperate concern with truth, proof and presenting all the facts that is built into every scientist. The consequences are texts of great length and weight, packed with exact and minute detail, that leave to the reader the task of selection, which ought properly to have been performed by the writer.

A less praiseworthy explanation which has also been given for this overloading with detail can be summed up in the words of one of our correspondents:

> Many reports are purposely made long and contain much irrelevant material . . . . (because) . . . . the scientist feels that only by doing so can he justify the time and money which has been expended on the particular research.

Even when the vocabulary of scientific texts is comparatively simple and the selection of material is ruthless, the writing often lacks directness and vitality, owing to the strain of following the tradition which requires the writer to refer to himself in the third person and to his activities in the passive voice. This is a result of scientists' anxiety to suppress emotion, prejudice, commitment, and every trace of personality in their work. "It would not be unreasonable to assume", they write, when what they mean is "I think". "It has been noted by the authors" appears instead of "We have found". Perhaps authors are not to be criticized for being non-committal, as a tentative manner of presenting ideas seems to be academically fashionable in most subjects. But why must they write "It became apparent at an early stage" when they mean "We soon saw", which is not only simpler but more accurate? There is no self-importance or prejudice in writing simply and personally.

Schoolteachers must take some blame

I have touched on what is probably the most important educational aspect of the problems of scientific writing—the harm done by teachers of English who have allowed their enthusiasm for Shakespeare, Shelley, and 'elegant variation' in essays to obscure the need for different styles of writing in different contexts.

A partial defence of these teachers—supported by many of our scientific colleagues—is that while this may have deprived students of experience in 'practical' writing, there is no reason why it should have led to incorrect use of the English language. However, English teachers cannot evade most of the blame for the characteristic attitude of scientists to anything that can loosely be termed 'English'. The attitude is one of sullen apathy or even open hostility. Many scientists, mindful of the irritation and sense of irrelevance they felt when criticized for not writing stylishly at school, now reject all such considerations as unimportant pedantry, even when they are writing about science.

Consequently, very reasonable protests about unmanageable length and content in sentences, rhythmically distressing groups of polysyllables, unvarying pace, tone and structure, and careless selection of words, all go unheeded. It is characteristic, too, for scientists to be impatient with criticisms of their imprecise use of everyday vocabulary and other ambiguities. They are unwilling to recognize that the whole responsibility for conveying meaning rests with the writer, and that even a factually correct presentation may completely fail in its purpose if the meaning is disguised by jargon or impenetrable style.

Reluctantly, but repeatedly, scientists have said that few of their colleagues are convinced that writing is an integral part of science: few will admit that careful reporting and effective presentation are skills that may be expected of them as essential features of their professional competence. In the words of one senior university lecturer in mechanical engineering:

> Underlying the failure of students, both undergraduate and post-graduate, to write good English is the widespread attitude of mind which regards communication (or anything pertaining to the arts) as a lower-grade activity, unworthy of comparison with scientific enquiry. It is hence relegated to a lower place in the students' esteem.
> . . . . The general attitude to English among scientific workers is betrayed by the fact that an error in mathematics tends to be an occasion for shame, an error in English an occasion for mirth.

Though it may well be the fault of English teachers that such a situation has arisen, its continuance should be disturbing to teachers of science and English alike.

Home-made barriers

In analysing the problems of scientific writing we have tried to give due weight to the very real difficulties that arise. For example, the need to integrate visual material and formulae into prose texts presents problems of choice and layout which call for insight and skill. Also, there are many scientific terms which are clumsy in themselves and will destroy the effect of the most skilfully balanced phrase. Even the time, space and peace required for writing up scientific or technical work is often lacking.

In addition, difficulties of communication are increased by the attitude of distrust or even hostility—so often a defence mechanism—adopted by 'innumerate' non-scientists. Perhaps, as Dr Bronowski suggests,[4] these are difficulties scientists have brought on themselves:

> They have enjoyed acting the mysterious stranger, the powerful voice without emotion, the expert and the god. They have failed to make themselves comfortable in the talk of people in the street; no one taught them the knack, of course, but they were not keen to learn. And now they find the distance which they enjoyed has turned to distrust, and the awe has turned to fear.

But it is equally true that non-scientists reinforce the communication barrier. They try to hide their embarrassed ignorance behind a façade of disinterest and false jocularity.

—even a factually correct presentation may completely fail in its purpose if the meaning is disguised by jargon . . .

They are not prepared to try to learn even the basic concepts of science and the basic vocabulary of scientific idiom, or to accept the fundamental attitudes of science. They assume that receivers of information can play a passive role. Scientists cannot be blamed if they sometimes give up in exasperation.

However, the tendency of scientists to look upon themselves as a superior group emerged often enough in our discussions to qualify as a real problem. In academic circles, to write something simply so that it can be understood easily by everyone is to invite the damaging description 'popular'. In industry, the desire to establish status frequently reflects itself in the way described to us by the manager of the operations research department of a steel works:

> . . . . there is often a feeling amongst scientific and technical staff that they are rather superior beings and (they) consider that to uphold this thesis they must present reports that cannot be understood by managerial, commercial and similar company officials.

Perhaps the most disturbing feature of this problem is the way intellectual arrogance is spreading among scientists themselves. Men in research laboratories feel superior to development men (an extension of the schism between pure and applied science, between scientists and engineers); and development men look patronisingly even on qualified men in the works.

Minority groups, in order to establish and preserve their identities, tend to exaggerate those characteristics which mark them off from the community at large. No doubt this has always been so; and in the case of scientists and engineers, the exaggeration is motivated in part at least, by very reasonable professional pride. But scientists are now becoming not simply marked off, but cut off, from the rest of the community by communication barriers—linguistic and emotional—that are partly of their own making. These barriers must be broken down before they become too massive to scale.

Conclusions

Many of the causes of current communication problems have been implied or stated in this discussion of our preliminary survey. The consensus of opinion has been that most of the problems are rooted in the school and university training of scientists and engineers.

Above all, this training has been criticised as requiring far too much mechanical absorption of pre-digested information in ready-made shorthand. Apparently there is little encouragement of critical appraisal and individual expression. Almost all the writing practice students get is note-taking in lectures. This is a helter-skelter business, and all too often degenerates into dictation or mere copying from a blackboard. Though many science and engineering departments pay lip-service to the importance of good report-writing, few give any

Mr John Kirkman *is a Senior Lecturer in the Department of English and Liberal Studies in the University of Wales Institute of Science and Technology (formerly the Welsh CAT) in Cardiff. He was educated at Barking Abbey School, Essex, and at the University of Nottingham, where he took BA, MA, and PhD degrees in English and a Certificate in Education. After two years national service as an Education Officer in the Royal Air Force, he taught for three years at Colchester Royal Grammar School before moving to his present post in September 1961. Dr Kirkman is an examiner for the Council of Engineering Institutions, and his interest in the use of English in communication has led to many articles and broadcasts.*

positive instruction in technique, and fewer still bother to correct expression in written work.

As a result of such training, students rapidly become unwilling to make personal evaluations of ideas; it is sufficient —even required—that they repeat facts in standard phrases and insert formulae in standard problems. They gradually lose any capacity they once had to find their own words for what they know.

This serious decay takes place at a time in life when positive growth should be occurring, when students should be developing wider linguistic and general abilities to cope with the mature thoughts, attitudes and problems of adult life. Training is replacing education, and horizons are becoming steadily narrower.

In the course of this training, many scientists and engineers develop a disastrous attitude to communication: a mixture of apathy and arrogance with which they cannot hope to succeed. If they approached the difficulties of communication with a painstaking humility similar to that which they bring to scientific difficulties, they could no doubt overcome many of the problems of self-expression which at present cause so much irritation to writers and readers alike.

Much depends on the teachers of science and technology: for unless, by precept and practice, they convince their students of the need for high standards of skill in communication, scientists and engineers of the future will, by imitation, become as apathetic and incoherent as, by an overwhelming majority of those with whom we have discussed this matter, they are accused of being at present.

REFERENCES

1. *Proc. R. Soc. Conf. on Scientific Information*, p. 43. 1948, London.
2. The consideration of philosophical problems in the investigation has mainly been the work of Dr P. R. Bridger, Lecturer in the Philosophy of Science in the Dept of English and Liberal Studies in the Welsh C.A.T., and the author is grateful for his co-operation in the preparation of this survey.
3. The notes on open-ended tests are based on Dr Hudson's accounts of his work in *Nature*, 10th Nov. 1962; *Where*, No 12, Spring 1963, and *Tongue-tied Scientists*, B.B.C. *Science Review* January, 1963.
4. BRONOWSKI, J., *The Common Sense of Science*, p. 142. 1951 Heinemann, London.

It Takes Two to Communicate

by E G Semler, BSc, CEng, FIMechE

Engineers are now quite aware of their difficulties in communicating with one another, let alone with non-technical people but many are under the wrong impression that good writers or speakers are born, not made. Technical communication is largely a matter of commonsense plus self-discipline. In this article the author writes about a few simple rules which everyone can follow.

Why should engineers have to take trouble over writing and speaking? Should they not get on with their jobs and leave the communicating to journalists and others who are paid for it?

One short answer is that communication is not just writing for the Press. One's own boss may well be non-technical but failure to communicate with him may be fatal to one's career. Again, one has to deal with other engineers who are non-specialists in the subject concerned and must therefore be regarded as approximating to a non-technical audience.

But can we afford to neglect the world at large? We are always complaining that the general public does not understand either ourselves or our work. Who is to blame for this? Leaving the job of interpretation to professional writers, however capable, has led to the present unhappy situation of our profession. Publicists have their place but they are likely to be non-technical people and, if we fail to get our message across to them, how can they pass it on?

No, whether we want to or not, we shall frequently have to communicate with non-engineers and if we learn to do this properly we shall then find it all the easier to communicate with our fellow specialists.

So how does one set about the presentation of technical information? There can be great art in writing brilliantly about nothing in particular but engineers are not usually artists and, the better the presentation, the more painfully obvious becomes the absence of useful information in a piece of technical writing. Conversely, much obscure writing is the result of an (often unconscious) awareness by the author that he has nothing much to say. Every article should therefore be preceded by careful heartsearching as to whether it is worth writing at all.

Visualise the customer

You may have a great deal to say but, unless you try to imagine the person with whom you are communicating, the chances are you will be wasting your time. At the very least you must guess how much of the subject the reader already knows, what is his principal interest in it and how much

"One's own boss may well be non-technical"

effort he is prepared to make. It is easier to write for someone you know, like your managing director, than for 70 000 readers of a journal who have nothing in common except the profession of engineering. But they must have *some* common denominator and it is therefore possible to construct in one's mind a model of the reader one is addressing.

When in doubt, you can get help from the editor of the journal concerned, the secretary of the learned society or whoever is in charge of the medium you will use. They are likely to know their customers much better than you do. Should you, however, be quite unable to visualise the typical reader, then it is best to scrap the whole idea: those who address all and sundry are likely to impress no one in particular.

First, then, we must determine the type of audience. Apart from one's fellow specialists, there are four main classes who, for practical purposes, are non-technical. Management, customers or users, production people and the general public, for instance, newspaper readers.

In each case we are essentially concerned with a sales effort: we are trying to get someone to absorb a message in which he is not vitally interested.

To forget this even for a moment, to write from the point of view that we are doing the reader a favour in sharing our precious knowledge with him, is a complete waste of time.

Take our manager, for instance: a busy man with a vested interest in the status quo. He is not too enamoured of new ideas, products or techniques, unless we can show him straight away how they will increase profits. The effect on profitability must therefore come at the beginning of every report for management. The practical applications and limitations will then make up the bulk of the report proper. The limitations will lead naturally to an outline of further work required to overcome them, accompanied by an estimate of costs still to be incurred, and other consequences of our proposals that affect management problems. Will they necessitate changes in staff structure? Will they interfere with other current work?

In short, those who want to sell their ideas to management must not only make the package attractive but should include a set of instructions for its use.

This is not a bad idea either when writing or speaking for customers. Presumably we are trying to sell them a product or are showing them how to use it.

In the latter case the approach will resemble that to our own management but, instead of specific managerial considerations, we shall stress such factors as price, delivery and other aspects of competitiveness.

Here again we are usually addressing someone who needs a very good reason for changing some well-established routine or a supplier he knows and understands. He is unlikely to be as fascinated by our product as we are and, while it does no harm to show some enthusiasm, the reasons for it must be clearly explained.

An instruction book is not a sales leaflet but few authors seem to appreciate this. Once we have sold something and the user is trying to operate it, irrelevant sales talk will merely infuriate him. Cut out all blurbs on the lines of:

Your Whipperty-gibbet has been designed with the greatest possible care and, if treated in accordance with these instructions, will last forever but . . .

It is much better to start with the BUT.

Instructions for use of a product are best accompanied by a fault-finding chart or check list which follows a logical sequence from the most frequent faults and easiest checks to those which require repair at the works. Operational instructions must similarly follow a logical sequence and avoid all jargon like the plague. A classification consultant once compiled a list of 78 different kinds of 'pin', not counting the two known to the housewife. So don't talk to her about a 'gudgeon pin'. Tell her what it does and where it is, preferably with the aid of a perspective diagram.

Those concerned with manufacturing our designs are likely to be engineers or at least skilled craftsmen who can read drawings. It is therefore tempting to fall back on technical jargon and orthographic drawings. But the works manager of an engine-block plant is not necessarily familiar with Schlieren photographs. So, even when addressing fellow engineers, please remember how far specialisation has progressed, how little they know about your particular bailiwick, and translate the jargon accordingly. If you are afraid of seeming too condescending—and this *can* put their backs up— it is always possible to do one's teaching unobtrusively. For instance:

"Schlieren photographs, taken with a high-speed camera through a window built into a cylinder wall . . ."

Just because a works manager, or even his fitters, can read orthographic drawings, they do not necessarily prefer them to simple perspective sketches which might be clearer. To turn Confucius base over apex, "one sketch can be worth three orthographic views and one sentence can be worth 1000 dotted lines." It all depends on what we are trying to show.

Finally, **the most difficult audience of all, the general public**: one cannot lay down the law here, so much will depend on the message. The best rule is to remember that all the axioms of good writing which we are about to discuss

"Tell her what it does and where it is—"

apply to the general reader even more than to the professionally interested types mentioned above. Style in particular must be clear and crisp, and no communication should be longer than necessary.

Remember, the general reader is is more likely to get bored than any other. This may hurt your ego but is very good for self-discipline. Not everyone thinks a well-oiled Geneva-cross mechanism is as beautiful as a Beethoven sonata.

Jargon, of course, is poison for the general reader and technical principles should be explained only where essential to the argument. One does not need calculus to compare the areas under two curves. Here, as always in writing, it is far more important to know what to leave out than what to put in.

Decide on a structure

Having discovered what type of reader we are concerned with, we shall find it easier to decide upon a suitable coverage and basic structure for our article or lecture. But it is essential to make a conscious decision for to start writing without a logical plan is as silly as trying to build a machine without first designing it. And a good designer constantly bears in mind both the user and the purpose of his product.

A slip of paper with section-headings will usually be sufficient to plan the structure but this slip of paper is vital. There are always several possible structures for any subject though some are easier to follow than others. The most important thing is that, once decided, the structure should be adhered to throughout. If you start off in the order in which a machine operates, do not confuse the reader by slipping into a different plan half-way through, just because you find it more convenient to deal with, say, all types of controls at the same time. You should have thought of that in the first place. Like any other product, a communication must be designed for the convenience of the user, not the originator.

Unfortunately the structure best beloved by engineers, the chronological order of the underlying work, is the least suitable for many purposes of communication. The notes in the laboratory report, or the order in which the job was designed, are usually quite irrelevant to the reader.

The main exception to this is where the time-sequence has a bearing on the results, such as in testing to destruction or in the operation of equipment. It obviously will not do to tell the reader of an instruction book on page 5 why he shouldn't have touched the button mentioned on page 2. But in that case the time sequence must be that of operation, not the order in which you thought of the various points!

How to begin

Whatever basic structure we choose, the sensible way is to start with something that will rivet the reader's attention or else he might not stay the course! If we have world-shaking results to report, by all means let's put them first. But for those of us who have not discovered America, the best way to begin is with the reasons for the work in question, and its objectives. If they do not rouse the reader's interest, the rest of the message is hardly likely to.

The danger here is to get bogged down in long descriptions of the background: it is best to state the problems briefly and give the history in the appendix, if you must. Similarly, all details, theories, methods, literature searches, and experimental failures are best banished to the appendices. What the reader really wants to know are the results and their applications to his own work.

The detailed emphasis will of course vary according to whether we are writing a paper for fellow specialists or a newspaper article for the *hoi polloy;* but the general principle is exactly the same: we must on no account bore the reader with what is of interest only to ourselves.

Even in those rare cases where no particular structure is dictated by the interests of the reader, the author will be well advised to choose a plan and stick to it, for his own benefit.

By keeping similar items under the same sub-headings and by avoiding large jumps in subject matter, he can marshal his own thoughts in an orderly manner and incurs less risk of omitting anything important. For example, in discussing materials, it may not matter too much whether plastics come before or after ceramics, but it is messy to insert them between ferrous and non-ferrous metals; and messier still to discuss fibreglass with sheet metal, just because both are used on motor car bodies; or, more likely, because one has only just thought of the former.

Style is commonsense

We now come to the interface between structure and style, the order of paragraphs and sentences. "The style is the man". If you have an orderly mind, and therefore found my remarks on structure mere commonsense, you will also arrange your paragraphs and sentences in an orderly manner. But in that case you are not, I am afraid, typical of most engineers.

However grammatical each individual sentence, if it is written down in the order of thinking, rather than in the order of 'use', the message will be obscure. This applies all the more to changes of thought half-way through a sentence, where the author was too lazy to start again at the beginning. In such cases the reader usually has to do it for him—perhaps two or three times before the penny drops!

The first rule of good style, then, is that each word, sentence and paragraph should be correctly placed within the overall design. The second is that they should all be short. The use of telegram style can be irritating but the opposite extreme is even worse. The reader hasn't got all day: he wants to get to the point, to glean the maximum amount of information from the minimum amount of reading matter.

True, 'idea-density' can become excessive where the subject matter is difficult, particularly in lectures. Written matter, however, can always be put down for a breather and most people prefer to do this rather than be smothered in padded sentences which confuse the mind more than they rest it; let alone the pernicious doctrine that you have to "tell them what you're going to tell them; then tell them; and then tell them what you've told them."

Recently I had occasion to analyse a number of articles for major faults in style and, not perhaps surprisingly

"Written matter, however, can always be put down for a breather—"

the great majority are the result of sheer waffle, ie, of redundant words, sentences and even paragraphs. A good many secondary faults, such as ambiguity, word-repetition and obscurities can be traced directly to this. Conciseness therefore is the greatest single virtue of good style.

Repetition and waffle

Some of the more obvious blemishes are the repetition of words or even syllables within a short space. The mind has a built-in defence-mechanism which discounts the second reading as 'already known'. One can usually discover in time where a word repeated does not mean that an idea has been repeated does not mean that an idea has been repeated* but this often means going back and is therefore doubly irritating. And irritation is the greatest single obstacle to understanding.

It is however permissible to repeat a word almost immediately for emphasis, as I have done above with the word 'irritating'. Not so for the equally annoying repetition of syllables, as in "irresponsible irradiation is harmful." Alliteration may be a form of poetry but it has no place in technical writing.

To avoid repetition, more use can be made of the personal pronouns, provided no offence against clarity is committed. Alternatively, here is a legitimate use for such dummy words as 'equipment' or 'unit,' mentioned below but to replace a name, not as an addition to it.

Redundant words and phrases come in all shapes and sizes, according to the context. Here are some particularly apt to crop up in engineers' English, and almost invariably for no good reason.

AVAILABLE as in: 'There are several types available'.
IN ADDITION TO: as in 'There are nuts in addition to bolts'.
IT WILL BE SEEN THAT/IT WAS FOUND THAT as in 'It was found that the gears were not meshing'.
ASSEMBLY/UNIT/EQUIPMENT/INSTALLATION as in 'The motor-generator assembly/unit/equipment/installation.
LEVEL as in 'The voltage level is higher'.
PROCESS/METHOD OF as in 'The process/method of metallising'.
IN CONDITIONS OF as in 'In conditions of severe usage'.
TYPE as in 'The solenoid-type valve'.
TEMPERATURE/PRESSURE, ETC as in 'At a temperature/pressure of $350°F/600$ lbf/in²'.

There are many more such words and phrases which should be pruned ruthlessly for, while doing little harm individually, they have a deadening effect in bulk.

Ambiguity and inversion

Ambiguity is the next most frequent fault and probably the most dangerous of all. Sometimes it stems from the sense itself, reflecting a basic confusion in the author's mind, or a refusal on his part to commit himself to a definite opinion. More often, however it arises from bad syntax. A good example is the following.

The structural improvement may be judged accurately by comparing the structure attenuation before and after treatment . . .

* *In reading the proof I find that the printer has obligingly provided an excellent example right here—I shall let it stand.*

We are here talking about noise and the word 'attenuation' is a piece of jargon which means 'thinning out' but is used by engineers to connote 'reduction'. But what exactly are we reducing—the frequency? the volume? the audible noise? the noise from the structure? or are we referring to the reduction of noise due to the structure?

A little thought and re-reading of the context will often provide the right answer. But why should the reader make even a *little* unnecessary effort? A little here and there can add up to quite a lot. The best thing is to write every sentence so that it is intelligible by itself.

A good way to invite ambiguity is the ghastly habit of using the passive voice where the meaning calls for an active verb. This is thought to originate in an anxiety to eliminate personal and subjective considerations in scientific writing. It does nothing of the kind. Instead, it irritates and introduces needless ambiguity. Take my own sentence above: "this is thought to originate . . ." Did I mean '*I* Think' or '*you* might think' or '*the experts* think'?

Again, how often do we read in a report that 'it was decided to abandon the project.' Who decided—the author? his co-authors? his boss? his customer? It makes quite a difference but we may never know . . . .

Frequently allied to the passive voice is the inverted sentence such as this one. It can be used for establishing the main facts quickly but it usually results in complicated and confusing split sentences. For instance:

> Given, as we were, only a few data and these not being too reliable, the results, which must therefore be taken with a grain of salt, are listed in Table 1.

By the time the reader has got to the end of this he has forgotten what is supposed to be in Table 1. The general rule is put your main message in the principal sentence at the beginning and do not split it. If there must be qualifications, they should follow in a subsidiary clause or, better still, in a new sentence. Thus:

> Table 1 lists our results; these must be taken with a grain of salt since we were given only a few data which were none too reliable.

Inversion also leads to excessively long sentences which should be avoided. So should multiple adjectival nouns which are criminally misused by engineers, often without even the courtesy of a hyphen. 'High speed by pass jet engine noise' at least has the principal noun at the end, thus giving you some clue as to whether the noise comes from the high speed, the by-pass or the jet engine. But what about 'High engine by pass speed jet noise'? This is the kind of shorthand 'up with which the non-technical reader will not put', nor is it calculated to please even those who can decipher it.

Jargon and maths

Since communication should be designed for the user and not for the author, mathematics, technical jargon and symbols can be justified only where the recipient is certain to understand them. Which is much more rarely than most authors seem to think. All this is mere shorthand which saves writing

"—only where the recipient is certain to understand them"

but it is not usually essential to the content of the message. You would not like your secretary to write you a note in shorthand, knowing that you cannot read it.

In technical writing one often feels that the opposite approach has been used: 'let's show them how much we know and perhaps, if they don't understand it, they will at least be suitably impressed.' They won't, you know: that is not how the human mind works.

With few exceptions, any jargon and symbolic shorthand, even pure maths itself, can be expressed in simple English. Of course, when you are sure the other fellow understands the jargon you can save space by using it. But much more frequently it will not be needed at all or can be left to the appendix. This is certainly true with non-technical audiences, for whom the mathematics is usually quite irrelevant.

Specialists tend to forget that mathematical notation itself, explains nothing. Writing $\eta \times kt$ merely states that efficiency increases in linear proportion to temperature but not why. Putting it in simple words may help to raise this question and also the need for an answer.

Related to jargon is the use of long, pretentious words of Latin origin instead of simple Saxon ones. Typical of these are 'commence' for 'start' 'in consideration of' for 'because', etc. It is not that the average reader fails to understand the sense but why *should* he have to wade through a swamp of meaningless verbiage to reach the small trickle of clear thought that can sometimes be found in it?

I think I can tell you why: as in so many other faults of technical writing, the root cause is pretentiousness. Look back over most of the points I have mentioned: very few of them would occur in ordinary pub talk. But no sooner do we get up on a rostrum or sit down to write a formal paper, than our self-esteem appears to demand all these aids to bad communication.

We are not unique in this: other professions have their jargons which may be one reason why we had to invent one of our own. But in our case the paperwork eventually comes home to roost in the form of hardware. And when it is a matter of Lever A pushing Lever B, the operation is not improved by 'Lever A commencing a displacement which impacts on the corresponding extremity of Lever B.'

Verbal communication

By and large a talk can be approached much like a written effort but there are one or two important differences. Ideally it should not be read out but be freely delivered. If slides are used, the talk can be planned around them, much more than a paper should be based on illustrations. The reader expects a more formal and less digressive approach from even the most casual article than from a talk or discussion piece. But this does not mean that verbal communication can be formless or without style. Even when one is speaking from a few notes, these should be very carefully prepared.

The style, however, can be a little less stringent. While there is no room for excessive waffle, the listener cannot turn back to what has been said and this makes it advisable to give him a little more time to think. To avoid lengthy pauses, therefore, a certain amount of partial redundancy is advisable after each important idea. Nothing, however should be obviously and completely redundant. Instead, give the impression that you are elaborating and explaining the basic idea.

Many good speakers use relevant and often humorous digressions for this purpose. But the operative word is 'relevant'! If the digression is too far-fetched it will increase the audience's resistance and take their mind off the very idea one is trying to get across. The same applies to humour which, I am afraid, is best left to the humourists. Some people have a natural flair for it but those (like me) who have not, had better not try! Nothing is feebler than the joke dragged in by the scruff of its neck, on the lines of: "And, talking of Scotland, have you heard about the Scotch egg that came out of the oven.... ."

Write in haste, polish at leisure

Having assembled one's data and decided the coverage and main structure of an article, it is best to write it as quickly as possible, while the overall plan remains in one's mind and before the white-hot enthusiasm has evaporated. To write well, one must have a strong urge to communicate. But it is usually fatal to leave a piece of technical writing in the form in which it comes off the typewriter. The treble-spaced draft should be put away for a few days and then reconsidered in the cold light of unfamiliarity. The great thing is to treat one's own brainchild as ruthlessly as one would the horrible efforts of others.

Having corrected the most glaring mistakes and improved any unfortunate phrasing, get it retyped and put away for a few more days. Then read it again, projecting yourself as far as possible into the mind of your model reader: will he understand *this* term? will *that* statement strike him as obvious? Does he really want those two pages on work which *you* found fascinating but which ultimately proved useless?

Recheck for all unnecessary jargon, maths, passive voices and ambiguities. If the structure no longer seems logical, do not spare scissors and *Sellotape*; if necessary, re-write the lot in a different sequence, for a good, logical structure is vital. Remember that even professional writers depend on conscientious and prolonged polishing.

The result may not be a masterpiece—no one has the recipe for genius; but it will be a meaningful message, clearly conveyed and, if the substance was worth passing on, the reader will feel its impact.

Illustrations

Confucius he say "A picture is worth a thousand words." Certainly technical writing should rely heavily on illustration. But, alas, many pictures are not even worth 10 words. How often does one see a graph which proves nothing but a straight-line relationship; or a photograph which shows nothing but a box-like casing. Even a table of results is redundant if it merely indicates that A is always bigger than B.

In writing for publication, the first consideration must be not to include too many illustrations in proportion to the text. A 'story told by pictures' may on occasion make sense but is difficult to take in and generally looks unattractive in print. A talk based on slides may sometimes be entertaining but is quite unsuitable for conveying serious technical information.

Next, what kind of illustration to use? The general rule should be drawings rather than photographs. Drawings can contain so much more useful information because they can be done with the specific purpose in mind. Photographs are often more decorative and editors of semi-technical journals like them if they are reasonably dramatic, so that a few should be included as a leavening. The ideal functional illustration of a technical article, however, is a drawing.

But what kind of drawing? The conventional orthographic engineering drawing was invented for the purpose of conveying detailed dimensional information but it is about the most difficult picture to visualise as a real object, even for engineers, and certainly for others. Unless the component illustrated is very simple indeed, a perspective or (if dimensions must be given) an isometric sketch is preferable. For instruction books exploded views are ideal, provided they do not become excessively complex.

Except for design and manufacturing purposes, detail dimensions are rarely necessary and merely confuse the illustration. So do any other inessential details: in order to explain how a device works it is much better to sacrifice complete accuracy to clarity and simplicity and this is of course equally true of the verbal description accompanying it. It is, however, necessary to indicate somewhere that a sketch explains the principle, rather than depicting the whole equipment.

If I may be allowed one special *bête noire*, it is the block diagram for which, in its modern form, we are indebted to the electronics merchants. 'Black boxes' are delightfully simple and can be labelled 'controller' or 'Fx' but they tell us precious little.

Block diagrams can be very useful, but always provided that (a) they do not enclose in a box the very object or function that needs explaining: and (b) that they elucidate a spacial or temporal or logical relationship, as for instance in critical-path diagrams. Most box diagrams, however, waste a great deal of space to tell the reader nothing more than that a number of related functions or components interact; which will hardly surprise him.

To plot or not to plot? Every engineer loves graphs, they are thought to add 'tone' to his report. But how much do they actually tell the reader? A graph has two functions: one is to show the trend of a relationship between two variables; the other is to provide a (usually inaccurate)

Mr E. G. Semler *was, until 1970, Editor of the CME and was responsible for its business management. He is now a Publications Consultant and freelance writer. Educated at Herne Bay College and University College London, he graduated in Mechanical Engineering in 1942 and served a college apprenticeship with the Metropolitan Vickers Electrical Co. at Trafford Park, where he also edited the house magazine, Rotor. For the next 12 years, he was a welding engineer with MetroVicks and other companies, his work ranging from technical development to sales promotion and technical writing. In 1956 he became editor of* Automation Progress, *a technical journal, and in 1960 he joined the staff of the Institution as Editor of the* CME. *He became a Student member in 1942 and was made a Fellow in 1970.*

means of scaling-off values as a substitute for extensive tables of results. The latter will hardly ever be necessary in general technical writing; after all, only the actual plotted points can be vouched for.

The proper function of technical illustrations in graph form is, therefore, to show a trend. But where the trend is irrelevant, the graph is redundant; worse still, where there is no trend, the graph is positively misleading. Thus, if results depend in some way on the time of the year, a graph against time may be significant; but if the graph simply records results taken at monthly intervals it is worse than useless because it implies a relationship that doesn't exist. To convey the actual values of results, tables are far more accurate and usually take up less space.

Reproduction

In writing for publication, the author should bear in mind the medium to be used and the method of reproduction. For illustrations this is quite essential. Without going into the various processes that exist, it is safe to assume that, in normal printing to the same scale, a photograph will lose up to 20 per cent of its definition and 50 per cent of its contrast; in the coarser reproduction processes it may lose up to 80 per cent of both. This is another reason for preferring line drawings, even to show only the general appearance of an object, for they lose much less in reproduction. It is amazing what can be done by a good technical artist.

However, in most publications all illustrations have to be reduced because space is expensive and, in any case, limited to the size of the page. It is vital to bear this in mind and to design each picture for the average size of illustration in the journal one has in mind. As a general rule this will vary from four to 25 square inches, the width hardly ever exceeding 7 in.

It is usually quite impossible to draw to such small sizes and the picture will therefore have to be scaled down in reproduction. This means that every single line and every space between lines will also be scaled down, in addition to any definition lost in the printing process. More important still is the size and line-thickness of lettering: words which are very legible on a drawing two feet by 18 in become indecipherable when width and height are each reduced to a quarter. Lines which were perfectly visible on the original may then almost disappear or run together.

As a general rule, any drawing that is likely to be reduced by more than half should therefore be specially drawn with extra-thick lines and (where possible) larger spaces between them, and several times the normal size of lettering. If this cannot be done for reasons of space, the whole illustration needs rethinking.

Very much the same considerations apply to slides, for while they are enlarged on the screen, they lose a corresponding amount of definition and, by the time they are viewed from the back row, they will appear rather like reduced illustrations in a magazine.

Although all 'figures' should always be explained in the text, this should be done again in the captions underneath, for many a paper has been read only because a casual browser got interested in the pictures and their captions. This need not mean slavish repetition of the same description: the general background and the methods used may be described in the text but in the captions the emphasis should be on the significance of the illustration itself.

It goes without saying that all letters, symbols and legends used must be either self-explanatory or defined in a key within, or underneath, the illustration.

If all this is done, the resulting pictures will indeed be worth a thousand words, for you can cram an enormous amount of information into a well designed illustration. But to achieve it needs just as much thought and planning as the text itself.

INDEX